EUROPE AND WORLD ENERGY

**Butterworths European Studies** is a series of monographs providing authoritative treatments of major issues in modern European political economy.

*General Editor*

François Duchêne — Director, Sussex European Research Centre, University of Sussex, England

*Consultant Editors*

David Allen — Department of European Studies, University of Loughborough, England

Hedley Bull — Montague Burton Professor of International Relations, University of Oxford, England

Wolfgang Hager — Visiting Professor, European University Institute, Florence, Italy

Stanley Hoffmann — Professor of Government and Director, Centre for European Studies, Harvard University, USA

Roger Morgan — Head of European Centre for Political Studies, Policy Studies Institute, London, England

Donald Puchala — Professor of Government and Director, Institute on Western Europe, Columbia University, USA

Susan Strange — Professor of International Relations, London School of Economics, England

William Wallace — Director of Studies, Royal Institute of International Affairs, London, England

*Forthcoming Titles*

Monetary Integration in Western Europe: EMU, EMS and beyond
European Political Co-operation
European Environmental Policy: East and West
The Defence of Western Europe
The Mediterranean Basin: A Study in Political Economy
The Making of the European Monetary System
The EEC and the Developing Countries
Pay Inequalities in the European Community

The research for this volume was undertaken at the Sussex European Research Centre (SERC) in the University of Sussex, England. The SERC carries out policy-oriented research on political, economic and social affairs in contemporary Europe.

# Europe and World Energy

Hanns Maull

**Butterworths**

In association with the
Sussex European Research Centre
University of Sussex, England

LONDON BOSTON
Sydney Wellington Durban Toronto

| | |
|---|---|
| United Kingdom<br>London | Butterworth & Co (Publishers) Ltd<br>88 Kingsway, WC2B 6AB |
| Australia<br>Sydney | Butterworths Pty Ltd<br>586 Pacific Highway, Chatswood, NSW 2067<br>Also at Melbourne, Brisbane, Adelaide and Perth |
| Canada<br>Toronto | Butterworth & Co (Canada) Ltd<br>2265 Midland Avenue, Scarborough, Ontario, M1P 4S1 |
| New Zealand<br>Wellington | Butterworths of New Zealand Ltd<br>T & W Young Building, 77-85 Customhouse Quay, 1, CPO Box 472 |
| South Africa<br>Durban | Butterworth & Co (South Africa) (Pty) Ltd<br>152-154 Gale Street |
| USA<br>Boston | Butterworth (Publishers) Inc<br>10 Tower Office Park, Woburn, Massachusetts 01801 |

All rights reserved. No part of this publication may be reproduced or transmitted in any form or by any means, including photocopying and recording, without the written permission of the copyright holder, application for which should be addressed to the Publishers. Such written permission must also be obtained before any part of this publication is stored in a retrieval system of any nature.

This book is sold subject to the Standard Conditions of Sale of Net Books and may not be re-sold in the UK below the net price given by the Publishers in their current price list.

First published 1980

© Butterworth & Co (Publishers) Ltd, 1980

ISBN 0 408 10629 8

---

British Library Cataloguing in Publication Data

Maull, Hanns
 Europe and world energy. — (Butterworths European series).
 1. Energy policy — Europe
 I. Title
 333.7    HD9502.E8    80-40488

ISBN 0—408—10629—8

---

Designed by Malcolm Young
Typeset by Scribe Design, Gillingham, Kent
Printed and bound by Redwood Burn Ltd, Trowbridge & Esher

# Preface

The idea for this book originated in 1974, and the bulk of the research and writing was carried out in 1975 and 1976 at the Sussex European Research Centre. I am very much indebted to the Deutsche Forschungsgemeinschaft for a generous grant, which made this research possible.

Since 1977, I have been heavily struck by a trouble known to most policy-oriented social scientists, and certainly to all those writing about energy: the rat race against the turbulent flow of events, in which the author is bound to lose. At the same time, this period has also provided ample opportunity to test the ideas, concepts and perceptions on which this study builds — and although I write this preface with some trepidation about the inevitable next twist of the world energy situation and its impact on the findings presented here, I am also reasonably confident that by and large the test has been passed. How well, I shall leave to the reader to judge.

The people whose advice and opinion benefitted this study are too numerous to be listed here — to all of them my thanks and my apologies if I have taken ideas from them without due credit. I was fortunate enough to profit from a wide range of expertise in many countries, and my thanks for this must go largely to the Trilateral Commission: as European Secretary and as one of the authors of a Trilateral Commission Report on energy I had the opportunity to travel and to discuss with experts from Europe, North America, Japan and the Third World.

The onerous task of keeping abreast of events was shared by many people. I would like to thank in particular the librarians of the Institute for Development Studies at the University of Sussex, of British Petroleum in London, and Mlle. Brigitte Levrier of the *Comité Professionnel du Pétrole* in Paris. Bill Martin and his colleagues at the International Energy Agency in Paris provided valuable help and information. Apart from all my friends who had to bear eternal complaints about the burden of a race against events, Terrie Russell, Catherine Morel and Susan Patterson helped me in coping with the burden. Editing my manuscript cannot have been easy — many thanks to Ann Duchêne and Colin Joy, as well as to the editors at Butterworths.

Finally, my gratitude goes to François Duchêne. To work with him has always been a pleasure, and his bountiful stream of ideas, insights and suggestions has been an invaluable source of inspiration and encouragement. Many of his ideas have found their way into these pages, and their final shape owes much to his efforts and assistance. Through this joint endeavour developed our friendship — and this has certainly not been the least of the rewards of this work.

# Contents

*Perspective* 1

Part 1    *Europe in the world energy market* 13

    1   Europe's domestic energy supplies   17

    2   Europe's prospects as an energy importer   51

Notes to Part 1   87

Part 2    *The trade relationship between the Nine and her energy suppliers: asymmetries*   99

    3   Trade patterns between the European Community and its main energy suppliers   103

    4   The power aspects of the international trade in oil   111

    5   The oil export sector in a developing country; the political economy of oil   118

    6   Co-operation and conflict   136

Notes to Part 2   143

Part 3    *Objectives for a European energy policy*   149

    7   Security of supply   153

    8   Long-term availability of supply   173

    9   Price formation and evolution   200

    10   Prices and control in the international energy system: the prospects   220

Notes to Part 3   250

Part 4    *Towards a global management of energy problems: elements and proposals*   261

    11   The North–South dialogue   267

    12   The Euro–Arab dialogue   284

13 The Mediterranean policy, the Lomé Convention, and trade relations with Iran  293

14 A global energy policy framework: some proposals  303

*Postscript*  323

*Notes to Part 4*  327

*Index*  333

# Perspective

The year is 4979 A.D. — a remarkable date for the future of world society. The most ambitious and far-reaching project as yet undertaken by mankind has been completed: the reconstruction of the solar system aimed at providing living space and energy for the dispersed settlements of what is still called (a nice touch of anachronism) 'global society'. After some last-minute complications (a lunar colony had programmed the automats in their space communication and navigation centre to strike, to protest against a conservation programme converting their settlements into a museum of pre-technological space colonization; the inhabitants wanted to keep their domes as quiet weekend and holiday refuges sheltered from hoards of interplanetary tourists) the resettlement programme was initiated. The whole population of the inner solar system was to be shuttled during the coming years to its new habitat, the giant shell constructed out of the matter provided by a pulverized Jupiter and the asteroid belt. This shell, with a diameter of two Astronomical Units (or about $300 \times 10^6$ kilometres (km)), and a thickness of only several metres, would be a new 'earth': provided with an artificial gravity through a complex rotation around the sun, and with a man-created biosphere, its whole inner side could be settled. The sun, at the centre of the shell, would supply the energy for the huge population, and the gigantic productive effort, of the global society: its whole radiation would henceforth be intercepted by the 'Dyson-Tsiolkovskii-Shell'[1] and thus be available for harnessing by mankind — ending a period of energy scarcity caused by the ever-increasing demand for energy of a developed technological civilization. This demand for energy had reached a level of about $3 \times 10^{29}$ ergs per second (erg/s) — considerably more than the $2 \times 10^{24}$ erg of solar radiation falling on the surface of the earth each second. This was the hour of the ultimate triumph of nuclear energy: a gigantic reactor with practically inexhaustible supplies of fuel, safely tucked away at a distance of $150 \times 10^6$ km[Note 2].

This look into an admittedly rather opaque crystal ball should help to underline what the energy problem (and this book) is about and what it is *not* about. Energy debates often present themselves as differences between 'optimists' and 'pessimists' about resources — in fact, the difference concerns something quite different: political and economic decisions. The energy resources of our planet are so abundant that their assessment hardly deserves much attention. At the same time, resources in themselves are without practical importance. They have to be harnessed through the mobilization of technology; as our science-fiction scenario has shown, the amount of energy available, given appropriate technologies, is practically limitless. At the same time, the example also serves to demonstrate that technologies cannot be introduced without social and political change. A fascinating example of this has been documented by the historian Lynn White: the introduction of feudalism, and thus of a strong central order, in northwestern Europe under the Carolingians depended on the invention of the stirrup. This turned the knight into a vastly superior fighting unit[3].

The energy problem thus essentially concerns the political and economic decisions dealing with the mobilization of technologies to produce and utilize energy, and with their positive and negative implications. Since this book is primarily concerned with the next 10 to 15 years, it will be more about the decisions needed to manage the consequences arising out of our present energy structures. A number of major studies[4] have been undertaken during the years 1976–1978; together they allow a realistic assessment of the present and the near future up to the end of this century. Although those studies differ in some respects, they appear to agree on two simple points:

(1) The world is undergoing a period of energy transition — a readjustment of energy production and consumption patterns. Basically, the transition will be characterized by (relative) scarcity, hence higher prices, hence (after the transition is completed) energy abundance (presumably at new, higher but relatively 'low' prices)*

(2) The driving element in the energy transition is the very high probability of world oil production reaching its peak sometime between 1985 and 1995; thereafter, it will decline. Much the same seems true for natural gas, although maximum production might fall somewhat later: world natural gas resources have not yet been tapped to the same extent as oil.

Such transitions have happened before: the change from wood to coal, or from coal to hydrocarbons. Historical experience suggests a timespan of about 50 years for such a transition; and present projections seem to confirm this.

As for the initial 10 to 15 years of this period, which is the focus of this book, we encounter again the familiar dispute between optimists and pessimists. There are the Cassandras who predict disaster at some near — though receding — point in the future. Their arguments are the continuing link between energy input and growth, the irreplaceably vital role of oil in total energy supplies, and the massive hidden energy demand in the form of widespread underdevelopment. The optimists point to lower than expected growth rates in world energy and oil demand over the last few years (due largely, it is true, to low economic growth which nobody thinks desirable), and to new large oil finds in Mexico and China.

---

* The apparent paradox dissolves if one relates energy price trends to general price trends and purchasing power, or, put differently, if one looks at the incremental cost of energy production.

Again, a closer look at the debate reveals that it really centres around political and economic decisions and choices, around the *management* of the energy problem. The position of the optimist implies *either* that existing arrangements and policies will suffice ('the market will do the job'), *or* it entails new arrangements and policies which are not spelled out. For example, the exaggerated hopes about Chinese or Mexican oil exports in the 1980s assumed (a) the capacity of these countries to step up production and exports sufficiently rapidly — a capacity which in turn depends on a number of major political and economic decisions such as the degree of foreign participation in the development of the oil —, and (b) their willingness to step up exports in line with expectations by the rest of the world. Both countries have made it clear that their oil production will be determined by their own national priorities.

The pessimists, on the other hand, assume that existing arrangements will not suffice to avoid transition crises — and their argument often ends with an appeal to governments for additional measures (evidently, this is not always a disinterested appeal — but the fact that the argument might serve some particular interests does not invalidate it *per se*).

Politics, not resource issues, are thus the clue to the future development of the energy problem. Political decisions might even strongly influence the graphs of energy transition now projected by the forecasters: for instance, perceptions about future scarcity have already led several producer countries to restrict oil and gas production with the objective of flattening out the eventual decline of production. We would be missing the point if we directed our attention to what is technologically and economically feasible; the substance of the transition process will consist of the interaction between political choices and energy constraints and opportunities.

Whatever political and economic choices are made, they will have implications beyond the energy situation itself. These choices will also — inevitably — involve value judgements. This requires the evolution of an energy strategy: namely a set of decisions and objectives which consider the short, the medium and the long term; it also implies a sense of overall direction, of the ultimately desired state of affairs. Since it is unrealistic to assume that our strategies will be able to anticipate *all* implications of the decisions taken, a vital ingredient of energy strategy must be the creation of flexibility and additional capacity for energy management — not in the sense of a bureaucratic reduction of future choices, but on the contrary through the opening of a wider range of options. It is only this kind of increased flexibility in national policies which will offer a realistic choice of reconciling the energy strategies of countries on the level of the international system.

That such mutual adjustment of national energy strategies on the international level is necessary follows from the present world energy structure. The major elements of this structure are:

(1) A very high level of centralization and interdependence. This is reflected in dense and complex networks of oil and gas pipelines and electricity grids

within and between states, and in the large intercontinental flows of energy trade (oil, coal, natural gas, uranium).
(2) The staggering size of world energy consumption. The figure for 1975 was 6155 × $10^6$ tons of oil equivalent (toe), as compared to 4121 × $10^6$ toe in 1965, and 2059 × $10^6$ toe in 1950. The three-fold increase indicates a qualitative difference in the dimensions of the adjustment problem compared to previous energy transitions — even allowing for the dramatic difference in available knowledge.
(3) The energy system is no longer determined by market mechanisms. The very long lead times in energy supply (4 — 15 years) and demand adjustment, and the enormous investment outlays required, pose serious problems for future profitability and involves major risks, which partly inhibits the proper functioning of the market; political interference further constrains its role. Quite apart from the fact that government regulation (or some other form of industry control, such as cartels) might be necessary to insure the smooth functioning of energy systems because of some of their inherent characteristics — it simply is unrealistic to assume that government can abstain from interfering in decisions with major implications for national security, economic growth, inflation, social justice and equality, and environmental protection.
(4) A high degree of dependence and vulnerability of many states as a result of the level of interdependence in the international energy system. The issue at stake is economic security in a very fundamental sense: dependence on international energy markets has now in many cases reached a point where the social fabric and the political system could be cataclysmically changed if this lifeline was interrupted. This holds true both for importers and exporters. New and difficult problems of international management are bound to arise out of this — much as governments are bound to interfere with national energy systems, so states will also assert national interests and objectives in the international energy system. The result will be a vast complication of a process dictated by global rationality: the exploitation of the cheapest energy sources first, and the phasing-in of new, more expensive ones as supplies of readily produceable cheap energy decline. Up to a point, this is what happened after World War II with the rapid advance of oil; the representatives of 'global rationality' were the multinational oil companies. The organization of the international system into nation-states strongly contradicts such a global evolution wherever it implies — as it does now — an infringement on vital national interests and objectives.

Because of these factors, the adjustment process which has begun will be difficult — much more difficult than indicated by historical precedents. It will pose a serious challenge not only to mankind's ability to design new technological responses to new problems, but also to our capacity for purposeful social and political action and organization. There can be little doubt that the true dimensions of the adjustment problem are not yet fully understood: the response to the energy crisis will ultimately have to be found in new lifestyles, new forms of social organization, and even new cultures, as much as in economic and political action. It is equally evident that the transition process will have a profound impact not only on the nation-state level, but also on the international level. Indeed, the international system will have to cope with the task of organizing and managing the major part of the transition process, if we assume that our societies will be sluggish and reluctant to implement the changes needed — as we would certainly be justified in concluding if we consider the responses to the first signs of an energy crisis in 1973/4. International management of the transition could imply a vast array of processes — and the danger of the present energy situation lies in the possibility that some forms of 'management' will be extremely damaging. Military action to secure access to supplies is one example here; preferential access for rich countries another. The importance of international management also becomes evident if we consider that the international system for a long time has been, and will continue to be, the ultimate regulator and residue for national energy policies. Nation-states have used this system to balance their energy demand with supply, or to balance their foreign exchange needs with supplies from exports other than oil. Although in the past this caused problems at times, the international system was generally flexible enough to play this role. Severe imbalances and constraints on this flexibility, which have to be expected in the future, would change the situation fundamentally, and ideally require that national energy policies be subjected to the capacity of the international system to balance export availabilities with import demands. The failure of any important actor in the international energy system to behave in accordance with international constraints would inevitably pose severe problems for international politics.

In addition to this challenge to the management of a difficult transition period, the international system will be faced with a second task: the evolution of a viable and just new world energy structure. There is a danger inherent in the present situation that solutions will be developed, and processes of change initiated, which will have severe, but as yet unforeseen repercussions on the future international system at large. The challenge is not only to construct a new global energy system, it is to design and enact a viable one. This implies objectives such as:

(1) Sufficient flexibility for the system to cope with a broad range of different possibilities of future development of energy demand, economic growth, and social organization.
(2) Consideration of the position of less developed countries and their different economic and social structures, with a view to create energy structures matched to their needs and possibilities.
(3) Minimization of risks of environmental damage, accidents, and political abuse (embargoes, nuclear weapons proliferation etc.).

As already indicated, the transition process might be complicated by the interference of politics with the management of the energy crisis. One example here is the very first warning of this crisis: the supply reductions and selective embargoes decreed by the Arab oil exporters in 1973/4. The link between the energy crisis and this particular political action is remote but exists: it is the degree of vulnerability in the present energy situation of the industrialized importers. Thus, we really could talk about two rather different kinds of problems with regard to energy: the long and drawn-out transition process, and the possibility that the vulnerabilities which will mark this period could at any given moment be used for political objectives entirely independent of the international energy situation, or only partly linked to it. An example of the first would be the use of the oil weapon in the Israeli–Arab conflict, or an attempt by exporters to secure acceptance of certain demands within North–South negotiations; examples of the second type of situation would be a cutback of production to enforce price increases, or a reduction of supplies to force the importers to conserve energy. Such acute political crises could appear at any point during the transition process, and again require international management – preferably before, as well as during the event.

Historical experience will be of little help in guiding this process of global management. In the past, the international oil market, which stood for the international energy market as a whole, given its enormous importance, was by and large managed by the large international oil companies. Their dominant role rested, in turn, on an underlying structure of control supported by Western political and military influence in the supplier countries. As the European powers gradually withdrew from their former colonial positions, control over oil came to depend increasingly on the power of the USA as the supreme defender of the management of oil supplies by the oil companies. In the late 1960s, Europe was relying on the USA for the security of its energy supplies, much as it did for military security.

This relic of colonial control over one important aspect of the international system was bound to become eroded, and its restoration appears impossible, for reasons which will be discussed later. Yet the challenge of international energy management persists, and indeed has grown dramatically, as the world has become aware that the hydrocarbon age will end some time in the not too distant future. The cost of failure to control oil supplies internationally will be dramatic.

Yet it was – among other factors – the entrance of the USA into the international oil market as an importer of rapidly increasing quantities of oil which undermined this structure of management. The energy crisis would have come anyway, but the creation of an incessantly growing supply gap in the USA from 1970 to 1973 helped to create the particular political and economic configuration which led to the events of 1973/4, and to intense competition for scarce energy supplies between Europe, Japan and the United States, the most important consumers of imported oil. This competition persists as a structural element in the present international system. Its elimination by the restoration of the quasi-colonial situation of the past, when Western control over the international oil market was undisputed, appears impossible. Yet the challenge of international management is only made all the more important by this hidden, and often not so hidden, competition: the end of the hydrocarbon era holds out unpleasant, and potentially extremely serious prospects for intra-Western conflict.

Where does Europe stand in all this? Very much at the receiving end – in terms of both the long-term transition crisis and the potential future political supply crises. The reason lies in the poor resource-endowment of Europe. Without the comparatively recent discovery of the North Sea hydrocarbon province, Europe's only major energy asset would be coal. But even with North Sea oil and gas Europe will be far from self-sufficient in energy; and this gift of nature appears to be too small to alleviate energy pressure on Europe for more than a limited number of years. In the longer term even the energy-rich countries of Europe will have to adjust to new sources of supply and new structures of demand. They have to begin this process now. Yet the political response of Europe to the energy challenge has so far been sluggish, reluctant, and reactive as well as national rather than European. Overall, the response falls far, far short of the challenge. This applies at the national and European levels, in domestic adjustment processes, and at the international level, with its need for a new management structure for the world energy system. Europe still searches for an international energy policy, and has only developed some elements of a domestic energy policy.

The following chapters will focus on the former deficiency, the absence of a European international energy policy, and neglect the domestic energy policy aspect. In concentrating on the problems of international management of the adjustment process we shall primarily consider the next 10 to 15 years, without, however, holding too closely to this period: energy problems have to be analysed with a very flexible time-frame so as to recognize long-term as well as medium and short-term trends and difficulties. The national distinctions and differentiations of Europe are often neglected, and sometimes there is an optimistic assumption about the cohesion and common will of the European Communities\*. Yet this rather abstract way of analysing a complex and long-range problem appears preferable to an approach which would be more realistic, but therefore also constrained by uncertainties about future membership of the Communities, about the evolution of their institutions, and the general development of European politics. To be bogged down in such difficulties would only hamper the broader view chosen for this book. Hopefully, a broad view will not falsify the major conclusions, whereas to follow the meanderings of European politics would inevitably prevent us from arriving at these conclusions. The element of utopianism sometimes apparent in the following pages

---

\* The analysis in this book is based on the period when there were nine member countries.

is happily accepted since it is the author's view that the energy situation will eventually demand 'utopian' solutions and actions as the only alternatives to catastrophe.

Thus, the book is about one non-entity (a European international energy policy), and about an 'entity', the European Community. Around this soft centre, we shall try to establish at least a fairly solid picture of the *setting*, the *objectives*, and the *processes*, which will be relevant for the formulation of a future European international energy policy. To do this, the level of analysis will be changed during the study. In the first part, the focus will be on Europe. The second part will concentrate on a vital segment in the international energy system: the relationship between Europe and its energy suppliers. The third part will consider yet another segment of this same system: the importing countries and their objectives. Here, it will become necessary to broaden the scope gradually, so as to allow analysis of the whole international energy system, which will provide the perspective chosen for the rest of our argument.

As the argument unfolds, several major hypotheses will become apparent; they are summarized below so as to provide a guideline through the following chapters.

(1) Europe will continue to be heavily dependent on imported hydrocarbons for the years ahead; these supplies will to an overwhelming extent come from Middle Eastern and African suppliers (Part 1).

(2) The relationship between Europe and its major energy suppliers is characterized by a complex structure of interdependence and asymmetrical dependence ties. Both sides are very dependent on the energy trade; the major vulnerability of Europe refers to supply shortfalls (both in terms of an acute crisis, and in terms of permanent constraints on energy availability, if a general supply shortage/high price situation develops); the main vulnerability of the suppliers lies in the difficulties of freeing their economies and societies from dependence on oil/natural gas exports, and of restructuring them towards alternative sources of foreign exchange earnings (Part 2).

(3) The major policy objectives for a future European international energy policy can only be obtained in co-operation with other countries and groups, and in particular with the OPEC countries, which play a crucial role both with regard to the future availability and the price of energy (Part 3).

(4) The organization of the transition process from hydrocarbons to alternative energy sources on the international level will best take the form of an international agreement on energy between all the major groups of countries involved in the world energy markets. Such an agreement would have to incorporate aspects of international management of the world energy markets, some transfer of national authority, and its own supra-national decision-making procedures, although these would not amount to an absence of national prerogatives (Part 4).

# Notes

1 So called after two outstanding scientists who had speculated about the eventual evolution of such a project, Constantin Edwardovich Tsiolkovskii, who put forward similar ideas in a book published in 1895, and Freeman J. Dyson, who suggested the project in an article written in 1960

2 This scenario is taken from I.S. Shklovskii and C. Sagan, *Intelligent Life in the Universe*, Picador, London 1977, pp. 469–470

3 L. White, *Medieval Technology and Social Change*, OUP, Oxford 1962. Lynn's point is that the stirrup had long been known before but that it was the Carolingians who recognized its military potential, and mobilized this technology for their purposes

4 They are usefully summarized in R. Lattès, 'La crise de l'énergie en perspective', in *Défense Nationale*, Nov. 1978, pp. 19–40.
In the same sense, see also T. de Montbrial, *Energie: le compte à rebours*, Lattès, Paris 1978 (published for the Club of Rome), and World Energy Conference 1978, published in several volumes by IPC Science and Technology Press, Guildford 1978

PART ONE

# Europe in the world energy markets

The unprecedented period of economic expansion since the Second World War has also been the 'golden age' of oil. In all major industrial economies, and in particular the two regions most devastated by war — Japan and Europe — it was oil, imported oil to a large extent, which fuelled the boom in industrial growth, international trade and living standards. As the most convenient general source of energy, advocated and promoted worldwide by the giant international industry, it took over to a large extent from other fuels, and attempts to resist the switch were by and large unsuccessful.

The evolution of Europe's dependence on imported oil is clear evidence of the rapidity of these changes. In 1950, Western Europe was still 94% self-sufficient in total energy consumption, but in 1973, this had declined to 39.5% (though it has now risen again to 43%). During the period between 1950 and 1975, energy consumption doubled to about $950 \times 10^6$ toe (see *Table 1.0*).

Thus, Europe is heavily dependent on oil imports: its energy security has become energy insecurity. To find ways to live with this insecurity, to organize it in a form which minimizes the risks, will be the task of a future European international energy policy. But to what extent will this dependence continue? Might Europe not return within a reasonable time to the degree of self-sufficiency enjoyed in the 1950s? The response to this initial question must condition all further inquiry. To answer the question, we must look at the prospects of Europe's indigenous energy production, taking the energy sources one by one.

Table 1.0 Europe's Energy Balance 1977 ($10^6$ toe)

|  | All sources | Solid fuels | | Oil | | Natural gas | | Other | |
|---|---|---|---|---|---|---|---|---|---|
| (1) Production | 433.52 | 178.80 | (41.2)[a] | 48.12 | (11.1) | 142.35 | (32.8) | 63.46 | (14.6) |
| (2) Net imports | 534.74 | 29.20 | (5.5) | 487.15 | (91.1) | 17.01 | (3.2) | 1.4 | (0.3) |
| (3) Consumption[b] | 928.81 | 204.57 | (22.0) | 507.30 | (54.6) | 157.70 | (17.0) | 64.86 | (7.0) |
| (4) Self-sufficiency (Production (1) as percentage of consumption (3)) | (46.6) | (87.4) | | (9.5) | | (90.3) | | (97.8) | |

[a] Figures in parenthesis represent percentage of total
[b] Including stock changes, excluding bunkers
Source: OECD Energy Statistics

CHAPTER ONE

# Europe's domestic energy supplies

## Outline

### *Solid fuels*

The basis of Europe's industrial revolution began to decline from the late 1950s. While there have been absolute increases in world coal production, absolute quantities of Community-produced coal have, by contrast, almost halved since 1957. During the 1960s there was a relatively dramatic swing away from King Coal to the new Emperor, oil. The question now is whether the drastic increases in oil prices from 1973/4 have changed this basic trend. It seems not, at least not for the next ten years — although the decline of coal will be slowed in relative terms (i.e. as a percentage of energy consumption) and absolute quantities produced will be halved. The part coal can play is restricted as much by factors limiting demand as by the difficulties of increasing production.

Solid fuels mark a division, which we shall encounter repeatedly, between the 'haves' and the 'have-nots' in the Community energy field. Solid fuel production is heavily concentrated in West Germany and the UK; in other Community countries it plays a diminishing role, or none at all. The two coal giants could only with great difficulty increase production to compensate for the expected decline in the rest of the Community — the fall in numbers of coal-miners would have to be checked and large investments made to develop new mines and expand old ones. Dramatic technological changes would be needed to increase productivity, which apparently has levelled out. Industrial relations pose a particular problem in the UK. Apart from economic costs, the social costs of increased coal production also have to be considered. Overall, these difficulties appear too great to allow much expansion of coal's contribution to Community energy supplies in the near future.

This does not mean that solid fuels do not have a future, but that the decline in coal production can be reversed only by new technologies, such as a major step towards automation, or on-site processes for turning coal into gas underground. Coal gasification, given economically attractive conditions, would open up vast new markets and give coal renewed importance.

The Community's natural gas is also fairly concentrated, mainly in the north, in the Netherlands and in Britain which possesses the largest and fastest-growing industry. Unlike coal, natural gas has expanded dramatically since the late 1960s and shows every sign of continuing as Europe's primary growth fuel.

The North Sea contains one of the world's major gas provinces, whose full potential may still not have been assessed. Natural gas has great value, for it is clean, cheap to produce, easy to handle and particularly useful for certain applications (e.g. as raw material for chemical industries). An elaborate distribution network has been developed, making it easily available throughout most Community countries. Furthermore, natural gas is interchangeable with oil in almost all sectors of consumption except transport.

But what of supplies? There is some uncertainty about the ultimate size of gas reserves in the North Sea, as well as the optimum extraction rate, particularly for the latter half of the coming decade. But the real ceiling to production is political. In all producer countries except the UK and Netherlands, and the newcomers in natural gas production such as Denmark and Ireland, whose North Sea fields will contribute small but nationally significant quantities in the future, demand exceeds supply and can be expected to grow faster than domestic production (should the latter grow at all). To increase self-sufficiency through natural gas would require more intra-Community trade. So far Britain has no plans to export natural gas, and the Netherlands, though having committed a large amount of production to export, has become increasingly reluctant to conclude additional agreements. The husbanding of this scarce resource, its reservation for premium uses and for future generations at home, are becoming increasingly important in the energy policies of the well-endowed Community countries. The result will be another growing energy gap between the Community's demand for natural gas and its indigenous supplies.

## Nuclear energy

For 25 years nuclear energy has been regarded as the source of energy with the most hopeful future, and this view was powerfully reinforced by the oil increases of 1973/4. Was this not a perfect illustration of the main argument for nuclear energy, the need for secure supplies, quite apart from the confirmation of the cost advantages supporters had always claimed for it?

From the pattern of supply and demand, however, it seems none too certain that consumption of nuclear energy — i.e. electricity — will rise as much as some predict. The gap between the (historically higher) growth rate of electricity consumption and that of total energy consumption has been closing in recent years, possibly indicating some limitations on electrification of our societies' energy systems. Moreover, electricity production is very wasteful, consuming some two-thirds of the energy input in the transformation process. Nuclear electricity production will have to compete with solid fuel, and even oil-fired power stations. Heavy fuel oil could be offered at very competitive prices, since it is an inevitable by-product of petrol and jet fuel production. Total electricity demand will therefore have to be shared between these different types of power plant.

The real problem of nuclear energy, however, is supply. The main arguments in favour of it are (a) cost advantage and (b) security of supply. The first argument is dubious, since cost comparisons normally exclude the huge government R & D expenditures and the still unknown costs of waste disposal and disaggregation at the end of a reactor's lifetime. The second argument is of greater relevance (although it often ignores the fact that there is still some dependence on imports of uranium ore and of enriched uranium for fuel elements). Nuclear energy might reduce energy import dependence – provided it can survive the present widespread public opposition in several member countries, particularly Germany and France, the two main proponents of reactor programmes. The major arguments of the opposition to nuclear energy include: the potentially disastrous results of a nuclear reactor accident; the risks of radiation; the problems of waste disposal, which nuclear supporters tend to hand over to future generations; the dangers of nuclear sabotage and terrorism; and proliferation to more and more states.

Some of these arguments have not been sufficiently rebutted to justify dismissal. The argument about nuclear energy thus seems to take on tragic dimensions. On the one hand, concern about the dangers of nuclear energy argues strongly against its expansion. On the other hand, the abandonment of nuclear energy could cause a dramatic shortfall in energy supplies, also with potentially very serious consequences for Western societies as a whole. Yet this 'tragic' perception of the problem seems to be overdrawn. A more realistic assessment would have to take into consideration:

(1) The huge uncertainties surrounding future energy, and in particular electricity demand, which might make nuclear energy much less necessary than is currently assumed.
(2) The inherent energy consumption of a nuclear programme, which in the case of rapid expansion of capacity (say, a doubling every five years) would actually become a net energy *consumer*, the energy requirements of the programme exceeding energy output.
(3) The enormous scope for conservation and the fuller utilization of existing energy production, especially in the case of electricity.

On balance, it should therefore be possible to move very cautiously on nuclear energy, and to decide against a risky, expensive and awkward technology until uncertainties are dispelled by further research, new designs, engineering improvements, and a clearer view about true demand.

The nuclear problem is compounded by the way nuclear energy policy has been formulated, both in some member countries and at the Community level. There can be little doubt that public opposition to nuclear energy draws its strength from a closed and bureaucratic decision-making process which confronts the citizen with *fait accompli* which he perceives as fundamentally affecting his life\*. The question of nuclear energy has not, until very recently, been discussed politically in any significant sense beyond a small group of people, let alone in a broad democratic process. As in other instances, the economic, social and political ramifications of energy choices have hardly begun to be considered, let alone analysed.

## Oil

At present oil is the Community's main source of energy and will remain so for the next decade (the 1980s). Apart from prices, its rapid expansion at the expense of coal was mainly due to its flexibility, and the broad range of uses to which it can be put will ensure that demand for oil will remain. The novelty in the Community's oil economy is the existence of sizeable indigenous reserves. Production in the North Sea is now building up rapidly and the UK has become almost self-sufficient. Technical problems apart, the main factor likely to influence availability of British oil to other Community countries will be political: the British government can decree production ceilings in order to conserve resources and if exploitable reserves should turn out to be larger than presently foreseen, this option might be taken up. An independent Scotland may be even more restrictive, and the possibility of this cannot be ruled out. The precise direction of producer energy policies, however, is hard to foretell. In general, it seems reasonable to see them as shaped by bargaining processes involving domestic as well as international pressures, with the latter tending to push up production levels. Similar considerations also apply to the complex relationship between the producer governments and the private oil industry.

What conclusions emerge from this brief survey? First, Europe's capacity to step up indigenous energy production is limited. Future import dependence will depend on the pattern of consumption, and domestic and intra-Community energy policies will need to focus on possibilities for reducing demand for imported energy. In practice, a very large import gap will remain for the foreseeable future and for this reason Europe cannot be spared the formulation of an international energy policy. Second, energy questions are becoming increasingly

---

\* The bureaucratic character of nuclear energy policies in at least some member states raises a further concern: the political implications of the nuclear reactor programmes of the Community. The dramatic case of the leading German nuclear scientist and manager, Dr Klaus Traube, may be an isolated case but should not be ignored. Dr. Traube was suspected of terrorist connections by the German Secret Service, which put him under surveillance, 'bugged' his home and tapped his telephone, in violation of constitutionally guaranteed rights. The suspicion against Traube could not be corroborated: nonetheless, he was dismissed from his job on the advice of the Secret Service. The violation of constitutional rights in this incident was defended by reference to the extraordinary danger posed by someone in a position to gain access to nuclear installations and make his knowledge available to outsiders. It seems legitimate to ask, what would be the social and political costs of a heavily nuclear-oriented energy economy? How far would it pose insurmountable dilemmas of reconciliation with the existing political culture in Europe? Would it imply the gradual erosion of civil liberties?

politicized. Resource conservation reflects one aspect of this, for it tends to focus at national level and could effectively block an intra-Community energy policy. Further, as the case of nuclear energy demonstrates, increased public debate on energy objectives complicates energy policy-making. The time for bureaucratic energy policies seems to have passed, and politicization of energy questions will considerably slow down decision-making and implementation. This will add to the need to secure necessary energy supplies from external sources.

## Survey

### *Solid fuel prospects: predictable but unexciting*

Coal was the energy base for the industrial revolution in Europe, and formed the base for the dramatic recovery of Western Europe after the Second World War. However, since 1958 'King Coal' has been replaced by crude oil from the Middle East, and later from Africa, as the dominant energy base in Europe: in 1960, solid fuels accounted for 69% of total energy consumption in the present nine member states of the communities, and oil accounted for 27% — 15 years later the consumption of coal had fallen to 19% whereas oil, nearly all imported, accounted for 57% of the European Community's energy consumption[1].

Oil met the new rapidly rising energy demand, and made inroads into sectors previously dominated by coal. Solid fuel consumption declined, absolutely as well as relatively. (This was true for hard coal, accounting for some 87% of solid fuel energy content, but not for lignite, production of which expanded somewhat.)

Solid fuels at present meet about 19% of the European Community's energy consumption. Most of this, as *Table 1.1* shows, comes from indigenous supplies, nearly three-quarters of imports being of coking coal, of which the Community has a substantial deficit. Trade in coal between the member states amounts to about 7% of total European Community solid fuel consumption.

There are marked differences in national consumption and production of solid fuels. The United Kingdom relies on coal for 34.6% (1976) of energy consumption, Germany 29.9%, Belgium 22.0%, with the rest of the Community dependent on less than 20%. Luxembourg has a much higher dependence, 47.8%. Of the total European Community coal demand, over 81% is for use in transformations, essentially in power stations and coke ovens. For most member countries this concentration is even more pronounced.

The position in future will clearly depend crucially on developments in the British and German coal industries, which account for about 90% of European Community solid fuel production, and which will certainly further increase their share. The coal industries of France and Belgium are in decline, and the others in the European Community do not have important coal production industries. France has found it difficult to meet even reduced coal extraction targets[2], the Belgian industry has been in decline for years and showed no sign

Table 1.1 Excerpts from European Communities Coal Balance Sheet (Hard and Brown Coal) 1975 ($10^6$ toe)[a]

| | European Community | Germany | France | Italy | Netherlands | Belgium | Luxembourg | UK | Ireland | Denmark |
|---|---|---|---|---|---|---|---|---|---|---|
| (1) Production | 206.9 | 94.1 | 17.2 | 0.3 | – | 5.2 | – | 90.1 | 0.04 | – |
| (2) Imports | 40.8 | 5.2 | 12.2 | 8.8 | 2.9 | 4.4 | 0.2 | 3.6 | 0.5 | 2.9 |
| (3) Exports | 12.5 | 10.3 | 0.4 | – | 0.2 | 0.3 | – | 1.3 | 0.04 | – |
| Gross consumption | 223.1 | 84.2 | 26.4 | 8.9 | 2.6 | 8.9 | 0.2 | 89.1 | 0.5 | 2.1 |
| of which: | | | | | | | | | | |
| (5) Transformation | | | | | | | | | | |
| coke ovens | 72.3 | 31.2 | 10.4 | 7.7 | 2.4 | 5.2 | – | 15.4 | – | – |
| power plants | 108.8 | 39.7 | 9.6 | 0.8 | 0.1 | 1.9 | – | 55.2 | 0.03 | 1.6 |
| (6) Industry | | | | | | | | | | |
| energy sector | 2.2 | 1.0 | 0.3 | 0.1 | – | 0.02 | – | 0.9 | – | – |
| other industries | 16.7 | 1.5 | 2.4 | 0.2 | Negligible | 0.2 | 0.2 | 6.8 | 0.04 | 0.4 |
| (7) Transport | 0.4 | 0.2 | Negligible | 0.1 | Negligible | 0.01 | – | 0.1 | – | – |
| (8) Other | 16.2 | 2.1 | 2.4 | 0.1 | 0.1 | 1.4 | Negligible | 9.6 | 0.4 | 0.003 |

[a] *Conversion factors used: 1 t hard coal approximates 0.7 toe; 1 t brown coal approximates 0.2 toe.*
*Source: OECD Energy Statistics*

of recovery after the oil price increases[3], while the Netherlands' mines have now closed down altogether[4]. Only Britain and Germany have substantial economically recoverable reserves; in the UK, a crash exploration programme since 1973/4 has resulted in dramatic additions to resources[5].

The decisive questions are, whether coal will be able to win new markets in sectors presently dominated by other sources of energy, and whether the sectors where solid fuel consumption is concentrated at present provide much scope for increased demand. Supply and demand have to be considered together: increased production without markets would obviously make little sense; expansion of demand beyond the capacity of indigenous production, on the other hand, would mean growing reliance on imports (a possibility to be discussed later, which clearly depends on international availability and prices).

To stabilize total Community output, British coal production would have to be increased slightly, given the decline in output from member states other than Germany. German production is thought to be capable of stabilization at around $91 \times 10^6$ toe (involving some decline in hard coal extraction, but an expansion of lignite output)[6], but the British National Coal Board hopes to expand production of hard coal from $82.8 \times 10^6$ toe (1973) to $85-90 \times 10^6$ toe in 1985. These targets have now been agreed with the National Union of Miners, but the production figures for the years 1974–1977 do not offer much hope that the targets can be reached[7*]. As for recent German estimates, they assume a slight decline in output, from $91 \times 10^6$ toe in 1980 to $88 \times 10^6$ toe in 1985[8]. Using official estimates, total European Community solid fuel production by 1985 would reach $184.2 \times 10^6$ toe, against the Commission's target of $210 \times 10^6$ toe[Note 9].

Marginal costs have risen over the last decade, even while the industry has contracted and been rationalized, and will continue to do so if it expands[10]. The pressure to subsidize coal to allow it to compete with oil and natural gas has not diminished much, in spite of energy price increases since 1973. Direct and indirect aid by European governments to coal producers in 1976 varied between 23.68 units of account per tonne (u.a./t) (Belgium) and 0.28 u.a./t in Britain, with France paying 7.22 u.a./t and Germany 2.37 u.a./t (figures for 1975 are significantly lower, indicating renewed pressure on coal's competitiveness)[11]. The limits to growth in underground mining productivity appear to have been reached, and with labour costs accounting for between 51 and 58% of total costs in the British and German coal industries, further cost increases appear inevitable[12].

Whether the Commission's production targets could be met without recruiting additional miners is uncertain, but the National Coal Board hoped to reach its original ambitious targets of $100 \times 10^6$ toe in 1980, rising to $115 \times 10^6$ toe in 1990, without enlarging its present labour force of 250 000[Note 13]. Existing manpower should suffice in Germany where, although production is not planned to increase overall, some shift will take place towards open-cast brown-coal mining[14]. However, there will be the additional cost of environmental protection and this may set the ceiling on the expansion of coal mining in Western Europe, which is concentrated in densely populated areas.

The Commission's targets require massive investment in mining equipment, transport and infrastructure, and there are signs that such investment is available: in 1975 there was an increase of 60% over 1974 in total European Community coal industry investment, and in the following years the trend continued upward[15]. Major changes in production patterns could not be expected before the mid-1980s, however, as lead times for investment to become effective are estimated at around 10 years. The elasticity of coal supplies to energy prices is thought to be well below unity: a trebling of energy prices might barely double coal production, even in the very long run[16].

The substitution of coal for other fuels over the next 15 years appears limited and would require massive adjustment costs: in *transport*, the main consumers are now jet and petrol engines, for which there are no close substitutes. Coal consumption in *industry* might be somewhat increased, as costs of converting oil/gas-burning equipment into coal-fired plants are fairly low, with lead times reasonably short. However, there would have to be a reliable long-term incentive for such additional investment, for the switch would entail additional expenditure to meet the environmental problems posed by burning coal[17]. In *households* the use of coal can be expected to decline further, since convenience still favours oil and gas[18]. Moreover, with the expansion of central heating, capital costs have become a larger factor than fuel costs. Overall, therefore, substitution outside the presently coal-dominated sectors will be marginal, until conversion of coal into oil and gas becomes economic. This will probably hold true even after recent substantial oil price rises.

Some scope exists for short-term substitution of coal for oil in the two main sectors of demand, power stations and coke ovens. A 1972 analysis for Germany showed that, theoretically, consumption of heavy fuel oil could have been cut by 27% ($7.1 \times 10^6$ tons) and of light fuel by 15% ($7.2 \times 10^6$ tons) by substituting coal. Actual substitution in 1973/4 was below this, but did substantially reduce the demand for fuel oil[19].

While no such data are available for other countries, the scope for substitution obviously depends, to a large extent, on the composition of the power station sector: the more genuine dual-standard systems there are, the broader the scope of substitution. In all Community countries, the capacity of dual-standard systems is planned to be increased in absolute, and, in some cases, in relative terms[20]. However, dual-standard systems are often more suited to one kind of fuel, so that their flexibility is somewhat theoretical.

Coal-fired stations are often the oldest, and therefore the least efficient stations: they tend to be placed at the end of the power station merit order, supplying mainly peak demand[21]. Moreover, there must be some doubts about the economic desirability of new coal-fired stations. From 1973 to 1975, coal costs to the CEGB increased by 170%[Note 22]. Some oil-fired generators receive their fuel supplies from nearby refineries, and so their conversion to coal could not be justified economically. Substitution for planned nuclear power stations is another matter, to which we will return later, but here too prospects for coal are uncertain, though not necessarily for economic reasons.

In industry, about 80% of European Community coal is consumed in coke ovens, most of which is already met by imports. This is likely to expand only by a little, if at all, owing to technical improvements and a slowdown (or reversal) of growth in the steel industry (a 2% annual increase in coke demand is predicted against a 4.1% rise in pig-iron production)[23].

Rising costs and inflexibility of supply may restrain future coal demand and investment. At the same time, oil companies are likely to continue to price fuel oils competitively and to rely on products which cannot be substituted, such as petrol and jet fuel, to make profits[24]. Only competition from natural gas in transformations and heavy industry may peak off, since the inherent qualities of this fuel support its restriction to premium uses[25].

To sum up, the future of coal, although undoubtedly brighter because of higher oil prices, remains fairly predictable and utterly unexciting unless there is a technological breakthrough. Constraints on supply and demand will enable solid fuels to achieve little more than stabilization in absolute terms and a slowdown of the decline in their share of total energy consumption. Coal reached a turning point in Western Europe around 1958, when oil began to take its place. Another turning point could come when coal gasification and other new coal technologies become economically viable propositions, thereby vastly increasing markets.

While it is unlikely that this will happen before 1985, such new technologies could make a very important contribution to a revival of solid fuels in the longer run. It might therefore be worthwhile to take a somewhat longer time-perspective, and reconsider the issue.

On the *supply side*, reserves and production capacity will hardly place any constraints for a long time to come — British reserves, for instance, are under present circumstances and at present production rates 300 times as large as annual production. In the long run, this will become more important than the constraints on both supply and demand, as technological change overcomes present difficulties.

To revive the production and consumption of coal in Europe, there will have to be cheaper forms of production and cleaner and more efficient ways of using coal. The most dramatic progress in production could be made through *in situ* gasification of solid fuels under ground. Several such ventures already exist in the USSR, and research is done in Europe itself. But large-scale application is certainly still some way off; until then, productivity increases have to be sought through increased mechanization and improvements of coal production infrastructure[26].

In the consumption of coal, coal liquefaction and gasification offer exciting possibilities, although their costs are still very high, compared with other sources of energy[27]. If synthetic natural gas (SNG) could be produced from coal on a large scale, it could replace Europe's declining natural gas production without much additional requirements in infrastructure — the system of pipelines and final use appliances would simply be fed by SNG instead of natural gas. Large-scale demonstration plants are operating already or are under construction in the

UK and in Germany, where work is being carried out on a project to synthesize natural gas from coal through the utilization of nuclear process heat[28].

Another promising development concerns fluidized bed combustion of solid fuels, a process in which coal is burnt in a churning bed of very hot particles such as sand, and kept bubbling like a liquid by hot air blown in. Chemicals can be added to fix impurities such as sulphur in order to prevent pollution[29]. The heat produced by this process can be used to raise steam very efficiently, or to synthesize oil and natural gas. Two different lines of development are presently under way in the UK (in co-operation with Germany and the USA): the British National Coal Board is constructing a large experimental power plant to come on stream in 1979, and Babcock and Wilcox is using NCB patents to operate a large fluidized bed combustion boiler; the firm offers a range of commercial boiler designs based on this process[30].

The fluidized bed combustion technology seems particularly attractive: it is relatively cheap, versatile and easy to fit, and can burn even very low-quality coal cleanly. Thus, new technologies offer not only interesting alternatives to traditional forms of using coal in power generation, but above all possibilities to reconquer markets long lost to oil and natural gas, and even open up new ones through SNG and synthetic oil products[31].

An example of the long-term hope for coal is the British project for coal development until the end of the century (*Plan 2000*), which projects production levels of $150 \times 10^6$ t (deep-mined coal) and an additional $20 \times 10^6$ t (open-mined coal) at the end of the period[32]. Exploration since 1973 has uncovered vast new coal reserves, and a series of new mines are planned during the next 20 years. Two new mines have already been opened, in 1977 and 1978, and the first big new development, the Selby project with a planned annual output of $10 \times 10^6$ t, is to come on stream in 1982; even bigger ones are to follow. The sum of £$10 \times 10^9$ will have been invested over 25 years by the end of the scheme[33]. In Germany, test drillings during the 1960s have also located large additional reserves, if at much greater depth than existing fields[34]. Thus, the future holds some possibility for the revival of coal in Western Europe.

## *Natural gas: the growth fuel*

Since 1960, natural gas has been the European Community's most rapidly expanding source of energy, accounting for 16.5% of total consumption in 1975, compared to 2% in 1960. In future this fuel will play an even more important role. Although the expansion of indigenous production may slow down, the sector as a whole is expected to continue to grow rapidly.

Natural gas appears in much the same geological circumstances as crude oil. It can be dissolved in crude oil deposits or form a cap on top of them (associated gas); or constitute a separate field (non-associated gas). As *Table 1.2* shows, this fuel has made large inroads into every sector of European consumption except transport. It is also used as a feedstock for the petrochemical and chemical

**Table 1.2** Natural Gas Balance Sheet of the European Communities, 1976 ($10^6$ toe)[a]

| | European Community | Germany | France | Italy | Netherlands | Belgium | Luxembourg | UK | Ireland | Denmark |
|---|---|---|---|---|---|---|---|---|---|---|
| (1) Production | 136.1 | 14.2 | 5.8 | 12.6 | 71.7 | 0.02 | — | 31.8 | — | — |
| (2) Supplies from the European Community | 39.4 | 18.1 | 8.8 | 3.7 | (−38.5) | 8.5 | 0.4 | — | — | — |
| (3) Supplies from outside | 12.4 | 2.5 | 2.6 | 5.8 | (−1.2) | — | — | 0.8 | — | — |
| (4) Consumption | 146.7 | 35.1 | 16.5 | 21.5 | 32.0 | 8.5 | 0.4 | 32.7 | — | — |
| of which: | | | | | | | | | | |
| Transformation | 31.4 | 11.9 | 2.4 | 3.8 | 9.3 | 2.2 | 0.2 | 2.1 | — | — |
| Industry | | | | | | | | | | |
| energy use | 45.2 | 12.4 | 5.3 | 9.2 | 5.8 | 3.2 | 0.2 | 9.1 | — | — |
| non-energy use | 9.2 | 0.8 | 1.6 | 1.8 | 1.3 | 0.5 | — | 3.2 | — | — |
| Transport | 0.3 | — | Negligible | 0.3 | — | — | — | — | — | — |
| Households etc. | 55.7 | 9.1 | 6.4 | 6.9 | 14.3 | 2.2 | 0.1 | 16.7 | — | — |

[a] Sums are not necessarily accurate because of rounding, losses etc.
Source: Eurostat

industries (e.g. the synthesis of ammonia, methanol and acetylene, and the production of hydrogen, all key processes in the chemical industry) and in the direct reduction of steel.

There are a number of reasons for this expansion[35].

(1) Natural gas is a readily controlled source of energy which can be used very efficiently.
(2) It poses few problems of pollution, though, as with oil and coal, there are difficulties with nitrogen emissions.
(3) Natural gas reduces capital costs through saving on storage, through lower construction costs and (through its purity) through the longer life of plants.
(4) To capture markets, it has been offered at prices lower than other fuels. Undoubtedly this has been the main reason for its growth.

Natural gas is extraordinarily versatile: it is clean, easy to handle, and able to produce stable high temperatures. These advantages give it a premium for both commercial/residential and industrial use[36]. Use for cooking and water-heating is approaching saturation point, but space heating and air-conditioning usages are still expanding quite rapidly[37]. In industry, natural gas serves both as a raw material and as a fuel (often together with oil) in the production of ceramics, bricks and glass, iron and steel, and in chemicals where automation and continuous processing make it particularly useful. Steam-raising and thermal power plants do not employ its specific qualities to the best advantage and such uses can be expected to diminish[38].

Obviously, gas can substitute for other fuels over a wide spectrum, and Community energy planners are looking to it as much as to nuclear energy for diversification from oil. Unlike coal, there are few constraints on demand: supply will determine the future role of indigenous natural gas in the Community. How far imports might help satisfy demand will be discussed later.

The infrastructure needed to transport and distribute natural gas is to a considerable degree already in existence or under construction. Production levels will therefore depend mainly on the scale of reserves, prices and national (or Community) energy policies. At present, the Netherlands dominates Europe's gas production with the one giant Groningen field, discovered in 1959, with estimated reserves of between 1800 and 2200 $\times$ 10$^9$ cubic metres (m$^3$)[Note 39]. The Netherlands also has several smaller onshore fields with total recoverable reserves of about 230 $\times$ 10$^9$ m$^3$ [Note 40], and recent offshore discoveries promise to make a substantial contribution[41].

The other major producer is Britain, which in 1965 discovered gas offshore in the southern basin of the North Sea. Several major non-associated gas fields were established, and production has reached very substantial levels from the Hewitt, Viking, West Sole, Leman and Indefatigable fields. Proven reserves of this gas province alone were given as about 721.6 $\times$ 10$^9$ m$^3$ in 1976[42]. It is now thought unlikely to contain major surprises, though findings are still being made and estimates upgraded. Interest has shifted to the huge amounts of gas, associated and non-associated, discovered and widely expected in the north and west. In 1978, total UK proven reserves were given as $821 \times 10^9$ m$^3$, while those of the Netherlands stood at $1698 \times 10^9$ m$^3$. Next were Germany ($206 \times 10^9$ m$^3$), Italy ($235 \times 10^9$ m$^3$), France ($136 \times 10^9$ m$^3$), Ireland ($28 \times 10^9$ m$^3$) and Denmark ($48 \times 10^9$ m$^3$). This leaves Belgium and Luxembourg as the only countries without indigenous natural gas reserves (see *Table 2.3*).

However, 'proven' reserves is a dynamic term, indicating amounts assumed to be recoverable on available information and in existing technical and economic conditions. These can change. Extension drilling provides new data leading to new reserve estimates, usually upgraded, and improvements in recovery prospects may also change the picture substantially. Reserve figures for a gas field normally rise rapidly at first, then flatten out six to eight years after discovery[43]. So predictions should take account of the history of European reserve estimates (see *Table 1.3*, which shows the dynamism of the British and Dutch industries, but also the latter's maturation).

**Table 1.3** *Reserve and Production of Natural Gas in the European Community, 1966–1977 ($10^6$ m$^3$)*

|      | Italy (A)[a] | (B)[b] | France (A) | (B) | Germany (A) | (B) | Netherlands (A) | (B) | UK (A) | (B) | Denmark (A) | (B) | Ireland (A) | (B) |
|------|------|------|------|------|------|------|------|------|------|------|------|------|------|------|
| 1966 | 143 | 8.8  | 249 | 5.1 | 198 | 2.8  | 1274 | 3.3  | 0.28 | —   | —   | —   | —   | —   |
| 1967 | 152 | 9.2  | 244 | 5.6 | 241 | 3.7  | 1784 | 7.0  | 283  | —   | —   | —   | —   | —   |
| 1968 | 152 | 10.5 | 297 | 5.7 | 241 | 5.8  | 1925 | 14.0 | 708  | 2.0 | —   | —   | —   | —   |
| 1969 | 178 | 11.5 | 300 | 6.5 | 342 | 8.2  | 2329 | 21.8 | 990  | 2.0 | —   | —   | —   | —   |
| 1970 | 141 | 12.4 | 204 | 6.9 | 337 | 12.8 | 2349 | 31.4 | 1019 | 11.4 | —  | —   | —   | —   |
| 1971 | 170 | 13.3 | 195 | 7.1 | 396 | 15.4 | 2349 | 43.6 | 1132 | 18.5 | 14 | —   | —   | —   |
| 1972 | 173 | 13.2 | 192 | 7.5 | 359 | 17.7 | 2490 | 57.7 | 1274 | 25.0 | 51 | —   | —   | —   |
| 1973 | 148 | 14.4 | 182 | 7.5 | 345 | 19.3 | 2576 | 70.3 | 1400 | 26.5 | 50 | —   | —   | —   |
| 1974 | 340 | 15.2 | 164 | 7.6 | 325 | 20.3 | 2683 | 69.7 | 1415 | 26.5 | 14 | —   | —   | —   |
| 1975 | 283 | 13.7 | 150 | 7.3 | 235 | 17.8 | 1981 | 90.9 | 1415 | 34.9 | 14 | —   | 28  | —   |
| 1976 | 187 | 14.8 | 142 | 7.1 | 212 | 18.4 | 1752 | 94.8 | 1400 | 37.6 | 20 | —   | 28  | —   |
| 1977 | 235 | 13.7 | 136 | 7.7 | 206 | 18.9 | 1698 | 96.9 | 1400 | 40.2 | 48 | —   | 28  | —   |

a (A) = reserves
b (B) = production
Sources: Comité professionnel du pétrole; UK Department of Energy

A field's history passes through stages[44]. First, there is a pause for a few years after discovery, during which time transportation, infrastructure, and markets are built up. Then production accelerates rapidly, to a plateau of maximum output. When some 70% of recoverable reserves have been extracted, output begins to decline. The time span, and in particular the duration of maximum production, depends on the depletion rate chosen, upper and lower limits being set by reservoir-specific factors and economic return respectively (between 3.5% and 5% for European fields currently in production).

The depletion rate is set by the operator (quite often an international oil company) and/or the government concerned. Gas distribution in Europe is largely handled by utilities which are partly or fully state-controlled, so governments have long taken a particularly active interest in natural gas policies, including pricing. In this way, they are both parties and mediators amid conflicts of interest over the distribution of the economic rent from gas production, and between short- and long-term considerations.

The development of government policies has reflected these conflicts. Thus, the discovery of natural gas in Britain's North Sea sector took place in a complex framework of competitive licensing arrangements, which ensured a rapid rate of development of reserves in the initial stage. In the Danish sector, on the other hand, only one consortium, comprising *inter alia* Shell and Gulf, was given exploration rights, which led to a very slow pace of development. Similarly, the original Dutch exploration consortium NAM (50% owned by Shell and Exxon, 40% by the Dutch State Mines, and 10% by the government) only speeded up exploration and development when new acreage was handed over to other companies. This reflects the desire of international oil companies to bring additional resources onto the market in an 'orderly', controlled fashion so as to avoid pressure on prices through over-supply[45].

The companies' desire for gradual integration of natural gas into the European energy market, rather than aggressive competition with themselves as oil importers, was accepted at least by some governments. The Dutch certainly went along with this cautious approach. While aiming at a maximization of economic benefits, they took account of conservation and the balance of the domestic energy market. At the same time, they initially supported attempts to secure markets for Dutch gas in other European countries[46]. The British government put greater emphasis on short-term gains and maximum economic benefits. The British Gas Corporation, with its powerful market position as monopoly buyer, forced low prices, and pushed natural gas expansion to the highest possible extent[47]. While both governments tried to direct natural gas to premium uses, this was somewhat offset in the case of the UK by the low price level.

Pricing policy will continue to have an important impact on future levels of supply, as well as of demand: the British experience in the southern North Sea basin seems to indicate that low prices caused a decline in exploration beyond the level normally associated with the shift from the exploration to the development stage[48]. The Dutch government, on the other hand, has recently pushed natural gas export prices close to the level of alternative fuels, and increased domestic prices[49]. Contracts for gas from northern fields in the UK North Sea sector have also been based on much higher price levels[50]. With markets now well established, indications are that future gas prices will be more or less in line with prices of competing sources of energy. Higher prices should only slightly restrain growth in demand, and will help maintain supply. Growth of demand will nevertheless increasingly outpace indigenous production.

Supply will therefore depend mainly on factors other than price. Trends in reserve evaluation (*Table 1.3*) suggest no dramatic production increase over the next decade, except by the UK. Projected expansion is modest for France and

Italy, and also for Germany, where official targets and other estimates point to a production level of about $18 \times 10^9$ m$^3$ in 1985 — the 1974 OECD forecast of $25 \times 10^9$ m$^3$ seems now much too optimistic[51].

The Dutch target for 1985 is $82 \times 10^9$ m$^3$, and although estimates of production capacity are somewhat higher, the Dutch government is likely to hold to this target for political reasons. It is very concerned about long-term energy prospects, and foresees natural gas falling from about 50% of the total energy balance in 1975 to 42.6% in 1985 and 34.6% in 1990. After 1978 sales to power stations and large factories were phased out and exports are expected to fall rapidly, and to cease altogether by 1996[52].

The evaluation of production potential from the North Sea is particularly difficult, and opinions diverge widely. Official estimates for the UK sector in 1985 now assume $45 \times 10^9$ m$^3$ which means total Community production could reach about $165 \times 10^9$ m$^3$, a very small increase over 1976 figures and about 28% over the level of 1973[53]. The EC Communission's lower target of about $200 \times 10^9$ m$^3$, therefore, seems now impossible to reach.

Odell judges total resources to be much larger and suggests a much greater rate of production for Western Europe (essentially the Community plus Norway), reaching $520 \times 10^9$ m$^3$ in 1985, more than a quadrupling of 1973 production[54]. However, de Vries and Kommandeur have shown that even with ultimate recoverable resources three times the present proven reserves, a moderate 4% depletion rate and a 35-year build-up of reserves, production would reach a peak of only $470 \times 10^9$ m$^3$ in 1989 and rapidly decline thereafter[55].

Odell's scenario also appears unlikely for political reasons: as he himself points out, such a development would require an essentially autarchic European energy policy including collective agreement in the European Community on:

(1) A floor price for energy.
(2) Joint mobilization of capital and resources.
(3) Joint assistance to countries facing socio-economic problems as a consequence of high production and revenue rates[56].

Unfortunately this programme presupposes a degree of politcal co-operation which is unlikely to occur. But even if the common will existed, major problems would still have to be faced. Even massive assistance might not alleviate some of the social, economic and environmental consequences of very high rates of production. If so, producer governments would be expected to sacrifice their people's interests to those of others, most evidently through rapid depletion of their energy resources. In any case, prudence dictates conservative assumptions about ultimately recoverable reserves. Even if this did not preclude an increase of production until reserve depletion began to exceed additions by new discoveries and field re-evaluation, the very high levels of consumption reached just when production began to decline rapidly would then require very large imports and/or alternative sources of energy with all the adjustment problems that would imply[57]. This could lead to major strains on the international energy market comparable to the entry of the USA as a large-scale importer of oil from the Middle East around 1970.

## Nuclear power: major uncertainties ahead

In 1973, total European Community installed nuclear power capacity was 12 300 megawatts (MWe). In 1974 governments and the Commission set targets which were to bring this up to 61 800–64 800 MWe in 1980, and to at least 160 000, 200 000 MWe if possible, in 1985[58].

This massive expansion would have meant a substantial increase in reactors: in Germany from 10 to about 50 power stations, in France 16 new 1000 MWe generators coming on stream between 1979 and 1981, and a further dozen by 1985, with a target of 50 reactors in that year. In Italy, with capacity of 550 MWe installed, 850 MWe under construction, and orders for an additional 4000 MWe, the central electricity authority, ENEL, planned to raise total installed capacity to 20 400–26 400 MWe by 1985[59]. Belgium has also developed a sizeable reactor programme in co-operation with France. The Netherlands and Britain, however, have moved very cautiously down the nuclear road, since both are much better endowed with indigenous energy resources than the other European Community countries. In the Netherlands, reactor siting is also particularly difficult, given the high population density, while for the UK technical problems with indigenous reactor designs have contributed to the low profile of nuclear energy. (For details on nuclear programmes and calculations about capacity build-up see *Table 1.4*).

At the end of 1978, installed nuclear capacity in the Community was 26 700 MWe – up by 115 from 1973; but it was clear that the ambitious objectives of 1974 could never be met. In fact the Commission now aimed at an installed capacity of 102 500 MWe in 1985, and expected rather less – about 90 000 MWe[Note 60].

The figures can be read to mean the high expectations for nuclear fission will not be realized or to confirm that fission power will become the predominant replacement energy. Clearly national energy planners still rely on it as the main alternative to import dependence. Nuclear energy is considered a cheap and secure way to solve Europe's energy problems: the energy source of the future. Yet there are those who warn that nuclear energy could easily become mankind's most gigantic *détour* in problem-solving. The world's energy future, they argue, will certainly not be dominated by nuclear energy, but by renewable resources[61].

A look at supply and demand structures might again help us to find our way through these arguments. At present, nuclear power is used almost exclusively to generate electricity. Electricity, though present in all major sectors of consumption, is heavily concentrated in two: the industrial (48%) and the household sectors (49.3%) – both figures for 1975[62]. Despite this concentration, growth in Community electricity demand has outpaced that of energy demand in general, with an annual average rate of growth of 8.35% compared to 4.85%

**Table 1.4** Present and Future Development of Installed Nuclear Capacity in the European Community (GWe) (estimates for year end)

| Country | 1977 | 1978 | 1979 | 1980 | 1981 | 1982 | 1983 | 1984 | 1985 | 1986 | 1987 | 1988 | 1989 | 1990 |
|---|---|---|---|---|---|---|---|---|---|---|---|---|---|---|
| | | | | | | | | Installed capacity | | | | | | |
| Belgium | 1.7 | 1.7 | 1.7 | 1.7 | 2.6 | 2.6 | 3.5 | 3.5 | 3.5 | 5 | 5 | 7 | 8 | 8 |
| Denmark | — | — | — | — | — | — | — | — | — | — | — | 1 | 1 | 2 |
| France | 4.7 | 6.5 | 12 | 15 | 19 | 23 | 27 | 31 | 34 | 38 | 42 | 46 | 50 | 53 |
| Germany | 6 | 9 | 10.3 | 12 | 14 | 16 | 18 | 22 | 25 | 29 | 34 | 38 | 43 | 47 |
| Italy | 0.6 | 1.4 | 1.4 | 1.4 | 1.4 | 1.4 | 1.4 | 2.4 | 5.4 | 7 | 12 | 15 | 20 | 25 |
| Luxembourg | — | — | — | — | — | — | — | — | — | — | 1 | 1 | 1 | 1 |
| Netherlands | 0.5 | 0.5 | 0.5 | 0.5 | 0.5 | 0.5 | 0.5 | 0.5 | 0.5 | 1 | 2 | 2 | 2 | 3 |
| United Kingdom | 6.6 | 6.6 | 10.3 | 10.3 | 10.3 | 10.3 | 10.3 | 10.3 | 10.3 | 11.6 | 12.8 | 14.1 | 15.3 | 15.3 |
| Ireland | — | — | — | — | — | — | — | — | — | — | — | — | — | — |
| Total | 19.5 | 25.7 | 36.2 | 40.9 | 47.8 | 53.8 | 60.7 | 69.7 | 78.7 | 91.6 | 108.8 | 124.1 | 139.3 | 154.3 |

*Source: OECD Nuclear Energy Agency, IAEA, Uranium, Resources, Production and Demand, Dec. 1977*

in the period from 1950 to 1972[63]. The gap has been closing, however, and while electricity demand is generally expected to expand further, predictions made after 1974 of annual increases of 6.8–7.2% until 1980, and 7.5–8% thereafter until 1990[64], now appear too high. The latest estimates assume a figure of 6% annual growth in electricity demand for the period 1976–1985 in the Community[65]. Even this assumption is based on relatively optimistic estimates about future economic growth, and about the relative role of electricity in the total pattern of energy demand.

Is there any scope for nuclear energy outside the electricity sector? Nuclear heat might be used directly for industrial applications – but so far there exists only one example of this in the Community: the BASF plant in Ludwigshafen. Present reactors operate at temperatures too low for direct industrial use. The high temperature reactor (HTR) could change this – but it has received a serious setback because of the difficulties of its main industrial proponent, General Atomics, and the cancellation of the OECD *Dragon* project. Only Germany now still pursues the HTR technology.

One interesting way of using nuclear heat in the future could be coal gasification. Another possibility is the direct use of reactor heat for space heating; but since reactors are normally sited far away from population centres for safety reasons, heat would have to be transported over long distances[66]. In sum, it appears that prospects for nuclear energy outside the electricity sector will be insignificant for a considerable period of time to come.

Therefore, we have to return to the electricity sector itself, and to its two main customers: industry and the residential/commercial sector. What are the opportunities for a substantial increase in overall electricity demand in these two areas?

Industry has already demonstrated its ability to constrain demand growth in the face of higher prices; besides, co-generation (the production of power from industrial waste heat on the site) could be significantly expanded to cover incremental electricity demand. As a study for the USA has shown, industry could co-generate about 50% of its total electricity demand by 1985[67]. Since co-generation has been developed further in Europe, the possibilities are less dramatic but still substantial; and from the point of view of energy efficiency, this seems much preferable to installing new electricity generating capacity outside the industrial sector. Finally, future electricity demand of the industrial sector will depend on overall growth of industry in Europe; and prospects here seem to be less than outstanding.

In the residential/commercial sector, additional electricity demand would largely have to come from space-heating – this, at least, seems to be the position of the European Commission, and it is difficult to see any other large additional source of demand[68]. Yet a dramatic increase of electric space-heating seems economically unattractive: electricity would have to compete with fossil fuels burnt directly for heating purposes, and this would involve different degrees of thermal efficiency (15–40% for electricity, compared to 50–90% for direct burning of fossil fuels), since losses in electricity generation and transportation are enormous[69]. At present, this is not reflected in the price structure, since installed capacity permits very cheap production of electricity during off-peak periods. With a huge increase of electric space-heating, this price difference would have to disappear as new capacity has to be installed, and demand evens out over the day: costs would then come to reflect adequately the real competitive strengths and weaknesses.

This shows that electrification is not necessarily a desirable policy. It would reinforce a heavily centralized, therefore fairly rigid and vulnerable, energy structure requiring massive investments in distribution. (About one-quarter of European Community energy investments during the period 1976–1985 are planned to go into electricity distribution)[70]. Moreover, electricity has a very low degree of thermal efficiency: about two-thirds of fuel input are transformed into waste heat in electricity generation, and further large losses occur during transportation[71]. In both cases, the opportunities and prospects for dramatic improvements are slim.

Ultimately, these problems of electricity are going to be reflected in costs: and since the non-substitutable demand for electricity (light bulbs, electronics) is fairly small, electricity seems bound to slow its dramatic advance in the past. Thus, we should be careful about future demand for electricity. But what about the relative position of fission reactors in electricity generation? Can they, and will they, catch incremental demand, and even replace existing capacity?

Ultimately, this will again be a question of costs. Some recent comparative estimates for nuclear and fossil fuel plants are summarized in *Table 1.5*. It shows that nuclear power enjoys a competitive advantage because of its lower fuel costs, but also that capital costs for nuclear plants have risen dramatically. The advantage in fuel cost will remain even if the price of uranium should go up further, since uranium ore accounts for only a minor part of the total cost of nuclear fuel, which in turn only constitutes a small part of total operating expenses. The problem for nuclear reactors' competitiveness lies in the high, and still increasing, capital costs, which reflect not only inflation, but also more stringent safety requirements. Investments planned in the Nine between 1976 and 1985 with regard to nuclear energy amount to $55.52 \times 10^9$ u.a., with about the same sum allocated for investment in electricity distribution; together this represents about 50% of the total scheduled energy investments in the European Community[72].

The comparisons presented in *Table 1.5* offer, however, only a limited indicator of actual costs. First, they do not include the huge R & D expenditures channelled into nuclear energy development in the past, nor the costs for ultimate waste disposal, nor disposal of abandoned reactors, etc. Second, the calculations assume similar standards of performance between different types of power plants. This does not appear to be fully justified: in the USA, the 'availability factor' (i.e. the percentage of total time during which a plant actually supplied electricity) was 5–10% higher for coal-fired power stations than for

Table 1.5  Comparative Capital and Operating Costs of Power Stations

| | Nuclear (LWR) | | | | Oil | | | | Coal | | | |
|---|---|---|---|---|---|---|---|---|---|---|---|---|
| | (1) | (2) | (3) | (4)[a] | (1) | (2) | (3) | (4)[a] | (1) | (2) | (3) | (4)[a] |
| Capital cost (m$/KWe) | 950 | 390 | 600 | 1085 | 520 | 260 | 347 | 452 | 805 | 356 | 404 | 517 |
| Fixed costs (m$/KWh)[b] | | | | | | | | | | | | |
| Capital | 23 } | 10.8 | 11 | 16 | 12.5 } | 7.2 | 6.0 | 7.4 | 20 } | 9.8 | 6.9 | 8.5 |
| Operating and maintenance | 2 | | 4.5 | 3.0 | 1.5 | | 4.0 | 1.8 | 5 | | 4.3 | 6.3 |
| Fuel costs | 9 | 3.2 | 6 | 9 | 41.0 | 11.5 | 20.7 | 25.2 | 19 | 11 | 16.9 | 27.8 |
| Total generating costs | 34 | 14.0 | 21.5 | 28.0 | 55.0 | 18.7 | 30.7 | 34.4 | 44 | 20.8 | 28.1 | 39.9 |
| Assumptions: | | | | | | | | | | | | |
| Plant operational | 1985 | 1985 | 1980 | 1985 | 1980–1983 | | | | | | | |
| Coal: sulphur removal | yes | | yes | ? | yes | | | | | | | |
| Fuel cost, oil | $12/b | | $9/b | $118/t of fuel oil | $116/t of crude | | | | | | | |
| Fuel cost, coal | $1/10⁶ Btu | | | $67/t | $85/t | | | | | | | |

[a] Figures in 1972 constant $. All other figures in current $.
[b] m$ = $ 0.001

Sources: D.C. Ion, *Availability of World Energy Resources*, Graham and Trotman, London 1975, p. 146; *Energy Prospects to 1985*, Vol. II, OECD, Paris 1974, p.185; Service des Etudes Economiques Générales, *Evolution du coût du KWh nucléaire et du KWh fuel*, Electricité de France, Paris, Oct. 1977; H. Michaelis, *Kernenergie*, dtv, Munich 1977, pp. 232–233

nuclear reactors; and recent figures about reactor availability in Europe are rather lower than the US figures[73].

Even collectively, all this will probably still not erode completely the competitive advantage of nuclear energy in the near future — but the gap is closing. But economics alone are not the only aspect to be considered: the availability of secure energy supplies might well be another important element in the decision for or against nuclear energy. Will it not be worth paying more money for such additional security of supplies?

The security of supplies argument is indeed of importance. First, nuclear power stations are much less dependent on continuous deliveries of fuel. While a fossil fuel power plant of 1000 MWe consumes about $2 \times 10^6$ t of fuel per year, a nuclear reactor of the same capacity needs only 35 t of uranium dioxide[74]. Nuclear reactors can easily stock one year's supplies within the plant — fuel elements have to be exchanged only every three years[75]. Even though much of Europe's uranium requirements would have to be imported, nuclear power obviously helps limit import bills and dependence on OPEC. However, uranium ore itself might pose problems of scarcity. The reactor type most widely used now, and certainly until the late 1980s, is the light water reactor (LWR), which burns the uranium isotope uranium-235; uranium-235 forms only about 0.7% of natural uranium ore. This makes the LWR a relatively wasteful way of burning uranium. The next reactor generation could do better — the HTR could eventually operate on the thorium—uranium-233 cycle (thorium is relatively abundant), and the fast breeder reactor (FBR) could 'breed' plutonium out of a mixture of plutonium and natural uranium fuel[76]. The evaluation of international availability of uranium will be given in the next chapter; suffice it here to say that production capacity should be sufficient until about 1985, but that problems of economic scarcity cannot be excluded afterwards, possibly even before, depending on production and export policies. In any case, Europe now seems firmly committed to the nuclear fuel cycle using reprocessing and the fast breeder reactor so as to cut down gradually and continuously on uranium import dependence. Whether the timing of the introduction of the FBR and the occurrence of difficulties in the availability of uranium will coincide, remains to be seen. Equally, it is difficult at the moment to evaluate the problems and advantages of a FBR electricity system. Costs and risks cannot yet be assessed properly; yet some scepticism seems appropriate in light of the comments made below about the desirability of avoiding too much emphasis on electricity, and of the environmental and social constraints on a linear extrapolation of past trends in economic and energy growth[77].

This takes the argument to the environmental impact of nuclear energy, and public opposition. Advocates of nuclear reactors point to the fact that they are practically free from emission problems, while coal power stations emit large quantities of sulphur, carbon and nitrogen dioxides, as well as huge amounts of ash particles[78]. They also emphasize the favourable record of nuclear energy in terms of actual health and accident risks, as compared with other energy industries. Yet these arguments have not prevented the opposition to nuclear energy in several European countries, which has successfully interfered with the work on nuclear plants. The specific risks and dangers seen by the opposition are: the possibility of a large-scale accident, of nuclear sabotage and terrorism, the radiation risk, and the problem of nuclear waste.*

Accidents and technical malfunctions do happen in nuclear reactors — LWRs have so far been out of action for about half their operational lifetime[79]. Nevertheless, the safety record of the nuclear industry has up to now been quite satisfactory: fatalities (caused by radiation and other accidents) per 1000 megawatts per annum (MWe/a) have so far been 0.492 for the whole fuel cycle, with the heaviest toll at the mining and milling stage. This figure of about one death per two reactors a year compares favourably with the record of other sources of energy, particularly coal[80]. Major accidents have up to now been avoided — and incidents such as the fire caused by a natural candle in the Browns Ferry reactor in Alabama or even the more dramatic accident at Three Mile Island are quoted both by opponents and by advocates of nuclear energy: the former pointing out that human failure can create bizarre, unforeseen situations, the latter that the safety precautions did prevent an accident of major importance[81]. A major disaster could only happen as a consequence of a reactor core melting — only then would huge amounts of radioactivity be released into the environment. After a failure of the cooling system and a loss of coolant, the fuel in the reactor might heat up to its melting point (about 5000 °F). If, then, there were additional and multiple failures in the emergency cooling system, the pressure vessel and the containment building would break, and the radioactive core would penetrate into the surrounding areas[82].

A study commissioned by the US Atomic Energy Agency (the Rasmussen Report) tried to analyse the probability of a major accident of this type, and the Ford Foundation/Mitre study contains a careful re-evaluation of this report[83]. The former estimates the probability of an extremely serious reactor accident (3300 immediate deaths, latent cancer fatalities over 30 years of 45 000, damage of $14 \times 10^9$) at one chance in two hundred million years of reactor operation, and the probability of a major accident (70 immediate deaths, 170 cases of acute sickness, $2.7 \times 10^9$ damage) at about 1:1 million reactor years. The Mitre study concludes that the Rasmussen Report somewhat underestimates the range of possible probabilities, but that the overall risks involved in nuclear reactors are acceptable, although this does not imply that no further research will be needed in the area of reactor safety[84]. Ultimately, however, it is the wide range of probabilities which is significant: they underline the difficulty of estimating risks related to human, as well as technical, fallibility. All evidence indicates that there is no absolutely foolproof system of safeguards against the determined idiocy or inattention of the human operator.

* Nuclear proliferation risks relate to exports rather than domestic energy politics; see Chapter 2.

The human element enters the equation in yet another way: as a political threat. Nuclear reactors already produce about 400–600 lb of plutonium annually, which at present is recovered in reprocessing plants and stored, separately or together, with the other nuclear waste. This involves the transportation and storage of large amounts of highly toxic radioactive substances; and the FBR would multiply the problem. Nuclear plants, reprocessing plants and storage facilities are all vulnerable to attack with a view to diverting plutonium, or to suicidal terrorist attacks, or to conventional explosives in times of war. Terrorists in a high-jacked plane did threaten to dive-bomb the Oak Ridge National Laboratory in the USA[85]. Illegally diverted plutonium or enriched uranium could be used to assemble a primitive bomb; while this bomb might not be as 'effective' as a military nuclear weapon, it would suffice to create a major disaster. Even without a nuclear bomb, the threat of releasing plutonium into the atmosphere could destroy a metropolitan area, or a major grain harvest[86].

Ultimately, then, the risk of nuclear disaster by accident or intent would appear to pose a serious dilemma — 'the problem of guaranteeing the necessary social stability without being forced to engineer society itself'[87]. As a Swedish Nobel Laureate of physics puts it:

'Fission energy is safe only if a number of critical devices work as they should, if a number of people in key positions follow all their instructions, if there is no sabotage, or hi-jacking of the transports, if no reactor fuel processing plant or reprocessing plant or repository anywhere in the world is situated in a region of riots or guerilla activity, and no revolutions — even a "conventional one" — take place in these regions. The enormous quantities of extremely dangerous material must not get into the hands of ignorant people or desperados. No acts of God can be permitted.'[88]

Another serious argument against nuclear power is the radiation risk under normal conditions, i.e. without any accident. Radiation risks can be divided into two groups: electromagnetic (X- and gamma-rays), and corpuscular radiation (alpha- and beta-particles, neutrons). All radiations can basically affect living tissue through ionization, but the effects differ from case to case[89].

The effects can be *somatic* or *genetic*; somatic effects appear immediately after exposure to radiation (acute), or at a later stage (chronic). Since the repercussions of small doses of radiation are extremely difficult to establish, the effects are subject to controversy; similarly, no agreement exists as to whether there is a minimum threshold dose, or whether all radiation in principle involves a risk. It is, however, well-established that radiation can cause cancers, and reduce life expectancy. Equally well-known is that radiation triggers genetic mutations: 99% of all mutations are harmful[90]. Since genetic change is triggered by comparatively low doses of radiation, permissible levels should be defined with regard to those risks. This is difficult since overall ceilings do not account for biological concentration, e.g. in plants and animals which are part of the human diet[91]. Assuming the radiation exposure levels used in the Mitre study, the conclusion about latent fatalities as a result of radiation among workers in nuclear installations and the general public can be estimated at 0.5 deaths per reactor year[91a]. Those levels would be substantially increased through radiation exposure as a result of reprocessing and recycling of plutonium.

Nuclear activities at present entail three types of radioactive waste: solid, liquid, and gaseous. Gaseous radioactive substances are the by-product of the fission process, and of impurities in the air and cooling water in the reactor. The most important is krypton-85, which is fully released into the environment[92].

Liquid waste is the cooling water, which normally contains isotopes formed by bombardment of water impurities and corrosion products from the cooling pipes. It is released after demineralization, which does not retain all radioactive substances. Some reach the environment, although the level of radiation is low[93].

The most serious disposal problem is posed by the necessary replacement of spent fuel elements every few years. This nuclear ash can be separated into fission products and actinides. Fission products (mainly strontium-90, cesium-137, and krypton-85) have half-life times of up to 30 years. An example underlining the problem is cesium-137, which forms about 3% of the nuclear ash: One-thousandth of the annual reactor production (60 lb) dispersed over one square mile would result in radiation levels about 1000 times as high as the usual natural radiation from space[94].

The actinides (uranium, plutonium, actinium, and others) are all very toxic and have long half-life times — about 24 000 years in the case of plutonium-239. This means that, although they are initially less radioactive than fission products, they pose a problem for hundreds of generations to come. Plutonium, which is particularly dangerous given its quantities, and the fact that minute doses inhaled by human beings cause lung cancer, is at present the main source of concern[95]. Large quantities would have to be transported over long distances, even on an international level. Uranium and plutonium are at present the only substances recovered from nuclear ash (although only 99.5% are actually removed; the last 0.5% cannot at present be recovered economically, thereby prolonging considerably the time-scale of the nuclear waste problem)[96]. The remainder is stored in liquid form in huge steel tanks which have to be exchanged every few years. Vitrification of this part of nuclear waste has been developed recently as a more satisfying alternative. The question of ultimate disposal has not yet been solved, although various suggestions have been made (e.g. storage in solid form in salt mines, in Antarctic rocks or under the ice caps). To put the task in perspective, one has to remember that half a kilogram of plutonium is estimated to possess a *potential* to cause 9 billion cases of lung cancer; it was estimated that by 1980 annual world production of plutonium could be 40 000 kg[Note 97]. In addition, a nuclear reactor of average size produces as much long-lived radioactivity as the explosion of 1000 bombs of Hiroshima calibre[98].

Given the weight of these objections to nuclear energy, public opposition is not surprising, and doubtless would increase dramatically if there were a major nuclear accident anywhere in the world. In any case, it already has gathered enough momentum to be the most serious obstacle to a rapid expansion of the national nuclear programmes — not altogether undeservedly, given the largely bureaucratic way in which energy policy decisions have been made in the past, particularly in the nuclear field. Opposition is, from this perspective, at least partly a backlash against the lack of openness in decision-making in national energy policies. Quite apart from whether the opponents are right or wrong (and there seems to be at least a very solid body of doubts which, given the potentially disastrous consequences of false assumptions, justifies more caution than has been displayed by governments in the past), the question of nuclear power now seems deeply rooted in politics, and can be solved only by political means.

Since 1977, the political debate has finally begun both on the national and on the Community level; and in Germany and Italy nuclear opposition seems to have subsided somewhat. Both countries are now politically committed to a cautious go-ahead with nuclear energy[99]. France had not yet been fully affected by the opposition, which still has to master the obstructiveness of a highly centralized and non-participatory nuclear policy[100]. In Britain, an endless squabble over appropriate reactor technologies, and over the role of nuclear energy in general, together with technical problems and lower electricity demand forecasts, had slowed down the nuclear power plant construction independent of nuclear opposition[101].

The likely future development, therefore, seems to be further expansion of nuclear energy without the kind of dramatic break-through expected in 1974. Caution seems to be the notion of the day, and this caution appears quite justified, given some persistent problems with nuclear energy:

(1) Nuclear reactors contribute to the establishment and perpetuation of a rigid, highly centralized, capital-intensive energy structure, which would seem to be badly equipped to adjust to a zero energy growth[102].

(2) Nuclear energy seems an inelegant and inefficient way of addressing energy problems: thermal efficiency is very low even by standards of the first law of thermodynamics (which refers to the relationship between useful and wasted energy), let alone by those of the second law (which concerns the match between the quality of an energy source, and the specific demand to be met). This problem is a problem of electricity in general, but concerns nuclear energy in particular, since it offers little other possibilities of markets[103].

(3) In terms of energy analysis (which compares energy costs with energy benefits), nuclear reactor programmes with a rapid pace of expansion consume more energy than they produce during the time of the build-up; only if this pace is slowed very much, or capacity remains stable, will there be a net energy gain from reactors. This, too, points to the need for a carefully timed nuclear strategy, which treats nuclear energy as a bridge into a differently based energy economy, without creating too many irreversible structures[104].

(4) Nuclear energy cannot be considered only from an economic and technological perspective; it raises important social issues about the role of technology in our societies, about the control of sophisticated techniques such as nuclear reactors, and about the social costs of this control. Decisions about energy structures are not unrelated to priorities about the kind of society one desires; and some scepticism with regard to nuclear energy seems appropriate from this point of view. Moreover, it seems as if such social concerns and priorities will feed back into the development of nuclear energy; one should, for this reason, also be careful about too optimistic predictions concerning the future role of nuclear energy.

(5) Nuclear energy would be particularly ill-suited as an energy technology for the Third World — it is capital-intensive, difficult to control, and probably much less competitive than in the industrialized world with its established electricity grids. Yet it would be dangerous to assume that the LDCs would be willing to follow a different energy strategy, if they see that the industrialized world is unwilling to foresake a further development of this source. The result could be a proliferation not only of nuclear energy, but also of nuclear weapons in the Third World, of much increased accident probabilities, and of growing discrepancies within the Third World — a development hardly suited to improve international stability.

Nuclear fission is about to become an important factor in the energy strategy of most European countries. Since it is one of the most easily accessible means of energy diversification, such a development has to be accepted. A rational energy strategy should aim at the replacement of hydrocarbons in the generation of base-load electricity. At the same time, and given the at least partly justified concerns about the social implications of the widespread use of nuclear energy, this should be treated very much as an intermediate solution, and expanded very cautiously. Above all, energy policies ought to aim at discouraging rather than encouraging growth of electricity consumption. This would also help in the overall efforts to reduce growth in energy demand — and it is energy conservation which should be given the highest priority, rather than increase of energy supplies. Third, it seems vital that the expansion of nuclear energy should be fitted into a long-term energy strategy with clearly defined social as well as economic objectives, and with the operational emphasis on increased flexibility and the creation of new options. Nuclear energy, in other words, should be treated as a far from optimal bridge, and great caution has to be taken that it does not develop to an extent where future alternatives are foreclosed by past decisions and their inherent force of predetermining the future. In the longer run, at the end of the transition period, nuclear energy might be neither as necessary nor as economic as is now thought.

## Oil: *a new indigenous source of energy*

Oil is at present the Community's most important source of energy, appearing in

Table 1.6 European Community Petroleum Products Balance Sheet, 1976 (all products, $10^6$ t)

| | European Community | Germany | France | Italy | Netherlands | Belgium | Luxembourg | UK | Ireland | Denmark |
|---|---|---|---|---|---|---|---|---|---|---|
| Available for inland consumption[a] | 469.941 | 127.492 | 104.939 | 86.383 | 34.429 | 22.439 | 1.411 | 81.832 | 5.052 | 15.964 |
| of which: | | | | | | | | | | |
| Power stations | 55.456 | 5.597 | 14.139 | 17.049 | 0.950 | 3.752 | — | 10.028 | 1.306 | 2.635 |
| Non-energy use | 55.293 | 16.052 | 9.089 | 9.151 | 8.394 | 2.261 | 0.027 | 9.708 | 0.137 | 0.474 |
| Energy use | | | | | | | | | | |
| Industry | 98.027 | 25.184 | 22.294 | 19.214 | 2.995 | 3.527 | 0.645 | 20.765 | 1.249 | 2.154 |
| Transport | 124.235 | 32.925 | 26.916 | 17.796 | 7.823 | 5.055 | 0.354 | 28.659 | 1.442 | 3.265 |
| Residential/commercial | 133.944 | 47.188 | 31.157 | 23.106 | 4.268 | 7.615 | 0.395 | 12.123 | 0.794 | 7.298 |

a This does not add up to total availability because of miscellaneous other consumption and statistical discrepancies
Source: Eurostat

all main sectors of consumption, and with a near-monopoly in transport. (See Table 1.6.) It represents the residual fuel, which brings supply in line with demand. Yet at the same time, indigenous production is presently the lowest of all sources of energy supplies. With the important exception of the UK, prospects for increases in indigenous production are very limited, as can be seen from Table 1.7. which shows the present oil reserves of the Nine. Although

**Table 1.7** *European Community Oil Reserves, Production and Future Production Estimates*

|  | Oil reserves$^a$ ($10^6$ t) | Oil production, 1978 ($10^6$ t) | Estimated production ($10^6$ t) 1980 | 1985 |
|---|---|---|---|---|
| Belgium | — | — | — | — |
| Denmark | 36 | 0.37 | 0.4 | 2.5–4.0 |
| France | 18 | 1.12 | 1.2 | 1.5 |
| FRG | 120 | 5.06 | 5.0 | 4.0–6.0 |
| Ireland | — | — | — | — |
| Italy | 49 | 1.50 | 2.0 | 3.0–5.0 |
| Luxembourg | — | — | — | — |
| Netherlands | 42 | 1.60 | 1.5 | 1.5–2.5 |
| UK | 3 200 | 53.38 | 105.0 | 130.0 |

Sources: *Comité Professionnel du Pétrole; Petroleum Economist, Aug. 1979*

world oil price increases may ultimately add somewhat to production by making economically attractive secondary and tertiary recovery techniques, this addition is unlikely to be more than about $3.5 \times 10^6$ t in the long run, and has so far failed to stop a decline in production in most countries[105]. Apart from Britain, the only recent oil discovery in the Community has been made in the Po Plain[106]. The field is unlikely to make more than a modest contribution to Italy's oil supplies. Denmark has established oil production from the Dan field in the North Sea, and traces of oil have been found off France. So far, however, Danish reserves are minor, and the prospects in French territorial waters uncertain; exploration also has been inhibited by a demarcation line dispute with the UK, finally resolved in 1977[107].

On present form, then, the only significant expansion of Europe's indigenous oil production will come from Britain. Several North Sea fields have already started production, or are under development (see Table 1.8), and exports have begun to flow to other Community members. North Sea oil is of high quality (light, and with few impurities); its low specific gravity means that refining will yield only about 20% of heavy fractions, compared with about 40% for other crudes[108]. Britain will, therefore, continue to import heavy crude oil, and export on a large scale crude oil or petroleum products, becoming a net exporter by 1980/81. Exports could well be as much as $2 \times 10^6$ barrels per day (b/d)[Note 109].

There are virtually no limits to the amount of British oil which the Community could absorb, since it would (provided prices are equivalent) simply replace

**Table 1.8** *Build-up of British North Sea Oil Production*

| Name of fields under production | Proven reserves ($10^6$ b) | Production ($10^3$ b/d) | | | | | | | | |
|---|---|---|---|---|---|---|---|---|---|---|
| | | 1977 | 1978 | 1979 | 1980 | 1981 | 1982 | 1983 | 1984 | 1988 |
| Argyll | 36–50 | 17 | 14 | 20 | 9 | – | – | – | – | – |
| Auk | 50 | 48 | 27 | 15 | 10 | 10 | 7 | 4 | n.a. | n.a. |
| Beryl | 400 | 61 | 54 | 70 | 80 | 80 | 75 | 70 | 65 | 65 |
| Brent | 2 000–2 200 | 26 | 77 | 220 | 350 | 475 | 475 | 560 | 520 | 465 |
| Claymore | 404 | 6 | 59 | 120 | 130 | 110 | 80 | 70 | 60 | 50 |
| Dunlin | 600 | – | 14 | 80 | 120 | 130 | 150 | 150 | 150 | 140 |
| Forties | 1 800 | 413 | 500 | 500 | 494 | 428 | 380 | 350 | 330 | 310 |
| Heather | 150 | – | 3 | 45 | 50 | 50 | 50 | 50 | 36 | 31 |
| Montrose | 110–155 | 17 | 25 | 36 | 40 | 40 | 35 | 30 | 30 | 25 |
| Ninian | 1 200 | – | 1 | 100 | 250 | 300 | 360 | 340 | 305 | 275 |
| Piper | 618 | 176 | 254 | 280 | 240 | 180 | 140 | 105 | 75 | 60 |
| Thistle | 450–500 | – | 54 | 125 | 160 | 200 | 175 | 150 | 135 | 105 |
| Buchan | 50 | – | – | 5 | 50 | 40 | 25 | 15 | n.a. | n.a. |
| South Cormorant | 110 | – | – | 20 | 50 | 50 | 60 | 42 | 30 | 20 |

| Name of fields under development | Proven reserves ($10^6$ b) | Production start-up | Peak production expected |
|---|---|---|---|
| Beatrice | 160 | 1981 | 80 |
| Beryl B | 300 | 1983 | 85 |
| Brae | 250 | 1983 | 100 |
| Cormorant North | 400 | 1982/3 | 180 |
| Fulmar | 500–525 | 1981 | 180 |
| Hutton[a] | 250 | 1983/4 | 80–100 |
| Hutton North West | 280 | 1982 | 100 |
| Magnus | 450 | 1983 | 120 |
| Maureen | 150 | 1981/2 | 75–80 |
| Murchison | 350–380 | 1980/1 | 130 |
| Stratfjord | 477 | 1979/80 | 60 |
| Tartan | 250–300 | 1980 | 65–90 |

| Name of other major fields | Proven reserves ($10^6$ b) | Peak production expected ($10^3$ b/d) |
|---|---|---|
| Andrew | 511 | 200 |
| Grand Alwyn | 511 | 100–150 |

a Development not yet approved by Department of Energy in late 1979
Sources: Petroleum Economist, April 1979; Financial Times, 16 Nov. 1979; Comité Professionnel du Pétrole

oil imports from elsewhere. The only question mark is British export policies – will London insist on a large share of refined products in its total exports?[110]

The most obvious factor determining production levels in the British North Sea in the longer term is recoverable reserves. At the end of 1976, proven reserves were officially given as $1405 \times 10^6$ t, with an estimated possible total of $3000–4500 \times 10^6$ t[Note 111]. Whether this will be the last word depends on three factors: new finds, appreciation of old fields and recovery rates. Great uncertainties surround all three, but the appreciation factor is central.

This factor relates the size of initially declared reserves to the final established figure. I.D. MacKay and G.A. Mackay[112] assume total recoverable amounts will be upgraded by not less than 25%; Odell's simulation model for North Sea oil assumes appreciation by 1.2 to 2.8, with a factor of 2 as the most probable result[113]. The oil industry accepts the appreciation factor but argues that it has been reduced by improved seismic techniques[114]. Nevertheless, substantial upgrading of reserve estimates has already occurred in several fields, and will probably occur again — given the extremely complex geological structure of the North Sea oil province, which appears to make initial assessments hazardous.

Discovery of new fields is another obvious factor. While a record 24 new discoveries were made in 1975, and a further 12 in 1976, none of these finds appears to be a major field of, say, more than $1 \times 10^9$ b (barrels). This seems to prove the consistent argument of the oil companies, that the major fields have already been established through improved exploration techniques. Odell, on the other hand, insists that new giant fields will be found.

The rate of extraction depends on geological and technical factors such as type and magnitude of expulsive forces, the permeability of the reservoir rock, the viscosity of the oil, and the nature of field exploitation. The normal recovery rate in the initial stage is between 30 and 40%[115] with the North Sea fields normally at the higher end of this spectrum since water injection will be used from the beginning. (Laboratory experiments on water injection have shown recovery rates of up to 69.9%[116].) The degree to which ultimately recoverable reserves will be increased by secondary and tertiary recovery techniques in the North Sea is uncertain; the importance of an increase of, say, 10 to 15% is, however, obvious.

Large parts of the British, and above all the Norwegian, sectors still have to be explored by drilling, and in addition exploration has now begun in the west Shetland basin and the Celtic Sea (which, however, is similar to the southern North Sea, and therefore more likely to contain natural gas than oil.) Both areas have so far proved disappointing[117]. Exploration drilling has undoubtedly moved towards less obvious targets, and the very high success rate of the past ten years (about 1 in 5)[118] will presumably decline.

The highest estimate of ultimately recoverable reserves is put forward by Odell at between 6 and $10 \times 10^9$ t[Note 119]. The 1974 OECD study gives a figure of $3562 \times 10^6$ t, with a further 50% as possible with tertiary recovery[120]. Oil company estimates range from 2850 to $5479 \times 10^6$ t, with I.D. MacKay and G.A. Mackay opting for a figure of $3836 \times 10^6$ t[Note 121]. Peak production rate estimates for the UK in 1985 fall between a conservative company estimate of 3.6 and Odell's figure of close to $9 \times 10^6$ b/d. This latter figure, and an optimistic company estimate of $7 \times 10^6$ b/d, both now appear unrealistic, and would have to be modified by political considerations, even if such production rates were technically feasible. The government would in that case almost certainly constrain oil output below theoretical capacity levels. But apart from this, Odell's computer simulations seem to be questionable since they use historical experience to assess the appreciation factors, somewhat modified, from other parts of the world. Against this, the companies argue that the North Sea province has very specific geological characteristics, that in any case offshore oil provinces differ from onshore production (in the former, original estimates are subsequently down-graded, and only ultimately up-graded again to close to the initial estimate), and that Odell's model neglects the vastly improved seismic techniques and know-how available, which change the appreciation factor[122].

The *economic* constraints on future North Sea production obviously relates to the price of world oil. The International Energy Agency's floor price of $7/b can already be considered purely hypothetical. The landed price of UK oil has not yet been disclosed, but given its quality and transport premiums, it should be considerably above the price of the world marker crude, Saudi Arabian light[123].

The geological and geographical circumstances of North Sea oil production push development and production costs to very high levels. Capital costs per barrel of daily installed capacity in established fields vary between £829 (Argyll) and £4437 (Dunlin). The average capital cost is estimated at around £2500, compared with £100–150 in Middle Eastern onshore production, and £150–250 in offshore production[124]. Moreover, North Sea costs have been rising rapidly (a consequence of inflation, delays and supply shortages).

Operating costs have been estimated at $2.10 (Argyll), $1.55 (Auk), $0.54 (Piper), and $0.93 (Forties)*. If one assumes a rate of return on capital outlays of 25% then net profits required by the companies (after royalties and taxes, but before operating costs) would have to average $4/b with a range from $2.91 (Piper) to $4.83 (Forties)[125].

Government revenues will come from three sources: a 12.5% royalty, a 45% petroleum revenue tax (PRT), and 52% corporation tax. The PRT, specially introduced to prevent windfall profits, can very flexibly differentiate between fields that are small and big, expensive and cheap: apart from a capital allowance of 175% with a maximum benefit in the early stages of production, there is a tax exemption clause for $7.3 \times 10^6$ b annually, up to a maximum of $73 \times 10^6$ b, and even a possibility of remitting PRT and royalties in certain circumstances[126]. In spite of these (from the point of view of the companies) comfortable taxation arrangements, some fields will remain unprofitable to develop, at least for the near future. Since conditions vary considerably, generalizations are difficult. But a field with total reserves of about $100 \times 10^6$ b was in 1978 at the bottom range of profitability, with this figure being dependent on the future relation between cost escalation in the North Sea and the evolution of international oil prices. Small finds might in the future be developed only if they are located

\* Exchange rate $2/£1.

close to existing pipelines, which would reduce development costs. Even a field as large as Hutton (reserve estimate 310 × 10$^6$ b) was originally of doubtful profitability[127]. In early 1978, however, it figured among the number of fields whose development was thought to start in the near future. Others in this category included Magnus, Beatrice, Fulmar, and North Cormorant. In addition, there were still some potentially economic finds already established but awaiting further confirmation[128]. From a purely economic viewpoint, a production level of about 3.0 × 10$^6$ b/d in 1985 appears feasible, although the British government estimates are at present still lower (about 2.5 × 10$^6$ b/d).

However, political considerations could well affect production rates, at least in the longer run. They can be considered under two headings — conservation policies by the UK government, and the secession of Scotland from the United Kingdom (or a form of devolution which gives Scotland full sovereignty over 'her' oil resources).

Both major parties in Parliament have advocated government control over depletion rates and the Petroleum and Submarine Pipeline Act has given government the necessary legal instrument. Whether the British government actually decides on depletion controls, and if so when, will depend on a variety of practical factors such as the capacity to absorb oil revenues, the size of production potential, and proven reserves. Such controls appear unlikely before 1985. The former Labour government accepted and supported the companies' drive for production rates at the maximum technically feasible level. It gave the companies assurances: that no cuts will be demanded before 1982; reductions will not exceed 20% of production capacity; no cuts will be decreed before 150% of the operators capital expenditure is recovered; and all eventual measures will give due consideration to commercial, economic and market aspects[129].

In principle, depletion measures would appear to be in the interest of a country only if international oil prices are expected to increase at a rate higher than income on capital investment. Strategic considerations (such as security of supplies and availability of oil in the longer run) and economic reasons such as the impact of large oil exports on the balance of payments, on the value of the pound, and on the competitiveness of industrial exports, might moderate such calculations. A government could also be inclined to try to minimize the side effects of high oil production in terms of economic and social distortions, and environmental damage[130].

The discovery of oil has undoubtedly provided a major stimulus to Scottish nationalism, and the Scottish Nationalist Party (SNP). If Scotland were to become independent, it would presumably inherit the bulk of the oil and gas deposits in the middle and northern North Sea basin. Since the direct impact of the oil industry (the creation of income and employment) is relatively small because of its high capital intensity, the high import share in oil installations and the relatively low share of costs in the selling price of oil, the benefits from oil operations are mostly indirect (i.e. government revenues). An independent

Scotland might by 1980 receive about 62% of her GNP in 1973 in the form of oil revenues[131]. Given her small population (about five million) she would rapidly accumulate huge foreign exchange reserves, and eventually have to consider reductions in output (after possibilities of increased domestic spending and/or reduction of taxes and investment abroad have been exhausted: I.D. MacKay and G.A. Mackay consider a production of more than $2 \times 10^6$ b/d for an independent Scotland as unlikely[132]. In addition, the uncertainties and legal complications accompanying a secession of Scotland might well imply delays in production expansion, since contracts would have to be renegotiated, and the legal framework for company operations re-established. During this period of uncertainty, oil industry activities could well slow down considerably.

That Scotland will secede now seems unlikely. In any case, Britain (and the Netherlands with her natural gas resources) is as unlikely as an independent Scotland to go for all-out production of natural gas and oil if the benefits do not far outweigh the disadvantages. Obviously, the absorption capacity of the whole UK is much greater than that of Scotland alone, but the risks remain real: of exacerbated regional problems, of economic distortions, of environmental damage, and of running out of a precious resource. On the other hand, neither the UK nor an independent Scotland could isolate itself from the European Community to which they are tied by multiple economic, social and political links. Some of the pressure from outside to increase production will, therefore, be effective, and there will be room for bargaining and *quid-pro quos*. The European Community could, for instance, make available capital for exploration and production, and the Council has already suggested rules for such investment aid[133]. Some aspects of a common European energy policy envisaged by Odell might, therefore come true — how much, will clearly depend not only on the mutuality of benefits but also on the emerging degree of common political will.

## *Conclusion*

The overall conclusion of this detailed survey of Europe's energy future seems clear: import dependence will continue on a high level, largely determined in its extent by the evolution of energy demand in Europe. While the build-up of North Sea oil and gas production will for some years help to reduce energy import dependence *in relative terms*, all indigenous energy sources seem to be faced with ceilings and constraints. To a large extent, these constraints are political — reflecting the rapid politicization of energy issues since 1973. In the energy-rich countries, the ceiling on oil and gas production is set by the concern for the long-term strategic energy position of those countries. The governments of the UK and the Netherlands seem inclined to take out an insurance policy: resources of hydrocarbons are to be conserved for future generations. This conservation policy is reinforced by worries about the economic, environmental and social impact of a very fast expansion of the hydrocarbon export sector. Obviously, such a policy involves a trade-off between present and future benefits;

but since the trade-off is relatively painless, given the benefits of oil and gas production even on a level lower than theoretically possible, it appears likely that conservation policies on UK and Dutch hydrocarbon production are here to stay.

Another type of constraint is environmental. This primarily concerns nuclear energy, but also involves coal. A likely development of the political bargaining between environmental pressure groups and governments is a cautious development of nuclear energy (and coal) under more stringent environmental and safety regulations — and at a higher price. It is as yet uncertain how this will affect the competitiveness of nuclear energy, which in any case rests on incomplete data. As in the case of coal, the real justification of nuclear energy appears to lie in the security of supplies it can provide, in the savings on foreign exchange for oil imports, and in the insurance against future structural shortages and/or dramatic price increases in the international hydrocarbon trade. This trade-off is politically more difficult to manage than the previously cited one in energy-rich countries; and it is compounded by worries about the social risks of a proliferation of nuclear reactors. Besides, nuclear energy expansion has — barring any technological break-throughs such as the super-battery — a natural ceiling: demand for base-load electricity, or roughly 50% of total electricity requirements.

Thus, the increasing contribution of indigenous oil and gas production in the Communities, as well as of nuclear energy development, will sooner or later reach its limits — sooner in the case of oil and gas, later maybe for nuclear electricity. Coal also does not offer any prospect of a substantial expansion of production. Thus, only for the next five years or so will the import gap not increase. Beyond that point, major technological advances and/or new massive hydrocarbon finds in the Communities will be necessary to foreclose an opening of the gap. Otherwise, increasing energy demand and levelling-off of indigenous oil and gas production could exacerbate problems of energy import dependence at precisely the time when many experts predict a real danger of structural oil shortages in the world.

CHAPTER TWO

# Europe's prospects as an energy importer

## Outline

Even the European Commission's very ambitious targets for Community energy production would still leave a substantial gap between supply and demand. Moreover, the failure, so far, to resolve the problems discussed in the previous chapter has already rendered these targets obsolete. Only the overestimation of energy demand, as a consequence of lower growth rates than predicted, offers some hope of reducing overall Community energy import dependence to about 50%, excluding imports associated with nuclear energy. But even this figure must be qualified as an optimistic assumption. Accordingly, Europe's import prospects for each of the four major fuels are a major consideration. This chapter first looks briefly, and then in more detail, at the import situation for each of the energy sources.

## *Solid fuels*

For solid fuels the general prospects for a substantial expansion of imports, and therefore a reduction of dependence on those of oil, appear unexciting, at least until 1985. In the longer run, coal might well experience a second spring, given its worldwide abundance. But any substantial swing back towards coal presupposes the advent of coal liquefaction and gasification on a massive scale, and thus cannot be expected before the end of the next decade. Until then, several interrelated constraints on the demand side will limit consumption increases to fairly modest amounts. Nevertheless, imports will have to be stepped up somewhat, since the best that can be hoped for is a stabilization of indigenous production in the Community. Additional supplies would mainly come from traditional exporters to the Community — North America, Australia, South Africa, Poland and the Soviet Union (see *Table 2.1*).

These countries all possess the potential to expand coal production, although the difficulties (such as the need for additional transport facilities required or environmental implications) might be greater than foreseen. The problem lies in the evolution of surplus production after allowing for increases in domestic consumption in these countries. Major uncertainties surround the future export capacity of the United States. Given the strong incentives, and massive scope, for more intensive domestic use of solid fuel, as well as the difficulties which might

**Table 2.1** *EEC Solid Fuel Imports, 1976* ($10^3$ metric tons)

| Importer | Imports total | of which: other EEC | South Africa | North America | Socialist bloc, total | of which: CSSR | Poland | USSR | Australia |
|---|---|---|---|---|---|---|---|---|---|
| Bel/Lux. | 7 830 | 4 370 | 400 | 2 020 | 750 | 130 | 270 | 350 | 250 |
| Denmark | 4 200 | 10 | 10 | 310 | 3 850 | 10 | 3 190 | 650 | — |
| France | 18 830 | 5 180 | 2 100 | 3 280 | 7 340 | — | 5 740 | 1 550 | 910 |
| Germany | 6 490 | 1 290 | 670 | 1 990 | 2 490 | 130 | 2 140 | 220 | 50 |
| Ireland | 600 | 110 | — | — | 490 | — | 490 | — | — |
| Italy | 12 130 | 2 420 | 430 | 3 820 | 4 550 | — | 3 290 | 1 260 | 890 |
| Netherlands | 4 910 | 1 090 | 20 | 2 420 | 920 | 190 | 660 | 70 | 460 |
| UK | 3 470 | 190 | 10 | 780 | 130 | — | 130 | — | 2 360 |
| Total EEC | 58 460 | 14 660 | 3 640 | 14 620 | 20 520 | 460 | 15 960 | 4 100 | 4 920 |
| % of solid fuel consumption | (18.8) | (4.7) | (1.1) | (4.7) | (6.6) | (0.1) | (5.1) | (1.3) | (1.6) |
| % of energy consumption | (4.9) | (1.2) | (0.3) | (1.2) | (1.7) | (0.03) | (1.3) | (0.3) | (0.4) |

*Source: UN Statistical Papers Series (Energy)*

well be encountered in attempts to step up production, US coal exports could actually peak out during the 1980s or 1990s at a level of 70—100 $\times$ 10$^6$ t compared to 60 $\times$ 10$^6$ t in 1975.

Attempts to utilize coal more fully will no doubt also have an impact in other traditional supplier countries, but some additional exports should become available from all of them: close to 60 $\times$ 10$^6$ t from Australia; up to 10 $\times$ 10$^6$ t in total for Canada; 20—40 $\times$ 10$^6$ t for South Africa; and 5—10 $\times$ 10$^6$ t altogether from Poland and the Soviet Union, with an export potential to the West for the two latter countries combined of about 40 $\times$ 10$^6$ t. All these countries are favoured by the geological structure of their fields: in some, open-cast mining plays a very significant role. However, supplies from South Africa might be affected by political disturbances.

Given the present and likely future structure of the international coal market, and the consuming countries low dependence on imports, this market can be expected to remain purely commercial. Politics will not enter the international coal trade in any significant way, and although there might be some attempt to reach a homogenous price structure among supplier countries, a producer cartel seems highly unlikely.

## *Natural gas*

As for natural gas, all the signs point to continued growth. There will certainly be a further rapid increase of the importance of this fuel within the Community's energy balance, and although, as we saw, a small increase of indigenous production can be expected over the next few years, demand is likely to increase even faster. As a consequence, Europe will play a major role in the nascent but rapidly growing world natural gas market. Imports to the deficit countries in the Community, which so far have almost exclusively come from the Netherlands, will now have to be bought from outside the Community, and even from outside Western Europe.

Within Western Europe, the major new supplier will be Norway, with a substantial contribution of about 30—40 $\times$ 10$^6$ toe of natural gas by 1985. However, fears of the effect on the economy of over-dependence on oil and gas production and exports, of inflationary pressures, and of harmful effects on the traditional way of life because of regional migration and environmental damage have led Norway to set a ceiling of natural gas production at 40 $\times$ 10$^6$ toe.

Outside Western Europe, the only non-OPEC source of supplies would appear to be the Soviet Union. Major contracts for pipeline exports of natural gas to the Community have been concluded, and deliveries started several years ago. The Soviet contribution to the Community's natural gas balance in 1985 will be substantial (see *Table 2.5*). Part of it, however, will come from a triangular deal involving the export of Iranian gas to the USSR in exchange for similar amounts of Soviet gas to Western Europe.

If we count these supplies as Iranian exports to Europe, OPEC would become the Community's major natural gas supplier in 1985, if present contracts firmed up or under negotiation materialize. Imports from Iran and Algeria could reach a maximum of 50 000–55 000 m$^3$ in that year. This is a high estimate, however, given that these two countries have introduced limitations on natural gas exports and massive investment required for pipelines or liquefied natural gas (LNG) carriers. These high costs make natural gas exports less attractive than oil experts, even if prices are similar in terms of heat equivalent. Iranian natural gas exports will also be affected by the new oil policy of the revolutionary regime.

This level of import dependence alone might not create unacceptable or even major political risks since the triangular character of the Iranian–Soviet–European import deal actually provides a significant element of additional security. Iranian interference with supplies would hit some areas in the Soviet Union directly, and therefore cause economic hardship or even severe damage. Similarly, Soviet interference with natural gas exports would deprive Iran of revenues, and presumably bring her closer to the West (unless the Soviet Union reimbursed Iran for any shortfall in revenues). In any case, import dependence on Iranian gas now looks likely to be much lower than assumed before the revolution.

The real security implication in this expansion of natural gas exports to the Community is a cumulative dependence on the same suppliers of natural gas *and* oil. While Algeria is unlikely to resort to natural gas cutbacks within the framework of OPEC action, this would appear less certain for Iran. The implications of this cumulative dependence will be considered later.

## *Nuclear energy*

Nuclear energy is regarded as the alternative energy source to enable industrialized countries to become independent of the OPEC countries – it also means that the Community will be considerably dependent on imports. The complex nuclear fuel cycle (see *Figure 2.1*) can involve international trade of several types: uranium ore, uranium hexafluoride (a gaseous substance), enriched uranium, and nuclear waste, including plutonium, as well as trade in nuclear installations (reactors, reprocessing and enrichment plants, etc.). Two aspects involve security of supplies: enriched uranium and uranium ore.

At present, the Community does not possess a significant enrichment capacity, and therefore the production of nuclear energy depends on the import of enriched uranium. The United States has been the Community's major supplier, but the reliability of this source has been undermined by the repeated stoppages of enriched uranium supplies since 1975. The Nine have, therefore, made an effort to diversify their sources of supply, in the short run by importing enriched uranium from the Soviet Union, which accounted for about 50% of demand in 1977. After 1980, the development of indigenous enrichment facilities will rapidly reduce European import dependence.

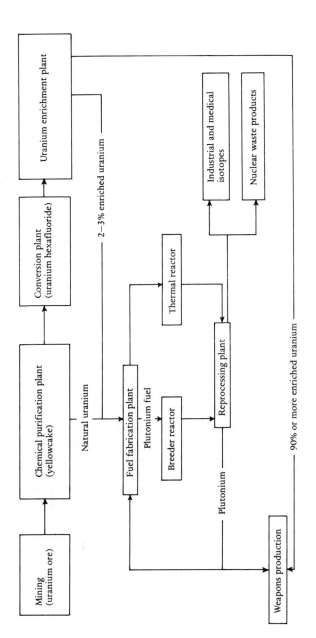

**Figure 2.1** *The nuclear fuel cycle*

The second aspect of import dependence is uranium ore requirements. Even now, the ore required for enrichment in the United States for European reactors has to be imported, since the United States has kept a very tight control on uranium exports. The indigenous uranium ore reserves in the Community (mainly in France) cannot sustain production sufficient to meet demand. About four-fifths of the requirements have to be imported, mainly from Canada, South Africa, Niger and Gabon. Australia should soon become another important supplier. The industrial countries in this group have all shown considerable interest in constructing their own enrichment industries, and can be expected to enter this stage of processing in the 1980s. This would mean some shift in trade patterns, and since the Community will by then have a substantial enrichment capacity of its own, it will need uranium ore supplies. The traditional suppliers might be reluctant to export them, because enrichment adds considerably to the value of the fuel. In future, therefore, uranium ore supplies will increasingly have to come from less developed countries.

The outlook for the international uranium market remains very uncertain, given the unpredictability of long-term demand, and of the restraints on investment by the multinational mining corporations. Presently known, easily accessible, uranium reserves could not meet a rapidly increasing demand much beyond 1990. Exploration is still in a very early stage, so this does not mean a physical shortage of uranium, but it could mean production bottlenecks, owing to insufficient investments. Uranium prices have already risen dramatically, though it is difficult to assess whether this is due purely to market forces, or is the result of collusion within the industry. A producer cartel of suppliers similar to OPEC in its relationship with the oil companies seems probable.

The German deal with Brazil — involving a complete nuclear fuel cycle — reflects such a concern over future uranium ore supplies: one aspect of the contract is a guarantee of future Brazilian deliveries of uranium to West Germany. This poses political problems for the Community which, in the long run, are more important than present import dependence on enriched uranium supplies. This dependence will be short-lived, and less dramatic than dependence on oil imports: stockpiling of nuclear fuel is easy, and reactors have to be refuelled only at fairly long intervals. Enriched uranium supplies, however, have already become enmeshed with the thorny issue of nuclear weapons proliferation, which has produced alliances and conflicts across traditional lines of alignment between East and West. It is not inconceivable that the two main suppliers of enriched uranium to the Community, the USA and the USSR, may jointly apply pressure to modify the nuclear export policies of Germany, although this possibility appears somewhat remote. Any unilateral attempt by the Soviet Union to use her supplier's position politically would take place within the framework of East–West interaction, where military and economic security overlap: such action would therefore be part of a general increase in the level of political and military tension.

Finally, the Community is a nuclear exporter as well as an importer. At present, this role is confined to exports of nuclear installations, but it might in the future also include enriched and reprocessed uranium (some reprocessing for foreign customers is already done by British and French reprocessing plants). To the extent that such exports go to energy exporters to the Community (i.e. OPEC countries) they create a measure of interdependence, and also add indirectly to the security of supplies through their export earning effect. These contributions to energy security are marginal; they create, however, dangers of nuclear proliferation.

This danger of nuclear proliferation, arising from the close links between peaceful nuclear energy and its military potential, has become a major area of conflict within the Western alliance, and in particular between Germany and France on the one hand, and the United States and Canada on the other. Divergencies of opinion have repeatedly interrupted the flow of enriched uranium ore from the two North American countries to the Community. The USA stopped supplies of enriched uranium within the framework of a revision of her nuclear export policy, and Canada did the same with regard to her uranium ore exports to Europe. These events have demonstrated the external vulnerability of the Community as regards nuclear energy. They do not totally invalidate, but do weaken, the argument in favour of nuclear energy on security grounds; at the same time, it strengthened European insistence on the reprocessing/fast breeder fuel cycle as a means to reduce nuclear import dependence.

The politically crucial dependence at present remains on *oil supplies* from OPEC countries. Oil imports will continue to balance energy demand and supplies from other sources — both indigenous and external. Is there any alternative to OPEC oil imports until 1985? Two candidates might be considered: Norway and the Soviet Union. In both cases, however, the possibilities are limited. Norway's contribution will be useful but restricted by the conservationist approach chosen. The production ceiling set by the Norwegian government is at present $50 \times 10^6$ t/a, with no indications that it will be substantially changed. On present form, this level will not in any case be reached during the 1980s.

The Soviet Union will presumably continue roughly its present exports to the Community, but large increases are unlikely. There are some doubts whether the Soviet oil industry can keep up the rapid expansion of the last 30 years. It faces substantial problems of adjustment towards Siberian-based activity, and has to meet increasing domestic demand, and also fast growing demand from the Comecon countries. They have been encouraged to import Middle Eastern and North African oil, but the Soviet Union will presumably want to remain the main supplier for political as much as economic reasons. Inevitably, then, Community dependence on OPEC will continue up to 1985, and well beyond. The implications of this dependence will be analysed more closely in the following parts of this book.

## Survey

### Solid fuels: limited scope for import expansion

At present, about 89% of total solid fuel demand in the Communities is covered by indigenous production; coal imports thus account for a mere 3.5% of total gross energy consumption in the Nine. In 1977, the Communities' biggest foreign supplier of coal was Poland (with $14.8 \times 10^6$ t — a decline of about 8% over 1976), followed by the USA ($10.9 \times 10^6$ t, minus 24% over 1976 levels), South Africa ($7.7 \times 10^6$ t, plus 122%), Australia ($6 \times 10^6$ t, plus 33%), the Soviet Union ($3.8 \times 10^6$ t, minus 6%) and Canada ($0.9 \times 10^6$ t, plus 35%)[134]. External trade of solid fuels is significantly higher than intra-Community trade of coal, which has been declining to such low levels that the EC Commission has now suggested steps to discourage imports of boiler coal from outside the Communities. This would not affect coking coal imports, of which the Nine have a substantial deficit (coking coal is used in iron and steel production)[135].

World coal reserves are abundant: it has been estimated that about 90% of global fossil fuel reserves are made up from solid fuels[136]. As *Table 2.2* indicates, these reserves are fairly widely spread, and Europe's share (represented mainly by Britain and Germany) is sizeable. However, the previous chapter has already shown that at least for some time to come this favourable resource-endowment is deceptive — for a variety of reasons, production cannot be expected to expand dramatically, if at all. Will imports, then, fill the possible gap between supply and demand of solid fuels — assuming such a gap will materialize?

Some scepticism as to the size of this gap has already been registered: the reasons for the scepticism are to be found in the constraints on coal demand. On

**Table 2.2** *Coal Resources and Reserves, Broken Down According to Continents and Countries*

| Continent | Country | Geological resources ($40^6$ tce) | | Technically and economically recoverable reserves ($10^6$ tce) | |
|---|---|---|---|---|---|
| | | hard coal | brown coal | hard coal | brown coal |
| Africa | Mozambique | 400 | — | 80 | — |
| | Nigeria | — | 180 | — | 90 |
| | Republic of Botswana | 100 000 | — | 3 500 | — |
| | Republic of South Africa | 57 566 | — | 26 903 | — |
| | Rhodesia | 7 130 | — | 755 | — |
| | Swaziland | 5 000 | — | 1 820 | — |
| | Zambia | 228 | — | 5 | — |
| | Other countries | 2 390 | 10 | 970 | — |
| Total | | 172 714 | 190 | 34 033 | 90 |

Table 2.2 *continued*

| | | | | | |
|---|---|---|---|---|---|
| America | Argentina | – | 384 | – | 290 |
| | Brazil | 4 040 | 6 042 | 2 510 | 5 558 |
| | Canada | 96 225 | 19 127 | 8 708 | 673 |
| | Chile | 2 438 | 2 147 | 36 | 126 |
| | Columbia | 7 633 | 685 | 397 | 46 |
| | Mexico | 5 448 | – | 375 | – |
| | Peru | 1 072 | 50 | 105 | – |
| | USA | 1 190 000 | 1 380 398 | 113 230 | 64 358 |
| | Venezuela | 1 630 | – | 978 | – |
| | Other countries | 55 | 5 | – | – |
| Total | | 1 308 541 | 1 408 838 | 126 839 | 71 081 |
| Asia | Bangladesh | 1 649 | – | 517 | 2 |
| | China (PR) | 1 424 680 | 13 365 | 98 883 | n.a. |
| | India | 55 575 | 1 224 | 33 345 | 355 |
| | Indonesia | 573 | 3 150 | 80 | 1 350 |
| | Iran | 385 | – | 193 | – |
| | Japan | 8 583 | 58 | 1 000 | 6 |
| | North Korea | 2000 | – | 300 | 180 |
| | South Korea | 921 | – | 386 | – |
| | Turkey | 1 291 | 1 977 | 134 | 659 |
| | USSR | 3 993 000 | 867 000 | 82 900 | 27 000 |
| | Other countries | 5 368 | 353 | 1 488 | 74 |
| Total | | 5 494 025 | 887 127 | 219 226 | 29 626 |
| Australasia and the Pacific | Australia | 213 760 | 38 374 | 18 128 | 9 225 |
| | New Zealand | 130 | 660 | 36 | 108 |
| | Other countries | – | – | – | – |
| Total | | 213 890 | 49 034 | 18 164 | 9 333 |
| Europe | Belgium | 253 | – | 127 | – |
| | Bulgaria | 34 | 2 599 | 24 | 2 179 |
| | Czechoslovakia | 11 573 | 5 914 | 2 493 | 2 322 |
| | Federal Republic of Germany | 230 300 | 16 500 | 23 919 | 10 500 |
| | France | 2 325 | 42 | 427 | 11 |
| | German Democratic Republic | 200 | 9 200 | 100 | 7 560 |
| | Greece | – | 895 | – | 400 |
| | Hungary | 714 | 2 839 | 225 | 725 |
| | Netherlands | 2 900 | – | 1 430 | – |
| | Poland | 121 000 | 4 500 | 20 000 | 1 000 |
| | Romania | 590 | 1 287 | 50 | 363 |
| | Spain | 1 786 | 512 | 322 | 215 |
| | UK | 163 576 | – | 45 000 | – |
| | Yugoslavia | 104 | 10 823 | 35 | 8 430 |
| | Other countries | 309 | 130 | 58 | 57 |
| Total | | 535 664 | 55 241 | 94 210 | 33 762 |
| World total | | 7 724 834 | 2 400 430 | 492 472 | 143 892 |
| | | | 10 125 264 | | 636 364 |

*Source:* World Energy Conference

the other hand, production is also facing serious difficulties, and shortfalls in the expected contribution of nuclear energy to electricity generation could also increase the indigenous supply—demand gap for solid fuels, since coal would be the most obvious replacement for nuclear energy — if not for economic, then for political reasons: a greater dependence on oil imports ought to be avoided as much as possible. Thus, additional coal import requirements in 1985 could be between 10 and $35 \times 10^6$ t[Note 137].

Will additional supplies of solid fuels be available? All the main suppliers of the European Community have a significant potential to increase production — most importantly the United States, but also Poland, Australia, Canada, South Africa and the Soviet Union. But whether this potential will be realized, and — if so — to what extent the resulting increase will go into exports, is another matter.

From this point of view, US coal exports are the most uncertain. True, President Carter's energy programme puts heavy emphasis on coal, production of which is to increase to $1 \times 10^9$ t by 1985. But some of the elements concerning solid fuels have already run into trouble in Congress; besides, the industry is facing severe problems in terms of labour relations, falling productivity, and slack demand[138]. The present level of production (about $665 \times 10^6$ t) indicates that the 1985 target will be impossible to reach under today's policies and trends. The achievement of an output of $1 \times 10^9$ t would, in fact, mean the opening of 270 new mines, the employment of 125 000 additional miners, and an investment of nearly $\$25 \times 10^9$ [Note 139]. Tough environmental standards of production and burning of coal will reduce coal's competitiveness, and the interplay of supply-and-demand constraints could keep production levels much below target. On the other hand, new coal technologies might in the longer run remove some constraints on the supply side, and thus lead to substantial increases in solid fuel consumption. This could mean that coal available for export could decline.

The picture is thus anything but clear, but it appears wise to assume only a moderate increase in levels of solid fuel exports from the USA in the 1980s — say, to about $70 \times 10^6$ t/a from the present $60 \times 10^6$ t/a. OECD estimates range from 78 to $109 \times 10^6$ t under various assumptions[140]. Some, but not all of the increase would become available for Europe.

Australia, South Africa and Canada are also favourably placed to increase output from large, low-cost, easily workable reserves; in these countries, coal gasification is less likely to affect long-term export availabilities. The greatest uncertainties concern South Africa, where wages of black miners have been kept artificially low, and the inherent instability of *apartheid* could lead to political violence affecting coal supplies. Altogether, export availabilities from Canada have been estimated at about $6 \times 10^6$ t in 1985 (as compared to $11.7 \times 10^6$ t in 1975), $57 \times 10^6$ t for Australia (in 1975 the level was $32.4 \times 10^6$ t), and $38 \times 10^6$ t for South Africa (in 1975 it was $3 \times 10^6$ t)[141].

As for the communist countries presently exporting coal to the Nine, Poland is thought capable of increasing its present exports of about $24 \times 10^6$ t: a large new coal field is at present being developed near Lublin in East Poland, and should reach a production level of $25 \times 10^6$ t by the turn of the century. A small increase of exports should therefore be possible by 1985, in spite of further domestic increases in demand for solid fuels[142]. The Soviet Union has achieved a steady, if undramatic annual increase in solid fuel output, and large new deposits are under development to make up for the eventual decline of traditional fields. Among the new developments is a brown coal field for open-cast mining, which will eventually produce no less than $700 \times 10^6$ t/a. This shift of production has already been quite successful, with 15 new mines being opened between 1968 and 1975. Recently, however, the Russian coal industry has run into serious difficulties in meeting targets. Over half of Soviet exports presently go to Eastern Europe, and this amount is likely to increase further. The share of solid fuels in Soviet energy consumption will probably decline further, and if this is so, exports to the West should increase slightly[143].

Obviously, prices will play a crucial role in the future development of the international coal market, which at present is quite limited ($194 \times 10^6$ t in 1975, up by 2.3% from 1974). Coking coal can be expected to be in heavy demand, although this depends on the evolution of the international steel industry, and on technological changes affecting coking coal input per unit of steel production. Since coking coal cannot be substituted, it will probably command a price premium. Steam coal, on the other hand, will have to compete with heavy fuel oil and nuclear energy, and price developments will therefore be crucial[144].

Of the factors influencing coal prices, labour costs still account for a major share: one-third of total costs, including depreciation, in Canada and Australia, 48% in US surface mining, and between 51 and 62% in US and European underground mines[145]. Once labour problems can be overcome, the trend of falling productivity in the USA might be reversed, but environmental costs will increase substantially. Finally, consumption is in some cases already tied very closely to oil or natural gas, and their advantages combined with existing capital investments in final consumption and competitive fuel oil prices seem to indicate that increases of coal demand in industry would require improbably large price advantages for coal — the assumptions of Carter's energy programme about the switch of industry from oil to coal are considered too optimistic. The fact that the oil companies can afford to make their profits on the non-substitutable light petroleum products, and so price the inevitable by-product of heavy fuel oils very competitively, could be an important factor here.

On balance, then, no dramatic changes can be expected over the next decade in the patterns and volumes of solid fuel imports to the Communities. Whether the world oil market becomes tighter or not, depends largely on the investments in the main exporting countries — given the timelags of 7—12 years for deep mines, and of 2—4 years for surface mines[146], investments might well be insufficient to cope with increased demand for coal from outside Europe, i.e. Japan and the LDCs. Even so, the international coal market will probably remain purely commercial, characterized by fairly long-term bilateral agreements with a relatively decentralized price-setting mechanism. In the longer run, the structure of the international market is likely to change in geographical terms, with new suppliers such as India and Indonesia playing a more important role[147].

One fundamental problem for European energy policies will be to address the large difference between world market prices for coal and indigenous production: in early 1978, the world market price was $30–35/t, while production costs in Europe ranged between $44 (UK) and $100 (Belgium), with the important German production at a high $72/t[Note 148]. This could undermine the position of European producers, but it seems likely that because of costs, national and Community subsidy policies will prevent a shift from internal to external sources of coal supply. In the longer run, the cost advantage of imports could, however, lead to a significant shift in coal supplies towards external sources, if the world coal markets expand, and if gasification of coal offers large new markets.

## Natural gas: growth story

Western Europe possesses, as pointed out earlier, relatively large reserves of natural gas — concentrated primarily in Britain, Norway and the Netherlands (see *Table 2.3*). Production is expanding rapidly, limitations being set more by

**Table 2.3** *World Reserves of Natural Gas 1978* ($10^9$ m$^3$)

| Europe | | Africa | |
|---|---|---|---|
| Germany | 206 | Algeria | 3 538 |
| Denmark | 48 | Libya | 728 |
| France | 136 | Nigeria | 1 217 |
| Ireland | 28 | Tunisia | 181 |
| Italy | 235 | | |
| Netherlands | 1 698 | Total | 5 781 |
| UK | 821 | | |
| Norway | 566 | *North America* | |
| Spain | 6 | USA | 5 943 |
| Greece | 113 | Canada | 1 642 |
| Total | 3 911 | Total | 7 585 |
| *Middle East* | | *Latin America* | |
| Abu Dhabi | 566 | Venezuela | 1 160 |
| Saudi Arabia | 2 476 | Mexico | 849 |
| Bahrain | 26 | Trinidad | 241 |
| Dubai | 43 | Argentina | 230 |
| Egypt | 91 | | |
| Iran | 14 150 | Total | 3 072 |
| Iraq | 862 | | |
| Kuwait | 891 | *Socialist Countries* | |
| Oman | 57 | USSR | 26 036 |
| Qatar | 1 132 | | |
| Syria | 88 | Total | 27 027 |
| Total | 20 457 | World total | 71 306 |

*Source: Oil and Gas Journal*

political than technical and economic constraints: problems of infrastructure installment and market penetration are nowhere as complex as in the case of solid fuels. The major source of supplies for the Communities is presently the Netherlands (see *Table 2.4*). The Dutch government, however, has not been willing to accept new export contracts; exports will be phased out during the 1980s, and will stop altogether at the end of the next decade.

**Table 2.4** *European Community Imports of Natural Gas, 1976* ($10^6$ m$^3$)

| Importer | World | Algeria | Libya | USSR | Netherlands |
|---|---|---|---|---|---|
| BeLux | 13 175 | – | – | – | 13 175 |
| France | 14 850 | 2 770 | – | – | 12 080 |
| Germany | 27 324 | – | – | 2 020 | 25 304 |
| Italy | 11 880 | – | 2 700 | 4 910 | 4 270 |
| UK | 1 055 | 1 055 | – | – | – |
| Total | 68 284 | 3 825 | 2 700 | 6 930 | 54 829 |

*Source: Comité Professionnel du Pétrole*

Another important source of natural gas imports in the future will be Norway, with the two large natural gas fields Frigg and Ekofisk. Frigg will eventually produce about 15 × 10$^9$ m$^3$ for export to the UK, and Ekofisk will supply about 19 × 10$^9$ m$^3$ by pipeline to Emden in West Germany, from where some of the gas will be passed on to other European Community countries[149]. Again, however, government policy seems likely to impose a ceiling on natural gas exports: the conservation-minded Norwegians have set a production maximum for natural gas of 40 × 10$^6$ toe, and so far, there are no signs that this ceiling will change. Export potential will depend on overall production of hydrocarbons, since it is clearly in the interests of Norway to utilize as fully as possible natural gas production associated with oil, and possibilities to do so domestically are limited. Even given present circumstances, however, Norway will make a major contribution to natural gas supplies of the Nine – but this contribution alone would only allow a modest increase in natural gas consumption.

Consequently, the Communities will have to look elsewhere for additional imports, if a further increase in supply is to be realized – and indeed, the Nine have already done so. Details of the Communities' natural gas imports, and of world reserves, are given in *Tables 2.3–2.6*. These indicate that the major sources of imports from outside Western Europe will be the Middle East and North Africa on the one hand, and the Soviet Union on the other. Overall, imports are likely to increase by a factor of 20 between 1973 and 1985; import independence could reach about 35% of natural gas consumption, and about 6% of energy consumption.

Natural gas is transported over long distances either by way of pipeline, or by

**Table 2.5** *Possible Natural Gas Imports to the Nine, 1985* ($10^9$ m$^3$)

| Importer | Algeria | Libya | Iran | USSR | Norway | (Netherlands) |
|---|---|---|---|---|---|---|
| BeLux | 5 | – | – | 4 | – | 11 |
| France | 9.15 | – | 3.66 | 4 | – | 11 |
| Germany | 8 | – | 5.5 | 8.5 | 18 | 30 |
| Netherlands | 4 | – | – | – | – | – |
| Italy | 12 | 2.35 | – | 10 | – | 6 |
| UK | 1 | – | – | – | 15 | – |
| Total | 39.15 | 2.35 | 9.16 | 26.5 | 33 | 58 |

*Source:* Table 2.6

special carriers (the gas can be liquefied under pressure, with a 600-fold contraction of volume, at a temperature of −161°C). Such long-distance transport poses a variety of problems, particularly with LNG carriers[150]. Because of the poor heat value/density ratio of gas, even pipeline transport of gas costs about twice as much as in the case of oil, and for liquefied natural gas (LNG) tanker transport is about six times as expensive as an equivalent amount of oil[151]. Liquefied natural gas also requires heavy investment in liquefaction and re-gasification equipment, and poses a variety of technical problems and safety hazards. Nevertheless, this method of transport has been established as a sound economic proposition since 1964, when the first LNG delivery from Algeria reached the UK. LNG systems are basically suitable for relatively small quantities; since the costs of liquefaction plants, storage facilities, and carriers are independent of distance, transportation in that case works out cheaper than long-distance pipelines, where costs are practically proportional to distance. Pipelines, however, provide rapidly decreasing costs with increasing quantities over a given distance, while the decrease is very slow with LNG after a certain minimum volume is reached, since new plants have to be added[152].

Both forms of long-distance transport already play a role in supplying the Communities' natural gas needs, and will continue to do so. In spite of further economies of scale which appear possible in LNG trade (the size of LNG carriers, for instance, has been growing rapidly), pipelines will presumably in the long run increase their importance as quantities of imports go up. They are used already, or planned, for imports from Norway, the Soviet Union, Iran and Algeria, while LNG arrives mostly from North Africa. Here, however, major opportunities still exist for pipelines — not only the recently agreed pipeline from Algeria via Tunisia to Italy, but also from Algeria to Spain, where a major contract is still under negotiation[153].

The growing dependence on natural gas imports will imply new technical and political risks. LNG is a highly explosive liquid, and major accidents could have catastrophic implications, and interrupt supplies. Technical problems with liquefaction plants could also cause difficulties, and indeed have already done so. The nature of LNG trade requires, however, storage facilities, since there is

Table 2.6 Details of Natural Gas Import Projects to the Communities (as of late 1977) ($10^9$ m$^3$)

| | Companies involved | Quantities supplied ($10^9$ m$^3$) | | | Initial date of delivery |
|---|---|---|---|---|---|
| | | 1975 | 1980 | 1985 | |
| A. Existing schemes / firm contracts | | | | | |
| (1) Algeria–UK (LNG) | CAMEL / British Gas Corporation | 1 | 1 | 1 | 1964 |
| (2) Algeria–France (LNG)[a] | SONATRACH / Gaz de France | 4 | 4 | 9.15 | 1965, 1972, 1980/1 |
| (3) Algeria–Belgium (LNG) | SONATRACH / Distrigaz | – | 5 | 5 | 1980 |
| (4) Algeria–Germany / Netherlands (LNG) | SONATRACH / Ruhrgas /Gasunie | | | 4 | 1984 |
| (5) Algeria–Italy | SONATRACH / ENI | | | 12 | 1983 |
| (6) Libya–Italy (LNG) | Esso Libya / ENI | 2.4 | 2.4 | 2.4 | 1972 |
| (7) Norway–Germany / Belgium / Netherlands / France[b] | Phillips / Ruhrgas, Gasunie, Distrigaz, Gaz de France | | 18 | 18 | 1977 |
| (8) Iran–Germany / France[b] | NIOC / Ruhrgas, Gaz de France | – | – | 5.5 and 3.66 | 1981 |
| (9) USSR–Germany | Ruhrgas (three contracts) | 2.2 | 8.5 | 8.5 | 1973, 1978 |
| (10) USSR–France | Gaz de France | – | 4 | 4 | 1976 |
| (11) USSR–Italy | ENI | 2.5 | 7 | 10 | 1974 |
| B. Projects under consideration | | | | | |
| (12) Algeria–Germany[c] (LNG) | SONATRACH–Thyssengas | | | 4 | 1981 ? |
| (13) USSR–Belgium | Distrigaz | | | 4 | ? |

a Several contracts
b Swap deal with USSR; now highly uncertain
c Conditional contract; depends on failure of negotiations of other contracts or new finds
Sources: Comité Professionnel du Pétrole; British Petroleum; Petroleum Economist; OPEC Oil Report (Petroleum Economist)

no continuous flow of deliveries. Depleted natural gas fields or other underground reservoirs can serve this function, and at the same time provide a cheap insurance against supply interruptions[154]. The levels of import dependence now predictable for the 1980s appear not excessive, and since much of them will come from Western European suppliers, they should pose no political problems. In the longer run, however, import dependence could reach significantly high levels that pose political problems, if supplies to replace declining imports from the Netherlands have to be met from OPEC sources, or from the USSR. The number of suppliers to the Nine is small — essentially, Algeria, Iran, and the Soviet Union. The triangular character of the imports from Iran which reach Europe via a switch deal with the USSR provides an additional element of security, since supply interruptions by one of the two would involve the other, and Iran and the USSR are unlikely to have common objectives in any disruptive action. The real problem seems to lie not in natural gas imports *per se*, but in cumulative dependence on OPEC for both oil and natural gas imports.

Several arguments could be made in favour of accepting this risk as limited: the natural gas suppliers of the Nine all depend quite heavily on income from hydrocarbons, and have so far shown no sign of 'mixing gas with politics': even Algeria, which cut back oil supplies to the USA in 1973/4, continued its natural gas exports to that country[155]. Besides, additional supplies which could increase import dependence on these countries might not materialize: both Algeria and Iran have recently indicated that they envisage no further export contracts beyond the ones already concluded. Algeria's export policy has come under attack within Algeria as too rapid, and in Iran growing domestic demand (which before the revolution was planned to reach about 35% of indigenous energy consumption in 1983), together with the reduction in gas production associated with the new oil policy of the revolutionary regime, will also limit exports[156]. Besides, cost escalation, technical difficulties, and more pessimistic expectations about future economic growth in Europe have generally dampened enthusiasm for natural gas imports, particularly in the form of LNG[157].

Since about 1977 the OPEC countries have shown considerable interest in co-ordinating their natural gas export policies, and a frequently mentioned objective is not only to bring the price of natural gas in line with current oil prices, but also to set up a unified price structure[158]. This might indicate a growing centralization of the international natural gas market, and to some extent this is likely to happen. However, this will be more difficult than in the case of oil: contracts cover long periods, they are negotiated between suppliers and consumers (the large gas utilities), and costs vary considerably from project to project. Thus, the establishment of a natural gas reference price appears difficult. What can be expected is that natural gas prices will keep closely in line with oil prices at the point of consumption — the net gains for the suppliers, however, are likely to be much lower, and to differ significantly from case to case. Nevertheless, OPEC is likely to increase its influence in world natural gas markets not only in terms of quantities supplied, but also in terms of its influence in natural gas policy.

## Nuclear trade: new problems on the horizon?

We already saw that official targets in the Community presume a very substantial expansion of nuclear electricity generation, and although one has to regard these targets with scepticism, further additions to the number of reactors in the Communities are certain. Worldwide, a similar trend is already firmly established, and barring unforeseen developments, the role of nuclear energy will grow substantially.

Consequently, international trade related to nuclear energy can also be expected to expand; it is interesting that about 125 utilities and firms from the United States, Canada, Japan, Australia, South Africa and Europe established in 1974 the World Nuclear Fuel Market designed to provide members with prompt and ready access to all stages of this market[159].

The nuclear industry, much like oil, but in a more complex way, comprises several stages from exploration to final conversion into energy. The nuclear fuel cycle (see *Figure 2.1*) spans uranium exploration, ore production and milling (the result of this first stage of processing is called yellowcake), then in most cases conversion into uranium-hexafluoride, a gaseous substance which is then enriched to increase the content of uranium-235, a uranium isotope which is fissionable. Some reactors, e.g. the British Magnox and the Canadian Candu types, can burn natural uranium, in which case gasification and enrichment are unnecessary. Enriched uranium, or yellowcake, is then processed to gain the reactor fuel rods; after this the uranium enters the reactor to be converted into energy. Used fuel can be reprocessed to gain fissionable material from the waste – the plutonium produced in burning uranium-235, and some remaining uranium.

International nuclear trade (to use a somewhat misleading, but convenient shorthand) essentially can be separated into two aspects: the nuclear fuel, and the plant and equipment to process and convert the fuel. The geographic distribution of uranium reserves, the technological complexities involved in some stages of the fuel cycle (particularly reactors and enrichment plants), and economies of scale obtainable by large units all favour international trade, sometimes of a complex nature. One example of this complexity is an agreement under which the United States sold West Germany $1.1 \times 10^6$ pounds (lb) of uranium ore. This ore is converted into uranium-hexafluoride in the UK, then enriched in the USSR, before it finally arrives in Germany[160].

The uranium ore requirements of the Community will largely have to come from imports – Western European reserves (mainly located in France, see *Table 2.7*) are only capable of covering about 20% of projected requirements. These requirements are at present extremely difficult to predict, although a substantial increase can be taken for granted. Where will the uranium ore for present and future reactors come from?

Looking at proven reserves, the main potential suppliers of the Nine are likely to be the USA, Canada, Australia, South Africa and some LDCs, particularly in Africa. The composition of imports will, however, depend on a number of other factors than reserves, both economic and political: most crucial will be government policies regarding uranium ore production and exports.

Table 2.7 World Production Forecasts and Reserves of Uranium Ore (excluding socialist countries) (t U)

| Country | Reasonably assured resources (up to $130/kg production cost) | Estimated additional resources (up to $130/kg production cost) | Production capabilities[a] | | | Production, |
|---|---|---|---|---|---|---|
| | | | 1980 | 1985 | 1990 | 1976 |
| Australia | 296 000 | 49 000 | 500 | 11 800 | 20 000 | 360 |
| Canada | 182 000 | 656 000 | 7 950 | 12 550 | 11 250 | 4 850 |
| France | 51 800 | 44 100 | 2 850 | 3 700 | 4 000 | 2 063 |
| Germany | 2 000 | 3 500 | 100 | 200 | 200 | 38 |
| Italy | 1 200 | 1 000 | 120 | 120 | 120 | — |
| UK | — | 7 400 | — | — | — | — |
| USA | 643 000 | 1 053 000 | 22 600 | 36 000 | 47 000 | 11 200 |
| South Africa | 348 000 | 72 000 | 11 700 | 12 500 | 12 000 | 3 412 |
| Gabon | 20 000 | 10 000 | 1 200 | 1 200 | 1 200 | 800[b] |
| India | 29 800 | 23 700 | 200 | 200 | 200 | 200[c] |
| Niger | 160 000 | 53 000 | 1 460 | 4 100 | 9 000 | 9 000 |
| World total (excluding socialist bloc) | 2 190 000 | 2 100 000 | 53 000 | 92 000 | 110 000 | 22 293 |

a Maximum obtainable capacity
b 1975
c Estimates
Source: OECD Nuclear Energy Agency/IAEA, Dec. 1977

The USA has in the past kept a tight control on uranium exports, and export agreements still have to be licenced by the Nuclear Regulatory Commission. Even fuel fabricated and enriched in the USA had to come from uranium ore bought elsewhere. At present, however, several hundred tons of uranium are committed for exports until 1979, and it appears as if domestic production capacity and imports contracted should be sufficient to allow a continuation of such exports during the 1980s[161].

Australia and Canada have had a political interest in their uranium ore exports. Policies are affected by:

(1) Environmentalist, conversationist and trade union resistance to nuclear exports.
(2) Concern about nuclear proliferation.
(3) The possibility to increase earnings from uranium exports by switching to exports of enriched uranium.

Australia's export policy or uranium tightened considerably under the Labour government which came to power in 1972. Apart from increasing the tax burden of foreign companies, and threatening to bar any foreign ownership of future mining projects, the government announced in 1974 a ban on uranium exports under pressure from trade unions and environmentalists, and commissioned a thorough assessment of the rights of the aborigines as to the uranium deposits in the far north-west. In 1976, under a new Liberal—Country Party government, the mining and export policy was gradually reversed; state participation in uranium mines was eliminated (although the new government also insisted on a minimum of 50% participation of Australian capital in any mining venture), and exports of uranium ore resumed. Exports were subjected to stringent controls and obligations, however, in order to satisfy critics worried about the possibility of Australian uranium becoming the raw material for the construction of nuclear weapons[162].

In Canada, as in Australia, resentment against the large share of multinational corporations in mining operations, and concern about nuclear weapons proliferation, led to considerable stiffening of the government's uranium export policy. After the failure to reach agreement with the Nine regarding more stringent controls over the uranium exports, the Canadian government decreed to halt all supplies to the Communities in early 1977; it took about a year to renegotiate the export agreements. The result was that supplies were resumed, but exports to French reactors were banned until the French government allowed IAEA controls[163].

Apart from these concerns about nuclear weapons proliferation, an important feature of future export policies could well be the desire to enrich as much of the uranium exports as possible in the producer country.

The typical fuel cost for a light water reactor consists of about 30% for mining and milling, 40% for enrichment, 15% for fuel fabrication, and the rest for reprocessing and miscellaneous costs[164]: therefore, exports of enriched uranium increase earnings substantially. Both Canada and Australia have been negotiating for enrichment technologies, and can be expected to enter the enrichment market in the 1980s[165]. South Africa has developed its own process based on the German 'jet nozzle' system[166]. In the long run, therefore, importers of uranium ore (such as the Community which is developing its own enrichment capacity) might well have to depend to a growing extent on new suppliers among LDCs — their uranium reserves are almost certainly vastly understated owing to lack of exploration.

Apart from national governments, a second group will continue to play an important role in the future uranium market: the multinational corporations. In 1975 sixteen of them founded the Uranium Institute to study joint problems of the industry[167]. Among the most important producers of uranium with large reserves are Anglo-American (South Africa), Rio Tinto Zinc (South and South West Africa, with one of the largest deposits in the world, and Canada, through a majority share in Rio Algom), United Nuclear / Homestake Partners (USA), Denison Mines, El Dorado Nuclear (both Canada, the latter state-owned)[168]. The French Atomic Energy Commission also holds huge reserves of about 100 000 t[Note 169]; and oil companies play an important part in the American uranium industry, with Kerr-McGee being the biggest uranium mining company in the USA[170]. Exxon also operates a uranium mine, fabricates fuel, and is involved in enrichment research; Gulf holds huge reserves, plans to open a uranium mine in 1978, and also launched a rather unsuccessful venture in the development of high temperature reactors, in which Shell later joined. Shell also has stakes in fuel fabrication and enrichment (10% stake in Ultra-Centrifuge set up by URENCO-CENTEC). Equally, CFP prospects for uranium, and holds a 10% share in a production venture in South Africa. Other US oil companies active in uranium production and exploration, and/or in fuel fabrication and reprocessing, include Continental and Getty Oil[171]. It will be interesting to see to what extent the large multinational enterprises in nuclear energy move further into downstream activities, particularly enrichment — the US government has repeatedly expressed its desire to turn over enrichment to private industry, at least for the additional capacity required[172].

The future interaction between producer governments and companies (through export licences and participation for mining) will be of crucial importance for the evolution of the international uranium market. These relationships have to be evaluated against the background of some general characteristics of the uranium industry. On the one hand, the industry is highly concentrated — four countries (Canada, Australia, South Africa and France, through her interests in Niger and Gabon), and a small group of multinational companies control almost all the present and projected trade in uranium ore. The elasticity of demand in the case of price increases is very limited in the short and medium term — uranium fuel accounts for only a small percentage of total operating costs of a nuclear power station, and price increases can be passed on easily to the final consumer because of the monopoly position of utilities. Supply elasticities are equally very low — the lead-time for the establishment of uranium mines is about 7–10 years, so that price increases would not bring additional capacity into the market, at least not for a considerable length of time[173].

These characteristics allow cartelization of the uranium market, but do not make it necessary: prices over a considerable spectrum can simply be raised in 'conscious parallelism' by producers without fear of market losses[174]. This situation invites the formation of a loose bipolar oligopoly of host governments and companies, and there are strong indications that such an oligopoly has already operated: in October 1976 a law suit by Westinghouse against 29 US and foreign uranium producers revealed that more than 20 companies, secretly encouraged by at least three governments, had set up a producer cartel in 1971. In June 1972, it was decided to impose a floor price on uranium, and within 20 months uranium prices rose by about 57%[175]. In mid-1977, the price per pound of uranium ($5 in 1971) had risen to about $40 in the spot market[176] The cartel was for some time backed by the Canadian government, which refused to grant export permits at prices below the floor set by the producers. This policy has since been reversed; whether the cartel still exists, is unclear, but given the present and projected shortage in mining capacity, some informal market organization would appear to be easily feasible. The uranium market can, therefore, be expected to continue as a highly centralized, homogenous system, with all the implications for consumer countries associated with such a structure.

Some recent calculations prepared jointly by the OECD and the IAEA have thrown light on the likely evolution of the international uranium ore market. Although major uncertainties persist, the conclusions of the report about the likely future supply and demand indicate that until about 1990 production capacity should be able to cope with increasing demand: capacity estimates for 1985 point to availabilities upward of 90 000 t/a, while demand in the same year could range from 71 000 to 88 000 t/a worldwide, depending on the future development of nuclear energy. Attainable production capabilities could reach 110 000 t in 1990, compared to a demand of between 102 000 and 156 000 t/a[Note 177]. The conclusion to be drawn from this is that during the 1990s shortages are possible; indeed, they could already happen earlier, if uranium policies of the major exporters keep export capacity significantly below the attainable maxima, and/or if continued uncertainty regarding the evolution of nuclear energy fails to stimulate the investments needed to bring about these maxima sustainable by presently known resources. (Details of supply and demand estimates for uranium can be found in *Table 2.7* and *Figure 2.2*).

Conversion of uranium oxide is a relatively cheap and simple exercise. Since a few conversion plants would be sufficient to produce the required feedstock for worldwide enrichment facilities, there would again be an incentive for the development of international trade in uranium hexafluoride. Such trade exists already: the USA, Canada, France and Britain supply all the demand of the non-Communist world: plants are operated by Kerr-McGee and Allied

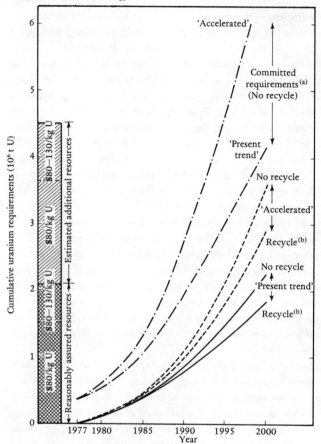

**Figure 2.2** *World cumulative uranium requirements (1977–2000) (a) Reserves needed by installed reactors during their operational lifetime (40 years); (b) recycle begins in 1985 (Source: OECD Nuclear Energy Agency, IAEA, December 1977)*

Chemicals in the USA, El Dorado Nuclear in Canada, British Nuclear Fuels (a state-owned enterprise) in the UK, and the French Atomic Energy Commission in France. The British company is the sole supplier of Germany and Italy, and also converts uranium for the USA; the French CEA will supply the French-dominated Eurodif enrichment plant(s)[178].

Reprocessing is a comparatively more complex operation, and capital costs for plants have increased rapidly in recent years, so much so that serious doubts exist as to its commercial attractiveness[179]. This, and the nuclear weapons proliferation risk associated with the plutonium production of reprocessing plants, have led the US government to postpone indefinitely this stage of the nuclear fuel cycle. In Europe, however, reprocessing already takes place on a commercial basis: BNFL has reprocessed spent nuclear fuel for foreign customers, primarily Italy and Japan (about 700 t up to 1975, compared to 17 000 t reprocessed for the UK), and hopes to expand its export markets to about 9700 t between 1975 and 1990[180]. An extensive public inquiry about the expansion of the Windscale reprocessing plant finally resulted, with considerable delay, in government authorization to proceed with the project, thereby opening the way for further agreements with other countries for the reprocessing of their spent fuel by BNFL[181]. France also operates a reprocessing plant, and the two countries have together with Germany established United Reprocessors — a trinational joint venture designed to ensure proper use of French and British facilities, and to build up a third large plant in Germany in order to help meet European demand and acquire some export business[182].

In the longer run — i.e. beyond 1985 — reprocessing of spent nuclear fuel is seen as an important way of reducing energy import dependence. The extraction of plutonium and uranium from the spent fuel can be used (a) in conventional reactors in the form of a fuel mixture of uranium and plutonium (as yet only done on an experimental scale, and involving higher fuel costs than for uranium fuel), and (b) in fast breeder reactors, which could reduce the requirements of uranium considerably, since this reactor type 'breeds' plutonium out of the relatively abundant uranium-238, and thereby produces new fissile material. However, the gain in economic security has to be seen in perspective: first, it will be limited in terms of overall energy dependence of the Community in the foreseeable future (and also involve a different set of suppliers). Second, this marginal increase of energy independence might have to be paid with a high economic and social price involved in the processing and transport of large quantities of plutonium. While the FBR seems essential to realize substantial gains in energy security (recycling of reprocessed nuclear fuel in LWRs will contribute only marginally to reducing uranium demand), this solution is still surrounded by major technical and economic difficulties. Eventually, the FBR technology may be mastered under economic and environmental safety conditions comparable to the LWR — but this would also seem to commit the Communities to an irreversible energy strategy based heavily on expansion of electricity, and thereby an inflexible, centralized system. The argument that the FBR would help to solve the nuclear waste problem associated with plutonium (which is also produced in LWRs), seems accurate only as long as the FBR programme expands; a slow-down or halt to expansion would accentuate the plutonium problem[183].

Most of the nuclear power stations operating or planned in the Community now depend on enriched uranium as fuel. Enrichment is, therefore, a crucial stage in determining stability and security of nuclear supplies; moreover, enrichment also constitutes a very significant element of total fuel costs. Until recently, the Community was almost totally dependent on US supplies for enriched uranium — only France and Britain had developed small enrichment capacities for their nuclear weapons programmes. This virtual monopoly of the USA has now been broken by the arrival of the USSR as an alternative supplier, and European efforts to build up their own enrichment capacity.

Enrichment is a highly complex and expensive process. In the past, the USA provided virtually all the requirements of the non-communist world within the framework of its Atoms for Peace programme through a number of bilateral agreements to supply enriched uranium for the operating lifetime of nuclear reactors. Prices were subsidized, and the US supplies were seen in Washington as part of a policy of preventing the spread of nuclear weapons by concentrating the sensitive enrichment technology within its own boundaries.

The USA turned out to be a supplier of limited reliability, however: in 1971 and again in 1973, prices were raised unilaterally, and the revision of the US nuclear export policy to develop better safeguards against nuclear proliferation affected supplies to Europe in 1975, and again in the period 1976–1978. The Carter government's nuclear policy, pronounced in April 1977, created severe differences between the USA and Canada on the one side, and Europe and Japan (as well as Third World countries) on the other. These differences were slowly and painfully narrowed[184].

Until recently the Soviet Union never exported uranium ore outside the communist bloc, but after the strategic embargo against socialist countries was lifted in the late 1960s it began to supply enriched uranium to the West. The first such agreement was signed in 1971 with France; further contracts were concluded with German, British and Italian utilities. Prices offered by the USSR have been competitive (about 5% below those of the USA), and its share in the European Community market in 1977 was estimated at about 50%. In the future, the share of external sources of supply will decline as European enrichment facilities become operational[185].

American enrichment plants use the gaseous diffusion technology developed for the US military programme. There are a range of other technologies, but only three seem so far to offer definite commercial possibilities: the gas centrifuge, the related 'jet nozzle' technology, and the still very much untested isotope separation process by laser bombardment[186]*. European enrichment projects have followed both the centrifuge and the gaseous diffusion technology, with the centrifuge technology largely developed by the partnership established by Germany, the Netherlands and Britain in 1970 — the Tripartite Centrifuge Agreement. The Agreement established two commercial operations, Centec (charged with the development and maintenance of the centrifuges) and Urenco, which operates the plants and markets the product[188]. Urenco has meanwhile started deliveries of enriched uranium[189]; the consortium plans to increase capacity to 10 000 tons of processing capacity ($10 \times 10^6$ separating work units (SWU)) in 1985, making the three countries theoretically self-sufficient, and practically exporters, since Urenco cannot hope to capture all markets in the three countries[190].

The French government has pursued a gaseous diffusion-based technology in Eurodif, which has also drawn some international support from Belgium, Spain,

---

* A fourth chemical process has been developed by the French CEA with special regard to the proliferation problem: it cannot produce weapons grade enriched uranium. However, it has not been tested commercially[187].

and Italy; Iran gave a $1 \times 10^9$ loan to the project in exchange for a 10% stake, and acquired a 20% share in another project by Eurodif, Coredif (which also includes the French Atomic Energy Commission)[191]. Together, those three projects will reduce the Community's import dependence on enriched uranium to relatively low levels, and possibly even provide it with some export capacity. The high degree of dependence on the Soviet Union will last only for a very short period and should therefore pose no political problems — besides, any political interference of the Soviet Union with her enriched uranium exports to the West would no doubt seriously affect the general climate of East—West relations. Also, the USSR will at that time still depend on trade with Western Europe, keeping enriched uranium trade within the framework of commercial exchange.

Thus it seems that the only sensitive area of European import dependence associated with nuclear energy production is uranium ore supply. Whether supply in the world markets will be sufficient will depend not only on reserves, but also on investments to bring uranium mines with sufficient production capacity on stream[192]. Projections about future supply and demand levels are hazardous, but for the next ten years it seems likely that shortages can be avoided.

In dealing with traditional suppliers, the Community will face a highly organized market; it would, therefore, seem desirable to secure some influence through the establishment of close relationships, e.g. the transfer of enrichment technologies in exchange for long-term supply contracts. The inclusion of consumers in the Uranium Institute also is a useful step in the same direction. Another obvious area of co-operation concerns investment by European mineral companies.

In the longer run, sources of uranium supply can be expected to shift towards LDCs, as traditional suppliers move into exports of enriched uranium, and as additional quantities of ore are needed. France and Germany already support efforts to locate and develop uranium resources in the Third World, and it is significant that the German-Brazilian nuclear agreement also provides for deliveries of uranium ore to Germany[193]. Attempts by European governments to secure uranium sources either directly (through state enterprises) or indirectly (through credits and other assistance) might lead to closer co-operation between host and consumer government, favouring some kind of commodity agreement of a regional or even global character[194]. The Lomé Agreement between the Community and African, Pacific and Caribbean countries might be expanded in this direction; it provides an interesting framework for possible co-operation in the uranium sector[195].

Some oil-producing countries have become more or less intensely interested in nuclear energy, and in several cases nuclear reactors have been ordered or are already under construction[195a] Countries in this group include Iran, Libya, Iraq, Saudi Arabia, Kuwait and Algeria; only Iran, however, tried so far to embark on a major nuclear programme designed to give a capacity of 23 000 MWe in 1992.

Iran had ordered four nuclear power stations from Germany, each with 1200 MWe (construction work has already begun), and two from France, each with 900 MWe. Iran is also said to have contracted 14 000 t of South African uranium oxide, and showed an active interest in Australian uranium ore and in a South African enrichment project[196]. The Iranian revolution seems to have all but killed the Iranian nuclear programme however.

Such a massive development of nuclear power would no doubt create some dependence on the suppliers of reactors, and thereby establish a mutual interest in uninterrupted relations: Iranian investment in European enrichment projects indicated that the Community might become a supplier of energy to Iran, as well as an importer of Iran's energy resources. In addition, nuclear reactors in developing countries will for some time depend on Western technological assistance. These factors contribute to European security of supplies, but only marginally. Similar considerations hold true in the case of nuclear exports such as the agreement to supply Brazil with a complete nuclear fuel cycle designed and built by German contractors: it is hoped that this will secure a stable source of uranium ore supplies for the Federal Republic. Expenditure on massive nuclear programmes (Iran's was expected to cost up to $\$25 \times 10^9$) would also help to balance the heavy energy import bills of some European countries (though not necessarily only the ones most in need of such a contribution) — again a marginal contribution to the security and stability of supplies of imported energy.

Against the marginal gains in import security have to be weighed the risks involved in the export of nuclear installations: i.e. risks involving the proliferation of fissionable materials such as plutonium-239 and uranium-235, which could be used by some governments as the base for nuclear weapons production, or as a political weapon (using the possibility of an easily available nuclear option as leverage). Plutonium could also be used by terrorists to fabricate a crude nuclear explosive, or simply a dangerous radiation weapon.

The temptation to proliferate seems not unrelated to the evolution of nuclear energy within the Communities — if nuclear power is given high priority, fluctuations in reactor demand would provide powerful incentives for Europe's nuclear energy to seek export orders. Whether under these circumstances the rules of the Nuclear Suppliers' Club survive in letter as well as in spirit, remains to be seem. The US government's emphasis on non-proliferation might yet lead to the establishment of international institutions and mechanisms to slow down the spread of nuclear weapons, and the search for alternative, less proliferating nuclear fuel cycles could still throw up new technical solutions. Yet the odds appear to be that Europe will become a major contributor to the spread of nuclear weapons[197].

## Oil: the crucial balancing fuel

Oil will continue to be the European Community's most important source of energy. Future oil import requirements are very difficult to predict, and estimates vary because of oil's balancing role; oil will have to fill the gap between energy demand and all other energy supplies. While there is now reasonable certainty about indigenous energy supplies in most respects, great uncertainties still surround nuclear energy and the evolution of energy demand in the Nine. In 1974, European Community Commission targets demanded a reduction of oil imports from $565 \times 10^6$ t in 1973 to about $420-470 \times 10^6$ t in 1985. In 1976, the Commission produced estimates which showed an import demand of $420-520 \times 10^6$ t, while in 1977 a further revised set of figures gave a level of $490-555 \times 10^6$ t[Note 198]. In 1977 the OECD forecasted $540 \times 10^6$ t[Note 199]. In the last two reports energy demand was estimated at about $1280 \times 10^6$ toe in 1985. The 1985 estimates have to be regarded with caution, however: it now appears as if this year will more or less coincide with peak levels of indigenous hydrocarbon production in the Nine; as demand for energy grows further, and indigenous oil and gas production stagnates, or even declines, demand for oil imports will increase further, even if nuclear energy substantially expands its contribution. First estimates prepared by the European Community Commission for 1990 indicate an increase in the importance of imported oil, both in relative terms (as percentage of total energy demand) and in absolute quantities[200].

All estimates, then, agree that there will be a very substantial import gap for the Nine during the 1980s; if one takes the Tokyo Summit targets as a guideline, the European Community would by 1985 still be the biggest oil importer ($472 \times 10^6$ t), followed by the USA ($421 \times 10^6$ t) and Japan ($315 \times 10^6$ t)[201].

In addition to OPEC, there are only two major sources of oil imports to the Communities: Norway and the USSR. In both cases, additional supplies appear insufficient to substantially alter present dependence on OPEC. Norway's currently established reserves are about $800 \times 10^6$ t, with ultimately recoverable reserves of about 3000 to $4000 \times 10^6$ t. Norway's own energy supplies come largely from hydro-power, with oil consumption in 1978 being just below $8 \times 10^6$ t. Thus, most of Norway's production would be available for exports. The government, however, has so far held firmly to its decision to set production ceilings of $50 \times 10^6$ t for oil and $40 \times 10^6$ toe for natural gas — levels which for oil will not in any case be reached on present form: it is thought that production will reach about $65 \times 10^6$ t in 1980, and then will stay at a plateau of about $60 \times 10^6$ t/a during the 1980s[202]. Further exploration licences are about to be prepared, but a modest production rate seems a firmly established principle of Norwegian oil policy. The objectives of this policy are to conserve a valuable source of energy for the future, and — more importantly — to control the impact of the oil sector on Norway's economy and society.

This impact will be *direct*, on industry and the labour market, and *indirect*, through increased government revenues. Directly, Norwegian oil has provided about 15 000 jobs in 1974, estimated to rise to 25 000 in 1980[203]. To this has to be added demand for oil-related equipment (worth about $1850 \times 10^6$ Kroner in 1974) and services, both of which have implications for employment.

Because of oil revenues, public revenues should increase considerably, to about 65—70 × $10^9$ Kroner in the period 1978—1980 (note, however, that oil income has been affected by technical delays and the Bravo accident, as well as by stable oil prices)[204]. This, in conjunction with spending by industry and consumers, will increase domestic demand, and also inflationary pressures. The government has calculated that every incremental 1000 × $10^6$ Kroner spent by the administration would lead to an additional labour demand of 7000—8000[Note 205]. A relative shortage of labour will tend to increase wages, which together with growing demand will push up prices. Since only industries producing goods and services which cannot be provided by imports (sheltered industries) would be able to absorb higher costs by increasing prices, there would be a transfer of labour from non-protected to protected industries.

The government has decided to use oil revenues in order to improve the quality of life of Norwegians; income tax has already been lowered, and increased investments in public services such as health, social welfare, education, and in regional development and environmental quality, are planned. The implication is that consumer goods will to a growing extent be imported, while labour will be transferred to sheltered industries and to particular service industries. The industrial sector can, therefore, be expected to shrink, while the service sector and the oil sector will expand. Assuming additional public expenditure of 6000 × $10^6$ Kroner up to 1980, together with developments such as reduction of working hours and expansion of the domestic labour force, sheltered industries will grow by about 95 000 man-years, while competing industries will decline by about 60 000 man-years[206]. Evidently, such a development would eventually threaten the diversified economic structure of Norway and create a heavily oil-dependent economy. Equally, a high rate of change would presumably involve a dramatic internal migration and destroy the present settlement structure of Norway, exacerbating regional problems. Further considerations are protection of fisheries, and the prevention of pollution and environmental damage. Such factors make the expansion of production beyond 50 × $10^6$ t unlikely. Quite understandably, 'considerations of what are the natural dimensions for an oil industry in a small society of four million inhabitants are the most significant factor in the shaping of Norwegian oil policy. The aim of this policy is that those highly valuable resources will be conserved for future generations'[207].

Another source of additional oil imports for Europe could be the Soviet Union. Total Soviet oil exports to the European Community in 1977 came to about 32 × $10^6$ t; oil production in that year stood at 546 × $10^6$ t, having grown by an annual average of about 10% since 1955[208]. The future development of exports to the West, however, is very uncertain, and on balance there seems little chance of additional oil supplies from the USSR in the longer run. More likely is a decline, and the appearance of the CMEA countries as competitors in world oil markets.

Future Soviet oil export levels will depend mainly on production levels, indigenous demand, and export policies. In recent years, Soviet oil production has shifted geographically from its pre-war centre in the Baku area on the

Caspian Sea (which provided 71.5% of total production in 1940) and from the Volga-Ural area in the two decades after the war (which accounted for about 70% of production in the early and mid-1960s, declining to 58.7% in 1970) to Siberia, with the huge Tyumen field alone providing 116 × $10^6$ t in 1974. Total production from this area is expected to reach 300–310 × $10^6$ t in 1980[209].

The reserves of Baku are largely exhausted, and production in the 'Second Baku' can now only be maintained by costly secondary recovery methods — there is little doubt that the Volga-Ural area has also peaked out, or will peak out soon[210]. Oil from the one new production area, Mangyslak, has a very high wax content (which makes transportation difficult and costly); besides, the desert character of the region poses problems of water supplies needed to maintain pressure in the fields[211].

Western Siberian oil is of very high quality, but even more difficult to produce and transport. The oil fields are in swampy areas plagued by mosquitos in the summer and strong frost in the winter. Western Siberia's oil province is almost unpopulated, and in spite of high wages it has proven difficult to attract labour — over one million people who came to Siberia between 1959 and 1969 left again[212]. The infrastructure for production and transportation is lagging behind, and hasty exploitation has led to repeated problems: difficulties in finding adequate water supplies, and integrating well, pipeline and pumping stations capacities[213]. At the same time, production has been hampered by the lack of technologically advanced equipment.

More fundamentally, the main condition for rapid advance of USSR oil production in the past — the existence of giant fields in each of the 'three Bakus' — no longer applies. In the future, the Soviet Union will have to draw production increases from a number of smaller fields in extremely adverse locations. Dependence on Western technology, capital and know-how will therefore be necessary if future targets are to be reached[214]. The signs of strain are unmistakable: growth rates in the oil industry have declined consistently, from 10.4% annually in the period 1960–1965, over 7.7% in 1966–1970, to 6.8% in 1971–1975[215]. From 1960 to 1975, production rose 3.4 times, while reserves showed only a 2.2-fold increase[216].

According to semi-official projections and serious Western studies, production of oil and condensates in the USSR was to reach 620–640 × $10^6$ t in 1980, 630–757 (with about 700 × $10^6$ t as the most probable figure) in 1985, and about 780 × $10^6$ t in 1990[217]. Domestic demand for oil has slowed its increase over recent years: growth rates of 11% between 1950 and 1963 have fallen to 7.5% for the period between 1963 and 1970, and less since[218]. Authoritative estimates indicate consumption levels of 470 × $10^6$ t in 1980, 603–650 × $10^6$ t (with about 635 × $10^6$ t as the most probable figure) for 1985, and about 650 × $10^6$ t for 1990. (See *Table 2.8*). On the basis of these figures, export availabilities (in net terms) could be between 27 and 107 × $10^6$ t in 1985 (with a central estimate of 78 × $10^6$ t), and around 130 × $10^6$ t in 1990.

Soviet oil export policy would then be faced with choices between a restriction

Table 2.8  The East European Energy Balance 1980–1990: Some Estimates of Three Major Fuels ($10^6$ toe)

|  | 1980 | | | 1985 | | | 1990 |
| --- | --- | --- | --- | --- | --- | --- | --- |
|  | USSR | Eastern Europe | Comecon total | USSR | Eastern Europe | Comecon total | Comecon total |
| Oil |  |  |  |  |  |  |  |
| Production | 901 | 26.4 | 927.4 | 900–1 081 | 25 | 1 043 | 1 175 |
| Consumption | 743 | 173.0 | 916.0 | 862–928 | 220–242 | 1 137 | 1 250 |
| Exports | 158 | −146.6 | 11.4 | 38–153 | −195 to −217 | −94 | −75 |
| Natural Gas |  |  |  |  |  |  |  |
| Production | 517 | 60.7 | 577.7 | 720–840 | 70 | 850 | 1 072 |
| Consumption | 428 | 99.0 | 527.0 | 550–580 | 122 | 702 | 1 026 |
| Exports | 89 | −38.3 | 50.7 | 170–290 | −52 | 148 | 46 |
| Coal |  |  |  |  |  |  |  |
| Production | 555 | 354.0 | 909.0 | 650–700 | 400 | 1 100 | 1 272 |
| Consumption | 540 | 328.0 | 868.0 | 600–650 | 377 | 1 027 | 1 247 |
| Exports | 15 | 26.0 | 41.0 | 50 | 23 | 73 | 25 |

*Sources:* GEPI, 'Situation et perspectives du bilan énergétique de l'U.R.S.S. de de l'Est Européen', *Le Courrier des Pays de l'Est*, No. 216, March 1978; UNECE, *New Issues Affecting the Energy Economy of the ECE Region in the Medium and Long Term*, ECE (XXXIII/2), Geneva 1978

Table 2.9  EEC Crude Oil Imports, 1975 ($10^3$ t)

| Importer | World total | Socialist countries | Algeria | Libya | Iraq | Saudi Arabia | Kuwait | Other Arab | Sub-total Arab | Nigeria | Iran | Venezuela | Sub-total, all LDCs |
|---|---|---|---|---|---|---|---|---|---|---|---|---|---|
| BeLux | 29 293 | 316 | 880 | 615 | 1 617 | 12 739 | 2 899 | 1 037 | 19 787 | 1 221 | 5 382 | 367 | 26 757 |
| Denmark | 7 915 | 436 | – | 138 | 166 | 1 672 | 624 | 790 | 3 390 | 889 | 3 045 | – | 7 324 |
| France | 106 081 | 1 191 | 5 873 | 2 183 | 12 018 | 33 482 | 2 723 | 14 030 | 70 309 | 8 738 | 13 290 | 752 | 93 089 |
| Germany | 91 023 | 3 383 | 10 214 | 14 900 | 1 404 | 18 811 | 2 692 | 6 406 | 54 427 | 10 105 | 14 189 | 2 307 | 81 028 |
| Ireland | 2 417 | – | – | – | 129 | 708 | 851 | – | 1 688 | – | 718 | – | 2 406 |
| Italy | 95 859 | 3 366 | 3 784 | 12 967 | 23 044 | 26 203 | 3 736 | 1 996 | 71 730 | 334 | 12 859 | 574 | 85 497 |
| Netherlands | 54 246 | 489 | 484 | 360 | 1 940 | 12 566 | 5 438 | 6 698 | 27 486 | 7 474 | 17 475 | 315 | 52 750 |
| UK | 90 412 | 882 | 1 531 | 3 024 | 3 078 | 23 320 | 11 803 | 9 235 | 50 460 | 6 100 | 20 635 | 3 502 | 80 697 |
| Total | 477 246 | 10 069 | 22 766 | 34 187 | 43 396 | 129 501 | 34 766 | 40 192 | 282 042 | 34 861 | 87 593 | 7 818 | 412 314 |
| % share of world total | (100) | (2.1) | (4.8) | (7.2) | (9.1) | (27.1) | (7.3) | (8.4) | (59.1) | (7.3) | (18.4) | (1.6) | (86.4) |

Source: Eurostat

of exports, or contraction of additional oil imports, and with allocation of exports between CMEA countries in Eastern Europe and other markets, including Western Europe. (See *Table 2.9*). Faced with growing import demands by Eastern Europe (whose demand had until recently been fully met by the Soviet Union), the USSR has encouraged its allies to turn to alternative sources of energy (such as nuclear), and to alternative suppliers of oil in the Middle East. Thus, Iraq, Iran, Egypt, Syria, Libya and Algeria have all supplied oil to Comecon countries[219]. At the same time, the USSR continued to be the main supplier of Eastern Europe, which has been assured that the Soviet Union will meet two-thirds of Eastern European oil demand in 1980[220]. Since Eastern European requirements will be in the region of $120 \times 10^6$ t[Note 221], and domestic production will be unlikely to exceed $20 \times 10^6$ t[Note 222], this means that up to $65 \times 10^6$ t of Soviet oil exports will go to those countries.

In the medium term, the Soviet dilemma will become greater. The Eastern European oil deficit in 1985 has been estimated at between 136 and $152 \times 10^6$ t, and in 1990 at $180 \times 10^6$ t[Note 223]. Soviet oil exports (net) could no longer even theoretically cover that gap — the CMEA area (USSR and Eastern Europe) will show an overall oil import demand of about $66 \times 10^6$ t in 1985 and about $50 \times 10^6$ t in 1990, according to two different studies for those years[223a]. Will exports to the West continue at all, given these circumstances?

The likely answer seems to be yes, but on a lower level. Oil exports constitute an important source of hard currency[223b]. Thus, the USSR could decide to maintain some of its exports to Western Europe, and to step up its own imports from the Middle East. Such imports have been contracted already in the past[224], and Moscow might find it economically attractive to continue and even expand such a policy. Some Soviet oil is also directly marketed in Belgium and Britain through Soviet distribution networks[225], and it appears unlikely that the Soviet Union would not attempt to maintain supplies to those outlets. Oil exports to Eastern Europe also profit the Soviet economy — prices have gone up steadily in recent years (although at levels still beyond world market prices)[225a], and Eastern European countries pay with industrial equipment and machinery which contribute to Soviet industrialization[226].

To sum up: the Soviet Union seems to face increasing difficulties in maintaining the past dynamism of its oil industry, and at the same time is confronted with growing demand from Eastern Europe, and even domestically. The Soviet government is unlikely to reduce by too much the exports of oil to Eastern Europe — to do otherwise would damage some of the firm links between Eastern Europe and the Soviet Union. The CMEA area is likely to become a net importer of oil during the early 1980s, and will thus compete with the Western world and other importers for oil from OPEC sources. To what extent such imports will be needed is still far from certain. During the 1980s, an average level of about $1-2 \times 10^6$ b/d seems a plausible assumption.

The political implications of this are difficult to foresee. Essentially, the Soviet Union has a choice between a purely commercial import strategy, and

Table 2.10  OPEC Oil Reserves, Consumption, Production, and Production Potential ($10^6$ t)

| Country | Reserves[a] 1977 | Reserves[a] 1978 | % of world total 1977 | % of world total 1978 | Reserve/production 1977[b] | Production 1977 | Consumption 1975 | Consumption 1976 | Production potential 1985 |
|---|---|---|---|---|---|---|---|---|---|
| Abu Dhabi | 3 956 | 4 229 | 4.5 | 4.8 | 53.1 | 79.7 | 0.534 | 0.619 | 150–175 |
| Algeria | 928 | 900 | 1.1 | 1.0 | 16.8 | 53.5 | 3.500 | 4.139 | 45–55 |
| Ecuador | 230 | 224 | 0.3 | 0.3 | 25.0 | 8.6 | 1.917 | 2.172 | 10 |
| Gabon | 290 | 280 | 0.3 | 0.3 | 25.0 | 11.2 | 0.255 | 1.000 | 10 |
| Indonesia | 1 433 | 1 364 | 1.6 | 1.6 | 16.4 | 83.2 | 10.485 | 11.694 | 80–105 |
| Iran | 8 595 | 8 458 | 9.8 | 9.6 | 30.6 | 276.4 | 17.581 | 21.506 | 275–325 |
| Iraq | 4 638 | 4 707 | 5.3 | 5.3 | 42.4 | 110.9 | 4.254 | 4.260 | 250–300 |
| Kuwait[c] | 9 625 | 9 563 | 11.0 | 10.9 | 101.7 | 94.3 | 1.190 | 1.498 | 150 |
| Libya | 3 479 | 3 411 | 4.0 | 3.9 | 34.1 | 100.1 | 2.557 | 2.711 | 100–150 |
| Nigeria | 2 660 | 2 551 | 3.0 | 2.9 | 24.5 | 104.3 | 3.400 | 5.503 | 100–150 |
| Qatar | 778 | 765 | 0.9 | 0.9 | 35.7 | 21.4 | 0.210 | 0.285 | 25 |
| Saudi Arabia[c] | 21 500 | 20 886 | 24.5 | 23.7 | 46.1 | 453.2 | 5.852 | 8.004 | 800–1 000 ? |
| Venezuela | 2 435 | 2 483 | 2.8 | 2.8 | 21.3 | 116.4 | 10.461 | 11.529 | 110 |
| Total | 60 547 | 59 821 | 69.0[d] | 67.9[d] | 39.5 | 1 513.2 | 62.196 | 74.920 | 2 105–2 565 |

a As of 1 January
b Reserves 1/1/78: production 1977
c Includes 50% share of Neutral Zone
d Figures do not add up because of rounding
Sources: Comité Professionnel du Pétrole; CIA

a strategy aimed at gaining a privileged relationship with certain suppliers, presumably only feasible through a measure of direct control. The former course seems more probable if the relatively modest estimate above about import requirements is correct — but the inconvenience here is the difficulty for Eastern European countries and the USSR to pay for oil imports in hard currency. The supply of arms might be a suitable substitute. However, this may involve more political risks: to establish control in Third World countries is anything but easy, and there would be the danger of confrontation with the West if the Soviet Union attempted to establish itself politically, economically and militarily in Gulf countries such as Iran or Saudi Arabia. Much will depend on the evolution of the situation in the Middle East in general; it might provide the USSR with openings which it could be tempted to exploit. Apart from securing oil imports for the requirements of its allies, it would increase the USSR's influence in world affairs. However, to speculate about the geopolitics of oil lies outside the scope of this book.

Apart from these dangers, Soviet oil exports to Western Europe would not create any security problems. They will be essentially commercial, as they were in the past: the USSR did not halt, for instance, supplies to Western Europe in the 1973/4 supply crisis[227]. Besides, the quantities are too small to create any problems in themselves. The contribution of Soviet oil to European energy security will therefore be positive, but minimal.

Over the next decade, OPEC suppliers will thus continue to be the main source of European oil imports. As *Table 2.10* indicates, OPEC countries have substantial reserves, sufficient production capacity, and relatively small domestic requirements. They could, in principle, continue supplies on present levels, and even higher ones if needed. This view is corroborated by substantial evidence pointing to massive additional oil and gas reserves in those areas[228]. The oil producers in the Middle East and Africa are the main partners of Europe in the international energy trade, and the Communities' international energy policy will, therefore, have to concentrate on securing supplies from these nations.

## *Conclusion*

The brief survey of prospects for indigenous energy production in the Community, and of patterns and volumes of imports required to balance supply and demand tried to answer a simple, but rather fundamental question: will the Community continue to be dependent on imports to an extent which exceeds levels amenable to purely commercial arrangements? In other words: will energy import dependence continue to pose political problems? Or, put another way: does the Community need an international energy policy?

The answer is affirmative — the analysis has demonstrated this quite clearly, no matter whether it is correct in all details (and, obviously, there are major uncertainties ahead which make it likely that *some* conclusions will not stand up to the test of time). More specifically, we can sum up our argument as follows:

(1) Indigenous production of coal can at best be expected to remain at present

Table 2.11  Europe's Energy Balance in 1985: Targets and Estimates ($10^6$ toe)

| | Consumption (1985) | | | | Indigenous production (1985) | | | | Imports (1985) | | | |
|---|---|---|---|---|---|---|---|---|---|---|---|---|
| | European Community estimate targets | OECD | Member state estimate | | Targets | OECD | Member states | | Targets | OECD | Member states | |
| | | | 1976 | 1977 | | | 1976 | 1977 | | | 1976 | 1977 |
| Solid fuels | 250 | 216 | 243 | 220 | 210 | 176 | 191–196 | 184 | 40 | 40 | 47–52 | 36 |
| Liquid fuels | 575–695 | 665 | 684–705 | 650–665 | 180 | 164 | 111–161 | 111–161 | 395–515 | 546 | 544–573 | 490–555 |
| Natural gas | 270–340 | 241 | 244–259 | 221–236 | 175–225 | 159 | 150–165 | 143–158 | 95–115 | 82 | 94 | 79 |
| Nuclear energy | 190–240 | 117 | 182–189 | 140 | 190–240 | 117 | 182–189 | 140 | — | — | — | — |
| Other | 45 | 37 | 35 | 35 | 45 | 37 | 34 | 31 | — | 1 | 1 | 4 |
| Total | 1 475 | 1 276 | 1 388–1 431 | 1 282 | 800–900 | 653 | 668–745 | 609–674 | 550–650 | 668 | 686–720 | 608–673 |

Sources: EC Commission; World Energy Outlook, OECD, 1977

levels. Constraints on the demand side, however, make a substantial absolute increase or, even more so, an increase of coal's share in the Community energy balance, unlikely: 'King Coal' will have to await the large-scale advent of gasification and liquefaction to resume his reign. Imports of coal will, therefore, expand only modestly, to about $50-70 \times 10^6$ t in 1985, depending on availability and demand. Beyond that date, the expansion of exports could accelerate.

(2) The growth of nuclear energy capacity is still surrounded by major uncertainties, although some fairly substantial expansion can be taken for granted. On the basis of a cautious assessment by the OECD in 1977, a figure of 85 800 MWe of installed capacity seems possible in early 1985. Even this estimate, however, included some reactors for which permission had not in 1976 been granted.

(3) Natural gas consumption will expand rapidly, and this fuel will be one of the balancing contributors to the Community's energy demand, given its wide applicability and environmental advantages. Indigenous production can also be expected to grow, but it will come increasingly under pressure from conservation policies. This holds true particularly for exports from the Netherlands to other member countries. Figures presented by member states on present form suggest a level of production of $140-160 \times 10^6$ toe. In spite of this development of indigenous resources, substantial imports will be needed, with about $79 \times 10^6$ toe already contracted.

(4) It can be assumed that indigenous oil production will rise to around $160 \times 10^6$ toe in 1985. Oil imports will, however, continue to be the main contributor to the Community's energy balance. The development of imports depends (as in the case of natural gas imports) crucially on the evolution of energy consumption. The Community estimates assume an annual growth rate in energy consumption of 3.3%, leading to a total energy consumption of $1282 \times 10^6$ toe in 1985. Other estimates project a higher energy consumption. In any case, an overall import dependence of about 50% appears to be the minimum obtainable; it could easily be higher. (A summary of several recent estimates of the Community's overall energy balance in 1985 is provided in *Table 2.11*).

# Notes to Part one

1. R.L. Gordon, *The Evolution of Energy Policy in Western Europe, The Reluctant Retreat from Coal*, Praeger, New York 1970; figures for 1975 from *Pétrole 1977*, CPdP, Paris 1978
2. *Le Monde*, 16 Nov 1976; J. Janiaud, 'Les charbonnages et leurs handicaps', *Energies*, 11 March 1977, pp. 6–7
3. *The Coal Situation in Europe in 1973 and its Prospects*, UNECE, Geneva 1974, p. 9
4. *Investment in the Community Coalmining and Iron and Steel Industries*, ECSC, Luxembourg 1975, pp. 21, 56
5. D.C. Ion, *Availability of World Energy Resources*, Graham and Trotman, London 1975, pp. 78–81; *The Financial Times*, 10 March 1977
6. *International Herald Tribune*, April 1977 (Survey: Germany Pt. II), p. 26 S
7. Bureau Européen d'Informations Charbonnières, *Nouvelles et Commentaires*, No. 4/1977, pp. 8–9
8. *Second Report on the Achievement of the Community Energy Policy Objectives for 1985*, EC Commission, Brussels 1977 (Com(77)395)
9. *Second Report on the Achievement of the Community Energy Policy Objectives for 1985*, EC Commission, Brussels 1977 (Com(77)395)
10. *Energy Prospects to 1985*, Vol. II, OECD, Paris 1974, p. 163; A.J. Bromley, 'Coal in Western Europe', Science Policy Research Unit, University of Sussex, 1973, Pt. I, p. 16 (unpublished paper)
11. *Agence Europe*, 1 Dec. 1977
12. *Energy Prospects to 1985*, Vol. II, OECD, Paris 1974, p. 164; *Agence Europe*, 21–22 March 1977
13. *The Financial Times*, 19 Feb. 1976
14. *International Herald Tribune*, 18 April 1975
15. *Investment in the Community Coalmining and Iron and Steel Industries*, ECSC, Luxembourg 1975, p. 18; *Agence Europe*, 22 March 1978. The year 1976 saw another 40% increase in investments in coal in the European Community, but since then investments have stabilized at the new high level
16. M.V. Posner, 'National energy policies: conflict and common interest', in Frans A.M. Alting von Geusau, (ed.), *Energy in the European Communities*, A.W. Sigithoff, pp. 155–165 (159)
17. A.J. Bromley, 'Coal in Western Europe', Science Policy Research Unit, University of Sussex, 1973, Pt. II, p. 8 (unpublished paper)
18. D. Schmitt, 'Alternatives to oil imported from OPEC countries: coal as a substitute', in Frans A.M. Alting von Geusau, (ed.), *Energy in the European Communities*, A.W. Sigithoff, Leyden 1975, pp. 89–104 (94)
19. D. Schmitt, 'Alternative to oil imported from OPEC countries: coal as a substitute', in Frans A.M. Alting von Geusau, (ed.), *Energy in the European Communities*, A.W. Sigithoff, Leyden 1975, p. 93 (94)
20. *Energy Prospects to 1985*, Vol. I, OECD, Paris 1974, p. 145
21. A.J. Bromley, 'Coal in Western Europe', Science Policy Research Unit, University of Sussex, 1973, Pt. II, p. 9 (unpublished paper)
22. *The Financial Times*, 19 Feb. 1976
23. *Investment in the Community Coalmining and Iron and Steel Industries*, ECSC, Luxembourg 1975, p. 14
24. A.J. Bromley, 'Coal in Western Europe', Science Policy Research Unit, University of Sussex, 1973, Pt. II, p. 9 (unpublished paper)
25. See for example *Activités de Gasunie, Rapport annuel pour l'éxércise 1976*, NV Nederlands Gasunie, Groningen 1977 (for Netherlands); *Erdoel / Erdgas Zeitschrift*, May 1977, p. 159 (for Germany); 'Natural gas in Europe: a survey', *Petroleum Times*, 28 Oct. 1977, pp. 31–37
26. *The Financial Times*, 7 Oct. 1977 (Survey: World Coal Mining); G. Foley, *The Energy Question*, Penguin, Harmondsworth 1976, pp 240–241
27. See for example, 'Mobil Oil dans le péloton de tête pour la récherche des nouvelles sources d'énérgie', *Techniques de l'Energie*, May 1977, pp. 18–20; R. Dumon, 'Le charbon et la charbochimie dans les prochaines décennies', *L'Industrie du Pétrole*, Nov. 1977, pp. 23–27; G. Foley, *The Energy Question*, Penguin, Harmondsworth 1976, pp. 238–240; N.J.D. Lucas, *Energy and the European Communities*, Europa Publications, London 1977, Pt. IV B; N.J.D. Lucas, 'Nuclear power in the EEC – the cost of security', *Energy Policy*, June 1976, pp. 98–108. A sceptical view on coal gasification can be found in A.B. Lovins, *Soft*

*Energy Paths: Towards a Durable Peace*, Penguin, Harmondsworth 1977, pp. 114–116
28 *The Financial Times*, 7 Oct. 1977 (Survey: World Coal Mining)
29 A.B. Lovins, *Soft Energy Paths: Towards a Durable Peace*, Penguin, Harmondsworth 1977, Chs. 2 and 7
30 *The Financial Times*, 7 Oct. 1977 (Survey: World Coal Mining)
31 G. Delannoy, 'L'avenir du charbon dans l'économie énergétique', *Revue Générale de Thermique*, Aug.–Sept. 1977, pp 627–633
32 *The Financial Times*, 7 Oct. 1977 (Survey: World Coal Mining)
33 *The Financial Times*, 7 Oct. 1977 (Survey: World Coal Mining)
34 *The Financial Times*, 7 Oct. 1977 (Survey: World Coal Mining)
35 *Energy Prospects to 1985*, Vol. I, OECD, Paris 1974, p. 118; *Impact of Natural Gas on the Consumption of Energy in the OECD European Member Countries*, OECD, Paris 1969, p. 62
36 E.K. Faridany, *LNG: 1974–1990, Marine Operations and Market Prospects for Liquefied Natural Gas*, Economist Intelligence Unit, London 1974, p. 67 (Quarterly Economic Review Special No. 17)
37 *Impact of Natural Gas on the Consumption of Energy in the OECD European Member Countries*, OECD, Paris 1969, pp. 61–64
38 See note 25
39 *Energy Prospects to 1985*, Vol. II, OECD, Paris 1974, p. 139
40 *Energy Prospects to 1985*, Vol. II, OECD, Paris 1974, p. 139
41 *Petroleum Economist*, Feb. 1975, pp. 45–47, Oct. 1977, pp. 320–321; *Petroleum Times*, 14 Nov. 1975, p. 4, 28 Nov. 1977, pp. 31–37. Production estimates for Dutch offshore fields in 1985 indicate a volume of 16 x $10^9 m^3$, as opposed to 3.5 x $10^9 m^3$ in 1976. Total proven offshore reserves in 1976 were 400 x $10^9 m^3$
42 *Petroleum Economist*, May 1976, p.177
43 B. de Vries and J. Kommandeur, 'Gas for Europe: how much for how long?' *Energy Policy*, March 1974, pp. 24–37
44 B. de Vries and J. Kommandeur, 'Gas for Europe: how much for how long?', *Energy Policy*, March 1974, pp. 24–37
45 P.R. Odell, *Oil and World Power: Background to the Oil Crisis*, 3rd edn., Penguin, Harmondsworth 1974; P.R. Odell, *Natural Gas in Western Europe, A Case Study in the Economic Geography of Resources*, De Erven F. Bohn, Haarlem 1969; L. Turner, 'State and commercial interests in North Sea oil and gas, conflict and correspondence', in M. Saeter and I. Smart, (eds.), *The Political Implications of North Sea Oil and Gas*, IPC Science and Technology Press, Guildford 1975, pp. 93–110; D.C. Ion, *Availability of World Energy Resources*, Graham and Trotman, London 1975, p. 36
46 *Impact of Natural Gas on the Consumption of Energy in the OECD European Member Countries*, OECD, Paris 1969, pp. 58–59
47 P.L. Cook and A.J. Surrey, *Energy Policy: Strategies of Uncertainty*, Martin Robertson, London 1977, Ch. 7
48 I.D. Mackay and G.A. Mackay, *The Political Economy of North Sea Oil*, Martin Robertson, London 1975, pp. 58–59
49 *Energy Prospects to 1985*, Vol. II, OECD, Paris 1974, p. 141; *Petroleum Economist*, Nov. 1977, pp. 354–356
50 P. Cook and A.J. Surrey, *Energy Policy, Strategies of Uncertainty*, Martin Robertson, London 1977, Ch. 7
51 *Second Report on the Achievement of the Community Energy Policy Objectives for 1985*, EC Commission, Brussels 1977, (Com(77) 395); *Energy Prospects to 1985*, Vol. II, OECD, Paris 1974, p. 139
52 *Petroleum Economist*, July 1976, p. 257, Nov. 1977, pp. 354–356; *Energy – situation in the Netherlands*, Nederlandse Middenstandsbank NV, Amsterdam 1978 (unpublished MS)
53 *Second Report on the Achievement of the Community Energy Policy Objectives for 1985*, EC Commission, Brussels 1977, (Com(77)395)
54 P.R. Odell, *National Gas in Western Europe, A Case Study in the Economic Geography of Resources*, De Erven F. Bohn, Haarlem 1969, pp. 58,67; P.R. Odell, 'The economic background to North Sea oil and gas development', in M. Saeter and I. Smart, (eds.), *The Political Implications of North Sea Oil and Gas*, IPC Science and Technology Press, Guildford 1975, pp. 51–75
55 B. de Vries and J. Kommandeur, 'Gas for Europe; how much for how long?', *Energy Policy*, March 1974, pp. 24–37

56 P.R. Odell, 'The economic background to North Sea oil and gas development', in M. Saeter and I. Smart, (eds.), *The Political Implications of North Sea Oil and Gas*, IPC Science and Technology Press, Guildford 1975
57 See P.L. Cook and A.J. Surrey, *Energy Policy, Strategies of Uncertainty*, Martin Robertson, London 1977, pp. 111–112
58 N.J.D. Lucas, *Energy and the European Communities*, Europa Publications, London 1977, Pt. IIIB; Guy de Carmoy, *Energy for Europe, Economic and Political Implications*, American Enterprise Institute for Public Policy Research, Washington 1977, Pts. III and IV; *Uranium, Resources, Production and Demand*, OECD, Nuclear Energy Agency, IAEA, Paris 1975
59 *Report on the Achievement of the Community Energy Policy Objectives for 1985*, EC Commission, Brussels 1976 (Com(76)9)
60 *Second Report on the Achievement of the European Community Policy Objectives for 1985*, EC Commission, Brussels 1977, (Com(77)395)
61 Amory Lovins is the best-known and most articulate of these experts. See his *Soft Energy Paths: Towards a Durable Peace*, Penguin, Harmondsworth 1977
62 *Overall Energy Balance Sheets, 1963–1976*, Eurostat, Luxembourg 1977
63 T. Leardini, 'Electric energy in the European Community: supply and demand patterns for the medium-term future', in Frans A.M. Alting von Geusau, (ed.), *Energy in the European Communities*, A.W. Sigithoff, Leyden 1975, pp. 33–54 (38)
64 T. Leardini, 'Electric energy in the European Community; supply and demand patterns for the medium-term future', in Frans A.M. Alting von Geusau, (ed.), *Energy in the European Communities*, A.W. Sigithoff, Leyden 1975, pp. 33–54 (38); *Energy Prospects to 1985*, Vol. II, OECD, Paris 1974, p. 171
65 *Second Report on the Achievement of the Community Energy Policy Objectives for 1985*, EC Commission, Brussels 1977 (Com(77)395)
66 *Petroleum Economist*, Aug. 1976, p. 311; M. Grenon, 'Alternatives to oil imported from OPEC countries; nuclear energy', in Frans A.M. Alting von Geusau, (ed.), *Energy in the European Communities*, A.W. Sigithoff, Leyden 1975, pp. 105–136 (106–107)
67 A.B. Lovins, *Soft Energy Paths: Towards a Durable Peace*, Penguin, Harmondsworth 1977; H.S.M. Keeney, Jr. *et al., Nuclear Power Issues and Choices*, Ballinger, Cambridge, Mass. 1977 (Ford/Mitre Report), pp. 152–153
68 N.J.D. Lucas, 'Nuclear power in the EEC – the cost of security', *Energy Policy*, June 1976
69 N.J.D. Lucas, 'Nuclear power in the EEC – the cost of security', *Energy Policy*, June 1976
70 *Second Report on the Achievement of the Community Energy Policy Objectives for 1985*, EC Commission, Brussels 1977 (Com(77)395)
71 G. Foley, *The Energy Question*, Penguin, Harmondsworth 1976, pp. 92–96. This finds its expression also in higher costs. For the USA, recent calculations show the cost of $1 \times 10^6$ BTU heat equivalent produced from electricity to fall between \$7.85 and 9.26, while the same amount of heat produced from OPEC oil imports will cost \$2.41 (*Petroleum Economist*, Aug. 1977, p. 316)
72 *Second Report on the Achievement of the Community Energy Policy Objectives for 1985*, EC Commission, Brussels 1977 (Com(77)395)
73 W.C. Patterson, *Nuclear Power*, Penguin, Harmondsworth 1976, p. 226; for the USA, H.S.M. Keeney, Jr. *et al., Nuclear Power Issues and Choices*, Ballinger, Cambridge, Mass. 1977, (Ford/Mitre Report) pp. 119–120; see also C. Sweet, 'Nuclear power costs in the UK', *Energy Policy*, June 1978, pp. 107–118
74 M. Benedict, 'Electric power from nuclear fission', in L.C. Ruedisili and M.W. Firebaugh, (eds.), *Perspectives on Energy, Issues, Ideas and Environmental Dilemmas*, Oxford University Press, New York 1975, pp. 196–208 (199)
75 M. Benedict, 'Electric power from nuclear fission', in L.C. Ruedisili and M.W. Firebaugh, (eds.), *Perspectives on Energy, Issues, Ideas and Environmental Dilemmas*, Oxford University Press, New York 1975, pp. 196–208 (199)
76 A good brief survey of reactor characteristics can be found in W.C. Patterson, *Nuclear Power*, Penguin, Harmondsworth 1976, Pt. I, and in T. Greenwood, G.W. Rathjens, J. Ruina,

*Nuclear Power and Weapons Proliferation*, IISS, London 1976 (Adelphi Paper No. 130), Ch. 3

77 A discussion of FBR issues can be found in H.S.M. Keeney, Jr. et al., *Nuclear Power Issues and Choices*, Ballinger, Cambridge, Mass. 1977 (Ford/Mitre Report), Ch. 12; for a supportive analysis, see H. Michaelis, *Kernenergie*, dtv, München 1977, pp. 51–52, 256 seq.

78 M. Benedict, 'Electric power from nuclear fission', in L.C. Ruedisili and M.W. Firebaugh, (eds.), *Perspectives on Energy Issues, Ideas and Environmental Dilemmas*, Oxford University Press, New York 1975, p. 199

79 W.C. Patterson, *Nuclear Power*, Penguin, Harmondsworth 1976, Ch. 8; for accident record, see Ch. 7.

80 D.J. Rose, 'Nuclear electric power', in L.C. Ruedisili and M.W. Firebaugh, (eds.), *Perspectives on Energy, Issues, Ideas and Environmental Dilemmas*, Oxford University Press, New York 1975, pp. 209–223 (216); H.S.M. Keeney, Jr. et al., *Nuclear Power Issues and Choices*, Ballinger, Cambridge, Mass. 1977 (Ford/Mitre Report), pp. 175 seq.

81 W.C. Patterson, *Nuclear Power*, Penguin, Harmondsworth 1976, pp. 214–215

82 US Atomic Energy Commission, 'Reactor safety study, an assessment of accident risks in US commercial power plants', in L.C. Ruedisili and M.W. Firebaugh, (eds.), *Perspectives on Energy, Issues, Ideas and Environmental Dilemmas*, Oxford University Press, New York, 1975, pp. 224–237 (Summary of Rasmussen Report)

83 H.S.M. Keeney, Jr. et al., *Nuclear Power Issues and Choices*, Ballinger, Cambridge, Mass. 1977 (Ford/Mitre Report), Ch. 7

84 H.S.M. Keeney, Jr. et al., *Nuclear Power Issues and Choices*, Ballinger, Cambridge, Mass. 1977 (Ford/Mitre Report), Ch. 7

85 D. Krieger, 'Nuclear power: a Trojan horse for terrorists', in *Nuclear Proliferation Problems*, SIPRI, London 1974, pp. 187–198 (193)

86 See for example, M. Willrich and T.P. Taylor, *Nuclear Theft: Risks and Safeguards, A Report to the Energy Project of the Ford Foundation*, Ballinger, Cambridge, Mass. 1974; *Nuclear Proliferation Problems*, SIPRI, London 1974. While the probability of terrorists or criminals acquiring nuclear weapons seems quite small, this does not remove the costs and dilemmas of precautionary measures against such diversion

87 D.P. Geesaman and D.E. Abrahamson, 'The dilemma of fission energy', in L.C. Ruedisili and M.W. Firebaugh, (eds.), *Perspectives on Energy, Issues, Ideas and Environmental Dilemmas*, Oxford University Press, New York 1975, pp. 273–278 (277)

88 H. Alven, 'Energy and environment', *Bulletin of the Atomic Scientist*, May 1972, p. 6

89 The following is based on R. Day, 'Nuclear power, possibilities and problems', Science Policy Research Unit, University of Sussex 1975 (unpublished MS); BEIR Committee, 'Report on biological effects on ionizing radiation, summary and recommendations', in L.C. Ruedisili and M.W. Firebaugh, (eds.), *Perspectives on Energy, Issues, Ideas and Environmental Dilemmas*, Oxford University Press, New York, 1975, Ch. 5; J. Picard, 'Some problems posed by the growing demand for energy', in S.A. Saltzman, (ed.), *Energy Technology and Global Policy*, Clio Books, Santa Barbara 1977, pp. 23–38

90 J. Picard, 'Some problems posed by the growing demand for energy', in S.A. Saltzman', (ed.), *Energy Technology and Global Policy*, Clio Books, Santa Barbara 1977

91 E. Gaul, *Atomenergie oder Ein Weg aus der Krise?* Rororo, Reinbek bei Hamburg 1974, pp. 34–35

91a H.S.M. Keeney, Jr. et al., *Nuclear Power Issues and Choices*, Ballinger, Cambridge, Mass.1977 (Ford/Mitre Report)

92 Environmental Education Group/ Environmental Alert Group, 'The clear and present danger: nuclear power plants', in L.C. Ruedisili and M.W. Firebaugh, (eds.), *Perspectives on Energy, Issues, Ideas and Environmental Dilemmas*, Oxford University Press, New York 1975, pp. 262–272 (265–266); R. Day, 'Nuclear power, possibilities and problems', Science Policy Research Unit, University of Sussex 1975, p. 13

93 Environmental Education Group/ Environmental Alert Group, 'The clear and present danger: nuclear power plants', in L.C. Ruedisili and M.W. Firebaugh, (eds.), *Perspectives on Energy, Issues, Ideas and Environmental*

*Dilemmas*, Oxford University Press, New York 1975, p. 265

94  D.P. Geesaman and D.E. Abrahamson, 'The dilemma of fission energy', in L.C. Ruedisili and M.W. Firebaugh, (eds.), *Perspectives on Energy, Issues, Ideas and Environmental Dilemmas*, Oxford University Press, New York 1975, p. 274

95  W.C. Patterson, *Nuclear Power*, Penguin Harmondsworth 1976, Ch. 9; Royal Commission on Environmental Pollution, *Nuclear Power and the Environment*, HMSO, London 1976 (Cmnd. 6618) (Flowers Report)

96  W.C. Patterson, *Nuclear Power*, Penguin, Harmondsworth 1976, p. 111

97  J. Picard, 'Some problems posed by the growing demand for energy', in S.A. Saltzman, (ed.), *Energy Technology and Global Policy*, Clio Books, Santa Barbara 1977; A.B. Lovins and J.H. Price *Non-nuclear Futures, The Case for an Ethical Energy Strategy*, Ballinger, Cambridge, Mass. 1975, Pt. I. Plutonium production estimate for 1980 based on data from *Uranium Resources, Production and Demand*, OECD Nuclear Energy Agency, IAEA, Paris 1977 on capacity and plutonium residue

98  Environmental Education Group/ Environmental Alert Group, 'The clear and present danger: nuclear power plants', in L.C. Ruedisili and M.W. Firebaugh, (eds.), *Perspectives on Energy, Issues, Ideas and Environmental Dilemmas*, Oxford University Press, New York 1975, p. 265

99  See for Germany, *Neue Zuercher Zeitung*, 16 Dec. 1977; for Italy, *The Economist*, 15 Oct. 1977, p. 93; *The Financial Times*, 4 April 1978

100  But see the incidents at the Maleville FBR site (*The Economist*, 6 Aug. 1977, pp. 37–38). On nuclear decision-making in France, see the provocative but fascinating book by P. Simonnot, *Les nucléocrates*, Presses universitaires de Grenoble, Grenoble 1978. An interesting nuclear 'dissident' is L. Puiseux, *La babel nucléaire*, Editions Galilée, Paris 1977 – he changed from an influential advocate and planner of nuclear energy within the French utility Electricité de France to one of the most articulate opponents – not without having his share of problems with his employer

101  P.L. Cook and A.J. Surrey, *Energy Policy, Strategies of Uncertainty*, Martin Robertson, London 1977, Ch. 8; on nuclear opposition in general, see A.J. Surrey and C. Huggett, 'Opposition to nuclear power: a review of international experience', *Energy Policy*, Dec. 1976

102  A.B. Lovins, *Soft Energy Paths: Towards a Durable Peace*, Penguin, Harmondsworth 1977, Pt. I

103  B. Commoner, *The Poverty of Power, Energy and the Economic Crisis*, Bantam Books, New York 1977, in particular Ch. 5

104  A.B. Lovins and J.H. Price, *Non-nuclear Futures, the Case for an Ethical Energy Strategy*, Ballinger, Cambridge, Mass. 1975, Pt. II

105  P.R. Odell, 'The economic background to North sea oil and gas development' in M. Saeter and I. Smart, (eds.), *The Political Implications of North Sea Oil and Gas*, IPC Science and Technology Press, Guildford 1975, p. 65. Production in the Community (6) has fallen from about $11.6 \times 10^6$ t in 1973 to about $8.3 \times 10^6$ t in 1973. For the Nine, however, the UK has made a big difference – an increase from $12.05 \times 10^6$ t (1973) to $47.2 \times 10^6$ t (1977)

106  *Petroleum Economist*, Nov. 1974, p. 418

107  *Petroleum Economist*, March 1976, p. 83

108  *The Financial Times*, 28 April 1976

109  *The Financial Times*, 28 April 1976

110  This issue has caused permanent conflict between the EC Commission and the UK; the Commission sees the British insistence on refining two-thirds of its own crude oil as a breach of the Treaty of Rome, and as an obstacle in its attempts to reorganize and streamline European refinery industries. See N.J.D. Lucas, *Energy and the European Communities*, Europa Publications, London 1977 pp. 121–123; *Agence Europe*, 11 Jan., 11–23 Feb., 1978

111  Department of Energy, *Development of the Oil and Gas Resources of the United Kingdom*, HMSO, London 1978

112  I.D. Mackay and G.A. Mackay, *The Political Economy of North Sea Oil*, Martin Robertson, London 1975, p. 62

113  P.R. Odell and K. Rosing, 'The North Sea oil province: a simulation model of development', *Energy Policy*, Dec. 1974, pp. 316–329

114  M. Saeter and I. Smart, (eds.), *The Political Implications of North Sea Oil*

*and Gas*, IPC Science and Technology Press, Guildford 1975
115 *Energy Prospects to 1985*, Vol I, OECD, Paris 1974, p. 95; C. Buckley, 'Oil in the North Sea', Science Policy Research Unit, University of Sussex, 1975, p. 2 (unpublished MS). BP hopes to recover 41% of the oil in place in the Forties field
116 *Petroleum Economist*. Aug. 1974, pp. 301–302
117 *The Financial Times*, 17 Dec. 1976, 2 Feb. 1978. In February 1978, oil was first struck west of the Shetlands, reviving new hopes for the hitherto disappointing area. The commercial viability of the find is as yet uncertain, however
118 Department of Energy, *Development of the Oil and Gas Resources of the United Kingdom*, HMSO, London 1978
119 P.R. Odell and K. Rosing, 'The North Sea oil and gas province: a simulation of its development', *Energy Policy*, Dec. 1974, pp. 316–329
120 *Energy Prospects to 1985*, Vol. I, OECD, Paris 1974, pp. 94–95
121 J. Birks, 'North Sea in 1973', a paper to the Financial Times North and Celtic Sea Conference, 1973 (mimeographed paper); BBC2 television programme 'Controversy', 29 Jan. 1975; M. Saeter and I. Smart, (eds.), *The Political Implications of North Sea Oil and Gas*, IPC Science and Technology Press, Guildford 1975, p. 77
122 M. Saeter and I. Smart, (eds.), *The Political Implications of North Sea Oil and Gas*, IPC Science and Technology Press, Guildford 1975, pp. 75–77. See also the controversy in *Energy Policy* in 1977 (see note 54)
123 *The Financial Times*, 28 April 1976
124 *The Economist*, 'Survey on North Sea oil', 26 July, 1975, p. 12; C. Buckley, 'Oil in the North Sea', Science Policy Research Unit, University of Sussex, 1975 (unpublished MS)
125 I.D. Mackay and G.A. Mackay, *The Political Economy of North Sea Oil*, Martin Robertson, London 1975, p. 97
126 *The Economist*, 'Survey on North Sea oil', 26 July 1976, p. 16
127 M. Saeter and I. Smart, (eds.), *The Political Implications of North Sea Oil and Gas*, IPC Science and Technology Press, Guildford 1975, pp. 76–77
128 *The Financial Times*, 19 April 1978
129 *The Financial Times*, 7 Dec. 1974
130 I.D. Mackay and G.A. Mackay, *The Political Economy of North Sea Oil*, Martin Robertson, London 1975, p. 44
131 I.D. Mackay and G.A. Mackay, *The Political Economy of North Sea Oil*, Martin Robertson, London 1975, pp. 168–173
132 I.D. Mackay and G.A. Mackay, *The Political Economy of North Sea Oil*, London 1975, p. 177
133 *Noroil*, April 1976, p. 27
134 *Agence Europe*, 25 Feb. 1978
135 *Agence Europe*, 23 Feb. 1978
136 D. Schmitt, 'Alternatives to oil imported from OPEC countries: coal as a substitute', in Frans A.M. Alting von Geusau, (ed.), *Energy in the European Communities*, A.W. Sigithoff, Leyden 1975, p. 91; see also M. Sakisaka, 'The present and future energy resources', in C. Gasteyger, (ed.), *The Western World and Energy*, Atlantic Institute, Paris 1974, pp. 15–36 (Atlantic Paper No. 1, 1974). For a detailed assessment of the reserve and resources outlook, see W. Peters, H.D. Schilling *et al.*, 'An appraisal of world coal resources and their future availability', in World Energy Conference, *World Energy Resources 1985–2020*, IPC Science and Technology Press, Guildford 1978, pp. 57–86
137 *Second Report on the Achievement of the Community Energy Policy Objectives for 1985*, EC Commission, Brussels 1977 (Com(77)395)
138 *The Financial Times*, 8 March 1978, 7 Oct. 1977 (Survey: World Coal Mining)
139 *The Financial Times*, 7 Oct. 1977 (Survey: World Coal Mining)
140 W. Peters, H.D. Schilling *et al.*, 'An appraisal of world coal resources and their future availability', in World Energy Conference, *World Energy Resources 1985–2020*, IPC Science and Technology Press, Guildford 1978, p. 81; *World Energy Outlook*, OECD, Paris 1977
141 W. Peters, H.D. Schilling *et al.*, 'An appraisal of world coal resources and their future availability', in World Energy Conference, *World Energy Resources 1985–2020*, IPC Science and Technology Press, Guildford 1978, p. 81
142 *The Financial Times*, 7 Oct. 1977 (Survey: World Coal Mining); W. Peters, H.D. Schilling *et al.*, 'An appraisal of

world coal resources and their future availability', in World Energy Conference, *World Energy Resources 1985–2020*, IPC Science and Technology Press, Guildford 1978, p. 81

143 J.H. Chesshire and C. Hugget, 'Primary energy production in the Soviet Union', *Energy Policy*, Sept. 1975 pp. 223–244; J. Russell, *Energy as a Factor in Soviet Foreign Policy*, Royal Institute for International Affairs, London 1976, pp. 71–79; W. Peters, H.D. Schilling et al., 'An appraisal of world coal resources and their future availability', World Energy Conference, *World Energy Resources 1985–2020*, IPC Science and Technology Press, Guildford 1978, p. 81; *The Financial Times*, 7 Oct. 1977 (Survey: World Coal Mining)

144 G. Foley, *The Energy Question*, Penguin, Harmondsworth 1976, pp. 115–128

145 *Energy Prospects to 1985*, OECD Paris 1974, Vol. I, p. 130, Vol. II, pp. 155, 164

146 A.J. Bromley, 'Coal in Western Europe', Science Policy Research Unit, University of Sussex, 1973, p. 44

147 *Energy: Global Prospects 1985–2000*, McGraw-Hill, New York 1977 (Report of the Workshop on Alternative Energy Strategies) Ch. 5

148 *The Financial Times*, 24 Feb. 1978

149 *Petroleum Economist*, Oct. 1977, pp. 320–321; *The Financial Times*, 9 May 1978

150 For a detailed discussion of hazards involved in LNG trade, see S. Mankabady, 'The affreightment of liquefied natural gas', *Journal of World Trade Law*, Nov./Dec. 1975, pp. 654–664; technical and economic problems are discussed in *Impact of Natural Gas on the Consumption of Energy in the OECD European Member Countries*, OECD, Paris 1969, pp. 40–43

151 *Briefing Service: Growth of Natural Gas*, Royal Dutch/Shell, London 1975, p.6

152 *Impact of Natural Gas on the Consumption of Energy in the OECD European Member Countries*, OECD, Paris 1969, p. 40

153 *Pétrole et Gas Arabe*, 1 Nov. 1977; *Middle East Economic Survey*, 31 Oct. 1977

154 *Gaz de France, Service de Presse*, No. 88, July 1977

155 E.K. Faridany, *LNG Review 1977, A Study of Marine Operations Ships Technology, Project Finance, LNG Prices and Market Prospects for LNG*, Energy Economics Research Ltd., 1977, p. 23. Although the Algerian government has repeatedly stressed that it wants to keep gas supplies free from politics, an interruption of supplies to the USA was actually considered (but rejected) in 1973

156 *The Economist*, 6 Nov. 1976

157 *The Economist*, 6 Nov. 1976; see also E.K. Faridany, *LNG Review 1977, A Study of Marine Operations Ships Technology, Project Finance, LNG Prices and Market Prospects for LNG*, Energy Economics Research Ltd., 1977; *Petroleum Economist*, March 1978, p. 123

158 *The Financial Times*, 9 Nov. 1977; *Petroleum Intelligence Weekly*, 31 Oct. 1977

159 J.A. Yager and E.B. Steinberg, *Energy and US Foreign Policy, A Report to the Energy Project of the Ford Foundation*, Ballinger, Cambridge, Mass. 1974, p. 344

160 *International Herald Tribune*, 23 April 1975

161 *Uranium, Resources, Production and Demand*, OECD Nuclear Energy Agency, IAEA, Paris 1977, pp. 119–121

162 See *The Economist*, 24 Sept. 1977, 15 April 1978 (Survey: Australia), p. 20; *International Herald Tribune*, 10 Oct. 1977; *The Financial Times*, 13 June 1975, 16 Feb. 1976, 14 April 1977, 23 May 1977, 16 Dec. 1977, 20 Jan. 1978

163 *Uranium, Resources, Production and Demand*, OECD Nuclear Energy Agency, IAEA, Paris 1977, p. 62; *Agence Europe*, 16–17 Jan. 1978; *The Financial Times*, 21 Dec. 1977; *Neue Zuercher Zeitung*, 11 Jan. 1978

164 J.A. Yager and E.B. Steinberg, *Energy and US Foreign Policy, A Report to the Energy Project at the Ford Foundation*, Ballinger, Cambridge, Mass. 1974

165 *The Financial Times*, 13 Oct. 1977, 14 April 1976; 12 April 1975. Canada, however, appears to have shifted back towards a preference for exports of uranium ore since its indigenous CANDU reactors do not require enriched uranium, and enrichment installations would therefore only serve for exports

166 See *The Financial Times*, 29 May 1975, 24 Aug. 1977. In 1975, a pilot plant started working, but a commercial

plant based on the jet nozzle technology planned for 1974 encountered major difficulties. *The Economist*, 25 Feb. 1978

167 *The Financial Times*, 13 June 1975, 6 Oct. 1975. The Institute does not include US companies which shied away from the cartel appearance of the organization. Discussions on prices are, however, explicitly ruled out by the statutes. Some consumers (utilities) have meanwhile joined the Institute. See *The Middle East*, Aug./Sept. 1975, pp. 38–43

168 C.-W. Sames, *Die Zukunft der Metalle*, Suhrkamp, Frankfurt 1974, p. 179; *Uranium, Resources, Production and Demand*, OECD Nuclear Energy Agency, IAEA, Paris 1977

169 *Le Monde*, 28 Nov. 1975; *Uranium, Resources, Production and Demand*, OECD Nuclear Energy Agency, IAEA, Paris 1977

170 C.-W. Sames, *Die Zukunft der Metalle*, Suhrkamp, Frankfurt 1974, p. 179

171 See J.-M. Chevalier, *Le nouvel enjeu pétrolier*, Calmann-Levy, Paris 1973, Ch. 5; C. Tugendhat and A. Hamilton, *Oil, The Biggest Business*, 2nd edn., Eyre Methuen, London 1975, Ch. 30; J.A. Yager and E.B. Steinberg, *Energy and US Foreign Policy, A Report to the Energy Project of the Ford Foundation*, Ballinger, Cambridge, Mass. 1974, pp. 233–234

172 *The Financial Times*, 17 July 1975; J.A. Yager and E.B. Steinberg, *Energy and US Foreign Policy, A Report to the Energy Project of the Ford Foundation*, Ballinger, Cambridge, Mass. 1974, p. 350

173 M. Willrich and P.A. Marston, 'Prospects for a uranium cartel', *Orbis*, Spring 1975, pp. 166–184 (173–178); see also J. Kostuik, 'The uranium industry, key issues in future development', *Energy Policy*, Sept. 1976, pp. 212–224

174 M. Willrich and P.A. Marston, 'Prospects for a uranium cartel', *Orbis*, Spring 1975, p. 182

175 *International Herald Tribune*, 26 April 1977; *The Financial Times*, 9 July 1976; *The Sunday Times*, 20 July 1976. See also M. Willrich and P.A. Marston, 'Prospects for a uranium cartel', *Orbis*, Spring 1975 and Z. Mikdashi, *The International Politics of Natural Resources*, Cornell University Press, Ithaca 1976, pp. 108–109

176 *International Herald Tribune*, 26 April 1976

177 *The Economist*, 3 May 1975

178 H.S.M. Keeney, Jr. *et al.*, *Nuclear Power Issues and Choices*, Ballinger, Cambridge, Mass. 1977 (Ford/Mitre Report), Ch. 11

179 *The Financial Times*, 15 Jan. 1976

180 *The Financial Times*, 28 Feb. 1978

181 *Uranium Resources, Production and Demand*, OECD Nuclear Energy Agency, IAEA, Paris 1977

182 J.A. Yager and E.B. Steinberg, *Energy and US Foreign Policy, A Report to the Energy Project of the Ford Foundation*, Ballinger, Cambridge. Mass. 1974, p. 349; M. Grenon, 'Alternatives to oil imported from OPEC countries: nuclear energy', in Frans A.M. Alting von Geusau, (ed.), *Energy in the European Communities*, A.W. Sigithoff, Leyden 1975, pp. 35–42; T. Greenwood, G.W. Rathjens and J. Ruina, *Nuclear Power and Weapons Proliferation*, IISS, London 1976, (Adelphi Paper No. 130), pp. 35–42; H. Michaelis, *Kernenergie*, dtv, München 1977, pp. 374 seq.

183 For a discussion of the FBR, see W.C. Patterson, *Nuclear Power*, Penguin, Harmondsworth 1976, pp. 75 *seq.*; H.S.M. Keeney, Jr. *et al.*, *Nuclear Power Issues and Choices*, Ballinger, Cambridge, Mass. 1977 (Ford/Mitre Report), Ch. 12; W. Haefele, 'The fast breeder as a cornerstone of future supplies', in S.A. Saltzman, (ed.), *Energy, Technology and Global Policy*, Clio Books, Santa Barbara 1977, pp. 59–80

184 *The Economist*, 19 April 1975, 12 July 1975; *The Financial Times*, 8 March 1977; *The Economist*, 25 Feb. 1978, 17 June 1978. By mid-1978, the difficulties were largely overcome; *The Financial Times*, 11 July 1978

185 *The Economist*, 19 April 1975, 12 July 1975

186 See T. Greenwood, G.W. Rathjens and J. Ruina, *Nuclear Power and Weapons Proliferation*, IISS, London 1976 (Adelphi Paper No. 130), pp. 21–27

187 *International Herald Tribune*, 17 Nov. 1975

188 H. Michaelis, *Kernenergie*, dtv, München 1977, pp. 359–366

189 *International Herald Tribune*, 17 Nov. 1975

190 T. Greenwood, G.W. Rathjens and J. Ruina, *Nuclear Power and Weapons*

*Proliferation*, IISS, London 1976 (Adelphi Paper No.130), p. 45; J.A. Yager and E.B. Steinberg, *Energy and US Foreign Policy; A Report to the Energy Project of the Ford Foundation*, Ballinger, Cambridge, Mass. 1974

191 M. Grenon, 'Alternatives to oil imported from OPEC countries: nuclear energy', in Frans A.M. Alting von Geusau, (ed.), *Energy in the European Communities*, A.W. Sigithoff, Leyden 1975, p. 129; H. Michaelis, *Kernenergie*, dtv, München 1977, pp. 361–363

192 For the pessimistic perspective, see W. Hager, *Europe's Economic Security, Non-Energy Issues in the International Political Economy*, Atlantic Institute, Paris 1975, pp. 55–57 (Atlantic Paper No. 3, 1975); for a more optimistic view, J. Kostuik, 'The uranium industry, key issues in future development', *Energy Policy*, Sept. 1976, and 'Future uranium supply and demand, industrial and commercial considerations', *Energy Policy*, June 1977, pp. 122–129. An authoritative assessment can be found in *Uranium Resources, Production and Demand*, OECD Nuclear Energy Agency, IAEA, Paris 1977

193 On the deal with Brazil, see *Der Spiegel*, 15 March 1976, and N. Gall, 'Atoms for Brazil, dangers for all', *Foreign Policy*, Summer 1976, pp. 155–201; on European measures to support uranium exploration, see C.-W. Sames, *Die Zukunft der Metalle*, Suhrkamp, Frankfurt 1974, pp. 180, 218

194 J. Maddox, *Prospects for Nuclear Proliferation*, IISS, London 1975, (Adelphi Paper No.113), p. 29

195 *The Economist*, 1 Feb. 1975

195a *Uranium Resources, Production and Demand*, OECD Nuclear Energy Agency, IAEA, Paris 1977

196 *Le Monde*, 2 Dec. 1975; *International Herald Tribune*, 13 Oct. 1975; *The Financial Times*, 3 May 1978

197 N.J.D. Lucas, 'Nuclear power in the EEC – the cost of security', *Energy Policy*, June 1976, Pt. IV F. One of the best accounts of the development of non-proliferation politics, and of the rift between Europe and the USA, is K. Kaiser, 'The great nuclear debate', *Foreign Policy*, Spring 1978, pp. 83–110. See also the documentation in *Europa Archiv*, No. 24, 1977, pp. D669–712

198 *Report on the Achievement of the Community Energy Policy Objectives for 1985*, EC Commission, Brussels 1976 (Com(76)9); *Second Report on the Achievement of the Community Energy Policy Objectives for 1985*, EC Commission, Brussels 1977 (Com(77) 395)

199 *World Energy Outlook*, OECD, Paris 1977

200 *Second Report on the Achievement of the Community Energy Policy Objectives for 1985*, EC Commission, Brussels 1977 (Com(77)395)

201 *World Energy Outlook*, OECD, Paris 1977

202 *The Financial Times*, 10 March 1978

203 Royal Norwegian Ministry of Finance, Parliamentary Report No. 25 (1973–1974), *Petroleum Industry in Norwegian Society*, Oslo 1974, p. 51

204 *The Financial Times*, 17 May 1977

205 Royal Norwegian Ministry of Finance, Parliamentary Report No. 25 (1973–1974), *Petroleum Industry in Norwegian Society*, Oslo 1974, p. 52

206 Royal Norwegian Ministry of Finance, Parliamentary Report No. 25 (1973–1974), *Petroleum Industry in Norwegian Society*, Oslo 1974, p. 8, pp. 55–59; see also A.C. Sjaastad, 'Norwegen als Erdoelproduzent, Aussichten und Probleme einer neuen Industrie', *Europa Archiv*, No. 8, 1978, pp. 241–250. The consensus about the need for a moderate production rate of oil and gas includes all political parties

207 X. Aamo, in M. Saeter and I. Smart, (eds.), *The Political Implications of North Sea Oil and Gas*, IPC Science and Technology Press, Guildford 1975 pp. 81–92

208 J. Russell, *Energy as a Factor in Soviet Foreign Policy*, Royal Institute for International Affairs, London 1976, pp. 42–47

209 See, for example, J.H. Chesshire and C. Hugget, 'Primary energy production in the Soviet Union', *Energy Policy*, Sept. 1975; J. Russell, *Energy as a Factor in Soviet Foreign Policy*, Royal Institute for International Affairs, London, 1976; W. Gumpel, 'Alternatives to oil imported from OPEC countries; oil and gas imported from the Soviet Union?' in Frans A.M. Alting von Geusau, (ed.), *Energy in the European Communities*, A.W. Sigithoff, Leyden

1975, pp. 137–151; *Oil and Gas Journal*, 17 Jan.1977, pp. 272–277 (275–276) (excerpt from CIA report to US Congress); UN Economic Commission for Europe (ECE), *New Issues affecting the Energy Economy of the EEC Region in the Medium and Long Term (Preliminary Version)*, Add 2, p. 23 (ECE(XXXIII)/2/Add. 1)
210 *The Economist*, 19 July 1975; W. Gumpel, 'Alternatives to oil imported from OPEC countries: oil and gas imported from the Soviet Union?', in Frans A.M. Alting von Geusau, (ed.), *Energy in the European Communities*, A.W. Sigithoff, Leyden 1975, pp.141–142
211 J.A. Yager and E.B. Steinberg, *Energy and US Foreign Policy, A Report to the Energy Project of the Ford Foundation*, Ballinger, Cambridge, Mass. 1974, p. 194
212 *The Economist*, 19 July 1975; W. Gumpel, 'Alternatives to oil imported from OPEC countries: oil and gas imported from the Soviet Union?', in Frans A.M. Alting von Geusau, (ed.), *Energy in the European Communities*, A.W. Sigithoff, Leyden 1975, pp. 141–142
213 *The Economist*, 19 July 1975; J.H. Chesshire and C. Hugget, 'Primary energy production in the Soviet Union', *Energy Policy*, Sept. 1975, p. 228; J. Russell, *Energy as a Factor in Soviet Foreign Policy*, Royal Institute for International Affairs, London 1976, Ch. 3; 'The Soviet economy', *Survival*, Nov./Dec. 1977
214 *The Financial Times*, 22 June 1978; *The Economist*, 30 April 1977
215 *The Economist*, 19 July 1975
216 *The Economist*, 9 July 1975; see also A. Wolynski, 'Soviet oil policy', in B. Crozier, *Soviet Objectives in the Middle East, Report of a Study Group of the Institute for the Study of Conflict*, ISC, London 1974, pp. 21–26
217 UN Economic Commission for Europe (ECE), *New Issues Affecting the Energy Economy of the EEC Region in the Medium and Long Term (Preliminary Version)*, Add. 1, p. 212; Groupe d'études prospectives internationales, 'Situation et perspectives du bilan énergétique de l'U.R.S.S. et de l'Est Européen' *Le Courrier des Pays de l'Est*, March 1978, pp. 3–16
218 *Petroleum Economist*, Aug. 1976, p. 303
219 *World Energy Supplies*, UN Statistical Papers, Series J, New York 1975
220 A. Wolynski, 'Soviet oil policy', in B. Crozier, *Soviet Objectives in the Middle East, Report of a Study Group of the Institute for the Study of Conflict*, ISC, London 1974, p. 25; see also J.A. Yager and E.B. Steinberg, *Energy and US Foreign Policy. A Report to the Energy Project of the Ford Foundation*, Ballinger, Cambridge, Mass. 1974, pp. 197–202; W. Gumpel, 'Alternatives to oil imported from OPEC countries: oil and gas imported from the Soviet Union?', in Frans A.M. Alting von Geusau, (ed.), *Energy and the European Communities*, A.W. Sigithoff, Leyden 1975, p. 149
221 Groupe d'études prospectives internationales, 'Situation et perspectives du bilan énergétiques de l'U.R.S.S. et de l'Est Européen', *Le Courrier des Pays de l'Est*, March 1978, pp. 13–16
222 Groupe d'études prospectives internationales, 'Situation et perspectives du bilan énergétiques de l'U.R.S.S. et de l'Est Européen', *Le Courrier des Pays de l'Est*, March 1978, pp. 13–16
223 Groupe d'études prospectives internationales, 'Situation et perspectives du bilan énergétiques de l'U.R.S.S. et de l'Est Européen', *Le Courrier des Pays de l'Est*, March 1978, pp. 13–16; UN Economic Commission for Europe (ECE), *New Issues Affecting the Energy Economy of the EEC Region in the Medium and Long Term (Preliminary Version)*, Add. 1, p. 212
223a Groupe d'études prospectives internationales, 'Situation et perspectives du bilan énergétiques de l'U.R.S.S. et de l'Est Européen' in *Le Courrier des Pays de l'Est*, March 1978, pp. 3–16; UN Economic Commission for Europe (ECE), *New Issues Affecting the Energy Economy of the ECE Region in the Medium and Long Term (Preliminary Version)*, Add. 2, p. 23 (ECE (XXXIII)/2/Add. 1)
223b In 1976 income from oil exports was estimated at \$5.5–6 x $10^9$ (*The Economist*, 30 April 1977)
224 Deutsches Institut fuer Wirtschaftsforschung (DIW), *Wochenbericht 21/1974*, 22 May 1974, p. 195
225 J.A. Yager and E.B. Steinberg, *Energy and US Foreign Policy, A Report to the Energy Project of the Ford Foundation*, Ballinger, Cambridge, Mass, 1974, p. 191
225a *The Economist*, 30 April 1977

226 W. Gumpel, 'The energy policy of the Soviet Union', in C. Gasteyger, (ed.), *The Western World and Energy*, Atlantic Institute, Paris 1974; J.A. Yager and E.B. Steinberg, *Energy and US Foreign Policy, A Report to the Energy Project of the Ford Foundation*, Ballinger, Cambridge, Mass. 1974, p. 202; J.A. Berry, 'Oil and Soviet policy in the Middle East', *Middle East Journal*, Spring 1972, pp. 149–160; *Petroleum Economist*, Feb. 1977, pp. 59–60

227 Deutsches Institut fuer Wirtschaftsforschung (DIW), *Wochenbericht 21/1974*

228 Z.R. Beydoun and H.V. Dunnington, *The Petroleum Geology and Resources in the Middle East*, Scientific Press, Beaconsfield 1975

PART TWO

# The trade relationship between the Nine and her energy suppliers: Asymmetries and dependencies

The conclusions of Part 1 were that the European Community will continue to be dependent on energy imports (mainly hydrocarbons) over the next decade, and probably well beyond. Since energy imports mostly come from a small group of suppliers in the Middle East and North Africa, this group is bound to be at the centre of Europe's future international energy policy.

Discussion of the shape of this policy requires a closer look at the implications of this relationship between Europe and her energy suppliers. Part I has established that hydrocarbons will continue to account for the bulk of energy imports and the main demand will, as now, be for oil and natural gas. The international coal market does not seem to pose any particular political problems in the foreseeable future, whereas in the case of uranium and nuclear technology such problems will persist, and potentially will affect supplies to Europe. However, the political difficulties with uranium lie not in the market itself, but in the Janus-like ambiguity of nuclear energy, which can be put to constructive as well as to destructive use. The politics of nuclear proliferation could affect — at least potentially — supplies of uranium to the Nine. Market considerations alone do not warrant the conclusion that the international uranium trade will become politicized in the way that oil is; if Europe follows her present policy of developing the plutonium cycle, dependence on uranium ore imports will diminish and thus further reduce any potential vulnerabilities. On the other hand, it is difficult to envisage a massive development of nuclear energy in Europe without the breeder: the greatest problems of nuclear energy — public opposition and limited demand for base-load electricity — will either affect both LWRs and the breeder, or will turn out to be superable. In either case, nuclear import dependence will be limited.

For these reasons, the following analysis is confined to oil and natural gas, and therefore to the Middle Eastern and African suppliers of energy to Europe. The emphasis will be on structural aspects in the trade between Europe and her energy suppliers, and — since the primary concern of this study is with the political aspects of Europe's energy dependence — it will be concerned above all with asymmetries inherent in this structure, of a kind which furnish political leverage for one side or the other, and with the elements of dependence and interdependence which will shape the evolution of this relationship.

The argument will be that Europe and her energy suppliers are linked to each other in a complex web of overlapping asymmetries and dependencies which work in opposite directions. More specifically, one can draw the following conclusions:

(1) The overall trade balance (i.e. considering the relative importance of Europe and her energy suppliers as trading partners for each other) favours Europe rather heavily — the Nine are much less dependent on trade with the selected group of OPEC suppliers than vice versa. But this overall picture does not adequately reflect the strategic importance of one element in this exchange: hydrocarbons. Nevertheless, this situation does point to a certain amount of leverage for Europe.

(2) The strategic importance of oil introduces another asymmetry in the relationship. Here, it is Europe which is the dependent and vulnerable partner, facing a considerable power potential of the suppliers. The suppliers' leverage differs, however, from country to country, and is concentrated in Saudi Arabia, Kuwait, Libya and the other low-absorbers.

(3) There is another aspect, however, to the trade in oil — this time an element of asymmetry favouring Europe, although in a diffuse and long-term way: the energy suppliers depend to a very large extent on this one raw material for their economic security — as underdeveloped countries they have little else to offer in exchange for European industrial goods other than raw materials. On the other hand, the oil sector as such is no guarantee for rapid development — problems of the development of oil economies are probably generally underestimated. These factors imply a considerable degree of long-term dependence on markets for oil to create export earnings — and these markets will have to be found to a large extent in Europe. This dependence on markets, however, differs from country to country, with accompanying differences in time horizons.

(4) Finally, there is the asymmetry resulting from the peculiar economic structure of most of the economies of the suppliers, though to varying degrees they all face the problem of creating alternative sources of wealth and income for their societies. For this they will need a long-term, successful development strategy — and international support not only in terms of technological assistance, but also, and probably soon, in terms of access to markets for newly constructed industrial production.

CHAPTER THREE

# Trade patterns between the European Community and its main energy suppliers

In our analysis, we shall proceed from quantitative to qualitative aspects, and from a general assessment to a more detailed scrutiny. The first perspective chosen to establish the validity of the above conclusions is that of the overall trade patterns between the two groups of countries — in other words: how important are the Community and the group of oil exporters to each other as trading partners?

The answer to this is provided by *Table 3.1*, from the point of view of the Community, and by *Tables 3.2 and 3.3* from the point of view of the suppliers. The main conclusions to be drawn from those figures are:

(1) In terms of overall trade, the group of energy suppliers are of minor importance to the Nine. Imports from this group accounted for 7.7% of all imports of the Nine in 1973, and 11.6% in 1975 — a change reflecting first, the increase in oil prices, and then the fall in demand in the Community in 1975 (the percentage share in 1974 had been 13% — higher than in 1975). Exports to the suppliers take an even smaller share of total Community exports: 2.9% in 1973, rising to 6.3% in 1975. Although the trend is clearly upward, the suppliers still constitute small, if rapidly growing, markets.

(2) Statistics for the suppliers point to a heavy concentration of trade in the relationship with the Community, if compared to total trade figures. As *Tables 3.2 and 3.3* show, this holds true for both imports and exports. The position differs substantially, however, from country to country, with a particularly high degree of trade concentration in the cases of Nigeria, Algeria, and Libya.

(3) There is a large difference between the value of Community imports from the suppliers and its exports to them. This imbalance reflects the 'oil deficit' of the Nine — underlining the strategic importance and peculiar role of oil in the trade relationship.

This crude quantitative assessment does not really give a very accurate picture of the reality. It does, however, draw attention to a number of crucial asymmetries in the overall trade patterns. First, there is the marked difference between the respective dependence on overall trade between Europe and her suppliers with each other. Second, the Community relies more on imports from the suppliers than on exports to them: the interest in access to supplies exceeds the interest in access to markets (in practical terms the two are connected: exports are the

Table 3.1  European Community—Oil Producers Trade 1973–1975 ($10^6$)[a]

| | European community imports | | | | | | European community exports | | | | | |
|---|---|---|---|---|---|---|---|---|---|---|---|---|
| | 1973 | % of total | 1974 | % of total | 1975 | % of total | 1973 | % of total | 1974 | % of total | 1975 | % of total |
| Algeria | 1 315.2 | 0.6 | 2 582.4 | 0.9 | 2 530.8 | 0.8 | 1 519.3 | 0.7 | 2 461.2 | 0.9 | 3 506.4 | 1.2 |
| Libya | 2 442.6 | 1.1 | 5 539.2 | 1.9 | 3 241.2 | 1.1 | 1 086.6 | 0.5 | 1 934.4 | 0.7 | 2 446.8 | 0.8 |
| Iraq | 1 668.8 | 0.8 | 3 055.2 | 1.0 | 3 475.2 | 1.2 | 272.5 | 0.1 | 942.0 | 0.3 | 2 305.2 | 0.8 |
| Saudi Arabia | 4 549.3 | 2.1 | 12 733.2 | 4.3 | 11 253.6 | 3.8 | 515.1 | 0.2 | 1 046.4 | 0.4 | 1 812.0 | 0.6 |
| Iran | 2 816.9 | 1.3 | 8 012.4 | 2.7 | 7 873.2 | 2.6 | 1 614.3 | 0.8 | 2 622.0 | 0.9 | 5 025.6 | 1.7 |
| Kuwait | 2 036.7 | 0.9 | 3 697.2 | 1.3 | 3 038.4 | 1.0 | 279.0 | 0.1 | 517.2 | 0.2 | 729.6 | 0.2 |
| Nigeria | 1 833.3 | 0.8 | 4 550.4 | 1.5 | 3 482.4 | 1.2 | 936.3 | 0.4 | 1 374.0 | 0.5 | 2 968.8 | 1.0 |
| Sub-total of oil exporters quoted | 16 662.8 | 7.7 | 40 170.0 | 13.6 | 34 894.8 | 11.6 | 6 223.1 | 2.9 | 10 987.2 | 3.9 | 18 794.4 | 6.3 |
| World | 216 308 | 100.0 | 294 490 | 100.0 | 300 052 | 100.0 | 213 115 | 100.0 | 276 085 | 100.0 | 295 988 | 100.0 |

[a] % figures may not add up because of rounding
Source: OECD Trade Statistics

Table 3.2  Oil Producer Imports from European Community Countries (in % of total imports)

| Importer | Total imports ($10$^6$) | European Community | Belgium | Denmark | France | Germany | Italy | Netherlands | UK | Ireland |
|---|---|---|---|---|---|---|---|---|---|---|
| Libya | | | | | | | | | | |
| 1974 | 2 763.0 | 56.7 | 2.5 | 0.3 | 10.4 | 11.5 | 24.8 | 2.2 | 5.0 | 0.0 |
| 1975 | 4 400.0 | 57.6 | 2.1 | 0.5 | 9.1 | 13.3 | 24.5 | 2.3 | 5.6 | 0.0 |
| Iran | | | | | | | | | | |
| 1974 | 5 425 | 39.4 | 2.7 | 0.4 | 4.0 | 18.3 | 3.3 | 2.4 | 8.3 | 0.0 |
| 1975 | 10 346 | 39.2 | 2.4 | 0.4 | 4.3 | 17.6 | 3.3 | 2.7 | 8.5 | 0.01 |
| Nigeria | | | | | | | | | | |
| 1974 | 2 773.9 | 57.8 | 2.0 | 0.9 | 6.6 | 15.1 | 5.4 | 4.6 | 23.0 | 0.2 |
| 1975 | 6 039.4 | 59.9 | 2.3 | 1.0 | 8.3 | 14.6 | 6.1 | 4.3 | 23.0 | 0.3 |
| Kuwait | | | | | | | | | | |
| 1974 | 1 550.1 | 33.3 | 2.6 | 1.4 | 3.9 | 10.9 | 4.0 | 2.2 | 8.2 | 0.1 |
| 1975 | 2 387.9 | 33.3 | 0.8 | 0.9 | 3.3 | 11.4 | 4.5 | 2.1 | 10.2 | 0.1 |
| Iraq | | | | | | | | | | |
| 1974 | 2 364.2 | 28.8 | 2.2 | 0.7 | 7.4 | 8.1 | 3.4 | 1.6 | 5.3 | 0.1 |
| 1975 | 3 557.2 | 36.1 | 1.5 | 1.1 | 6.6 | 16.5 | 3.4 | 1.8 | 5.2 | 0.0 |
| Algeria | | | | | | | | | | |
| 1974 | 4 405.4 | 61.4 | 3.2 | 0.5 | 32.4 | 12.0 | 8.1 | 2.0 | 3.2 | 0.002 |
| 1975 | 6 383.2 | 60.2 | 2.7 | 0.4 | 32.6 | 10.5 | 9.6 | 1.3 | 3.0 | 0.1 |
| Saudi Arabia | | | | | | | | | | |
| 1974 | 2 858.3 | 22.6 | 3.1 | 0.1 | 1.8 | 6.0 | 2.8 | 4.0 | 4.8 | 0.001 |
| 1975 | 7 176 | 27.8 | 1.8 | 0.5 | 3.1 | 8.7 | 4.9 | 2.0 | 6.8 | 0.04 |

Source: IMF Directions of Trade

Table 3.3  Oil Producer Exports to the European Community (as % of total exports)

| Exporter | Total exports ($10⁶) | European Community | Belgium | Denmark | France | Germany | Italy | Netherlands | UK | Ireland |
|---|---|---|---|---|---|---|---|---|---|---|
| Libya | | | | | | | | | | |
| 1974 | 8 260.8 | 77.0 | 2.0 | 0.3 | 5.9 | 22.0 | 33.4 | 0.9 | 12.5 | — |
| 1975 | 5 706.9 | 51.3 | 0.9 | 0.3 | 3.1 | 21.7 | 20.3 | 0.8 | 4.2 | — |
| Iran | | | | | | | | | | |
| 1974 | 19 167 | 38.0 | 1.6 | 1.7 | 3.4 | 5.9 | 5.3 | 14.2 | 5.7 | 0.2 |
| 1975 | 18 283 | 39.1 | 2.4 | 1.4 | 6.3 | 7.3 | 5.7 | 8.0 | 7.7 | 0.3 |
| Iraq | | | | | | | | | | |
| 1974 | 6 028.6 | 46.0 | 0.5 | 0.4 | 18.7 | 4.6 | 17.6 | 0.4 | 3.8 | — |
| 1975 | 7 123.4 | 43.8 | 1.7 | 0.2 | 13.8 | 1.6 | 21.2 | 2.3 | 2.9 | 0.1 |
| Nigeria | | | | | | | | | | |
| 1974 | 9 218.5 | 50.4 | 0.6 | 0.5 | 10.0 | 7.1 | 1.6 | 13.7 | 16.8 | 0.06 |
| 1975 | 7 993.3 | 46.5 | 0.8 | 1.2 | 10.9 | 6.8 | 1.2 | 11.3 | 14.2 | 0.07 |
| Kuwait | | | | | | | | | | |
| 1974 | 10 961.2 | 38.0 | 1.3 | 0.2 | 7.9 | 2.1 | 5.0 | 3.6 | 15.6 | 2.3 |
| 1975 | 8 740.4 | 32.8 | 1.2 | 0.2 | 6.0 | 1.1 | 3.9 | 8.2 | 9.1 | 3.1 |
| Algeria | | | | | | | | | | |
| 1974 | 4 537.4 | 50.6 | 1.4 | 0.6 | 19.2 | 21.8 | 5.4 | 0.5 | 1.7 | 0.001 |
| 1975 | 4 438.0 | 51.8 | 2.2 | 0.04 | 15.2 | 21.0 | 8.3 | 1.1 | 3.9 | 0.02 |
| Saudi Arabia | | | | | | | | | | |
| 1973 | 7 695.8 | 42.9 | 1.8 | — | 9.2 | 3.3 | 10.0 | 9.3 | 8.0 | 1.4 |
| 1974 | 30 992.3 | 42.4 | 3.4 | 0.002 | 11.5 | 4.4 | 10.3 | 2.6 | 9.3 | 0.7 |

Source: IMF Directions of Trade

direct way of paying for imports). Third, the suppliers depend on the Nine both in terms of supplies and in terms of markets for their exports, though with significant differences in degree. Overall, the level of this dependence on imports from, as well as exports to, the Community is certainly high enough to render the suppliers very sensitive to economic developments in Europe — an aspect which deserves emphasis.

A comparison of the figures given in *Tables 3.4 and 3.5* for the two years 1973 and 1975 provides some interesting insights. The most striking feature of these figures is obviously the massive expansion of trade in both directions. But almost as remarkable is the fact that this expansion has led to quite significant distortions of the second set of figures in comparison to the first: non-oil exports of Algeria, Nigeria and Iran (the only countries of the group with significant non-oil exports) have at best stagnated, and in most instances have declined in value terms (the quantity reduction must be even more pronounced). This applies particularly to the food sector, where the decline was marked; at the same time, there are disproportionately high increases of food imports in most members of the group. The conclusion is that the agricultural sector in these countries (the traditional mainstay of the economy in Iran, Algeria and Nigeria) is in trouble. The small and budding manufacturing industries also do not seem to be doing well, judging by the figures: only Algeria achieved some modest growth in exports in this sector, while Iran and Nigeria registered some decline in dollar terms. Obviously, these figures cannot be extrapolated directly, but they do reflect general trends in trade patterns of the oil exporters. In Chapter 3 we will take a closer look at these trends, and will attempt some explanation.

Apart from fuels, Community imports from the energy suppliers are marginal: they include mainly crude materials listed under SITC category 2 (palm kernels from Nigeria, unmanufactured textile fibres and crude animal and vegetable materials from Iran); SITC category 0 (agricultural products such as fruit and vegetables from Algeria and Iran, and cocoa and animal feedstuff from Nigeria); and SITC category 6 (manufactured goods chiefly classified by material, such as non-ferrous metals from Nigeria, iron and steel from Algeria, and textiles, mainly carpets, from Iran). The vast majority of these goods have, like oil, only a low degree of processing and value added.

Community exports to the suppliers are heavily concentrated in SITC categories 7, 6, and 5 — in that order. Together, these categories comprise 81.9% of total exports of the Nine to these countries (1975), with machinery and transport equipment alone accounting for almost half of total exports. On the other hand, these exports are only between 2.9 and 4.1% of total European Community exports in these categories. Again, trade in these goods seems much more important for the suppliers than for the Community. These categories contain many export items of vital importance for the development strategies of the suppliers, and are particularly needed by the high-population suppliers, where the need for development is greatest. In that sense these figures indicate the importance of European exports for these economies — an importance which

Table 3.4  European Community Imports from a Group of LDC Oil Exporters, by Major Classes of Goods ($10^6$)

| SITC group | Algeria | Libya | Nigeria | Iraq | Saudi Arabia | Kuwait | Iran |
|---|---|---|---|---|---|---|---|
| (A) In 1973 | | | | | | | |
| 0 Food and live animals | 49.3 | 0.04 | 140.9 | 8.1 | 0.15 | 0.51 | 41.0 |
| 1 Beverages/tobacco | 56.2 | — | 0.004 | — | — | — | — |
| 2 Crude materials, inedible, excluding fuels | 13.8 | 4.3 | 151.4 | 4.1 | 3.4 | 0.47 | 69.6 |
| 3 Fuels | 1 157.8 | 2 439.2 | 1 432.9 | 1 148.0 | 4 527.2 | 2 023.9 | 2 443.4 |
| 4 Animal/vegetable, oil and fats | 7.0 | — | 6.1 | Negligible | Negligible | Negligible | 0.06 |
| 5 Chemicals | 2.1 | — | 1.0 | 0.05 | Negligible | 5.7 | 3.9 |
| 6 Manufactured goods, classified by material | 24.8 | 0.1 | 41.8 | 0.55 | 0.65 | 0.21 | 242.8 |
| 7 Manufacturing and transport equipment | 8.8 | 1.1 | 2.6 | 4.9 | 10.9 | 4.9 | 8.9 |
| 8 Miscellaneous manufactured goods | 1.0 | 0.2 | 0.5 | 0.2 | 2.9 | 0.7 | 5.1 |
| 9 Goods not qualified by kind | 0.7 | 0.6 | 0.5 | 0.9 | 4.1 | 0.3 | 2.1 |
| Total | 1 321.5 | 2 446.2 | 1 777.7 | 1 166.8 | 4 549.3 | 2 036.7 | 2 816.9 |
| (B) In 1975 | | | | | | | |
| 0 Food and live animals | 36.0 | 0.08 | 128.5 | 5.4 | 0.2 | 0.7 | 0.07 |
| 1 Beverages/tobacco | 10.7 | Negligible | 0.01 | Negligible | 0.13 | Negligible | 0.04 |
| 2 Crude materials, inedible, except fuels | 21.8 | 1.9 | 90.5 | 2.1 | 1.7 | 0.5 | 96.1 |
| 3 Fuels | 2 425.5 | 3 240.2 | 3 198.5 | 3 476.6 | 11 215.9 | 2 990.8 | 7 529.7 |
| 4 Animal/vegetable oil and fats | 1.1 | Negligible | 18.4 | Negligible | Negligible | Negligible | Negligible |
| 5 Chemicals | 2.8 | 0.015 | 2.2 | 0.2 | 0.4 | 12.6 | 3.9 |
| 6 Manufactured goods, classified by material | 29.4 | 0.2 | 39.7 | 0.35 | 2.9 | 1.0 | 209.4 |
| 7 Manufacturing and transport equipment | 4.7 | 2.4 | 6.3 | 3.1 | 25.2 | 12.3 | 23.9 |
| 8 Miscellaneous manufactured goods | 1.5 | 0.4 | 1.1 | 0.7 | 2.0 | 1.6 | 9.5 |
| 9 Goods not qualified by kind | 4.1 | 2.7 | 2.2 | 0.2 | 0.5 | 1.0 | 4.4 |
| Total | 2 537.6 | 3 247.9 | 3 487.6 | 3 488.7 | 11 248.8 | 3 020.5 | 7 879.4 |

Source: OECD Trade Statistics

**Table 3.5** European Community Exports to a Group of LDC Oil Exporters, by Major Classes of Goods ($10$^6$)

| SITC group | Algeria | Libya | Nigeria | Iraq | Saudi Arabia | Kuwait | Iran |
|---|---|---|---|---|---|---|---|
| (A) In 1973 | | | | | | | |
| 0 | 109.4 | 116.9 | 82.9 | 19.5 | 62.8 | 44.7 | 34.8 |
| 1 | 1.5 | 2.1 | 8.3 | 0.4 | 19.2 | 6.2 | 1.8 |
| 2 | 23.0 | 9.9 | 10.8 | 4.0 | 1.5 | 1.0 | 20.0 |
| 3 | 27.1 | 32.0 | 9.6 | 3.5 | 4.7 | 3.8 | 7.2 |
| 4 | 10.3 | 0.8 | 0.8 | 0.5 | 3.6 | 0.6 | 12.4 |
| 5 | 157.6 | 48.2 | 131.1 | 41.3 | 44.5 | 25.9 | 192.3 |
| 6 | 395.8 | 250.8 | 195.6 | 81.9 | 112.8 | 50.8 | 358.3 |
| 7 | 708.4 | 534.3 | 430.7 | 109.6 | 220.0 | 97.8 | 900.5 |
| 8 | 80.0 | 84.5 | 61.5 | 11.2 | 38.6 | 46.5 | 68.0 |
| 9 | 6.1 | 6.1 | 5.1 | 0.5 | 7.6 | 1.8 | 18.9 |
| Total | 1 519.3 | 1 086.6 | 936.3 | 272.5 | 515.1 | 279.0 | 1 614.3 |
| (B) In 1975 | | | | | | | |
| 0 | 340.0 | 204.0 | 206.9 | 41.6 | 141.1 | 54.8 | 161.7 |
| 1 | 2.1 | 3.8 | 68.3 | 3.9 | 35.6 | 9.4 | 6.2 |
| 2 | 37.8 | 18.9 | 23.8 | 10.4 | 5.5 | 2.9 | 103.2 |
| 3 | 99.6 | 122.4 | 72.3 | 7.3 | 8.1 | 4.3 | 19.1 |
| 4 | 33.0 | 1.8 | 2.4 | 2.2 | 10.5 | 2.7 | 21.5 |
| 5 | 248.6 | 102.8 | 350.1 | 118.2 | 111.9 | 47.8 | 375.5 |
| 6 | 759.2 | 575.2 | 660.4 | 436.4 | 316.9 | 111.3 | 1 009.6 |
| 7 | 1 811.1 | 1 148.7 | 1 371.6 | 1 583.4 | 950.3 | 374.2 | 2 983.3 |
| 8 | 139.5 | 195.7 | 197.3 | 58.6 | 174.3 | 117.8 | 186.1 |
| 9 | 32.1 | 67.6 | 22.0 | 42.2 | 61.7 | 5.2 | 156.7 |
| Total | 3 503.0 | 2 441.0 | 2 975.2 | 2 304.3 | 1 816.0 | 730.4 | 5 023.2 |

Source: OECD Trade Statistics

means increased interdependence and sensitivity to economic developments in Europe, and a growing structure of links and relationships which will be increasingly difficult to sever. However, it should again be stressed that this relative importance of Europe to the suppliers varies from country to country. Also, while this introduces an element of asymmetry in favour of Europe, the asymmetry is somewhat diffuse, long-term in character and difficult to transform into political leverage on specific issues. As a background factor, however, its importance seems to be considerable.

*Tables 3.4 and 3.5* indicate a further asymmetry between the two groups: that between raw material exporters and exporters of highly complex equipment with a large share of added value. This asymmetry – a crucial aspect of development and underdevelopment – is the starting point for all development planning by the oil exporters. To overcome the draw-backs of an economy based on the export of one raw material, these countries plan to develop industrial production on a fairly substantial scale, implying (a) the loss of traditional European markets by OPEC for non-oil exports; (b) competition from new industries in OPEC countries in third markets (chemical and petrochemical industries, and to a lesser extent iron and steel could become examples here); and (c) ultimately, also competition in European markets from newcomers (the most obvious example here is refining). Despite this caveat as to the future there is a clear complementary element in the patterns of trade between the two regions, which augurs well for an increasingly close economic symbiosis, especially if new forms of division of labour offset the contrast between raw material and manufactured goods producers; the sectors in which the energy suppliers could move up the ladder of industrial development (or, for that matter, agricultural exports) should allow European industries to adjust in an orderly way into new forms of industrial and non-industrial specialization. OPEC's new industries will only be an element in a much wider adjustment process in the reallocation of world industry from the North towards the South (and from the West towards the Asian East?); the management of this transition will pose difficult problems. But the pressure from OPEC will be only minor on present form: industrialization is still in the beginnings, and it will take at least 5–10 years before new industries have some impact on world markets. Demands for markets should be relatively easy to meet, with one exception: refined products. This could become a major area of disagreement and trade conflicts between Europe and her energy suppliers.

CHAPTER FOUR

# The power aspects of the international trade in oil

Europe's very heavy dependence on oil (and natural gas) imports now and in the future is, so far, the only asset of advantage to the energy suppliers. Oil, and oil revenues derived from its sale, are their only strength, and today it is a very potent weapon indeed.

This section will consider exclusively the power implications inherent in the exchange of oil against goods, services and money. Other factors, which influence the power relationship the other way, such as the asymmetries noted in the previous chapter, or the larger framework of interaction defined by the political economy of oil to be analysed in the following chapter, are for the moment set aside, though the limitations of oil power should always be kept in mind throughout this assessment of its importance.

A power analysis of trade relations (such as suggested by A.O. Hirschmann[1]) has to consider two possibilities. First, the *increase* of power by trade, if the supply of goods leads to improved economic production, higher growth, or more effective social organization. Clearly, oil is significant in this respect; for instance it played a role in the outcome of World Wars I and II. Second, power can be *reduced*, at least potentially, by dependence: if supplies of goods, or of a single item such as oil, become vital to the political, economic and military vitality of an importing country — such as the industrialized countries now importing oil in great quantities — , but derive from only one or a few suppliers, then trade interruptions, or even the threat of them, weaken the power of the importer and enhance the power of the supplier. This latter aspect played an important part in the Middle East crisis of 1973/4, and has since become an important element in international relations.

The *ability* to interrupt a trade relationship is based on a nation's sovereignty in three respects (or in three different meanings of the word): *legally*, the concept of sovereignty covers the *right of a state to decide freely on the use of its natural resources* (and its industrial products). In *practical political terms*, sovereignty as a prerequisite to interrupt trade also implies a leadership which enjoys a *considerable degree of independence* from external influences exerted by the importer states, and a roughly *symmetrical political relationship* in the sense that the importers are unable to use political and, ultimately, military pressure to prevent the supplier from continued interruption of trade, or deter the supplier from interference with trade.

Sovereignty in these practical terms has been a fact of international relations of the LDC oil exporters for a few years only, and indeed the year 1970 can be considered as the turning point: it was then that a new, radical Libyan government (which had deposed one year earlier the conservative monarchy, in itself a sign of the changing times) demonstrated that an LDC oil exporter could successfully dictate terms to the international oil companies[2]. From then onward, international opportunities and internal pressures were stronger than consideration for the interests of the Western industrialized countries, and even led to conservative regimes increasing their independence of Western influence, and pursuing more autonomous policies. At the same time, American setbacks in Vietnam, the structure of superpower relations and the declining credibility of military intervention had removed the risk of countervailing military pressure, thereby enabling the Arab oil producers successfully to apply the oil weapon in 1973 without any military retaliation.

The *capability to exert influence* through the threat and/or implementation of trade interruptions is based on the dependence of the importers on the continuation of the trade — in the case of oil on dependence on continued oil deliveries. This implies:

(1) No alternative suppliers are available to fill the gap produced by interruptions of supply from a particular group of suppliers. This is the case if and when trade patterns show a high degree of concentration (which means that the gap to be filled would be large) and stand-by capacities outside the group of countries taking part in the interruptions or at least refusing to increase their production to make up the shortfall is insufficient. *Figure 4.1* indicates the disappearance of surplus capacity outside OPEC by 1970, while the enormous quantities of energy imports required by the European countries have already been pointed out in Part 1. The disappearance of surplus capacity largely explains why the Arab oil

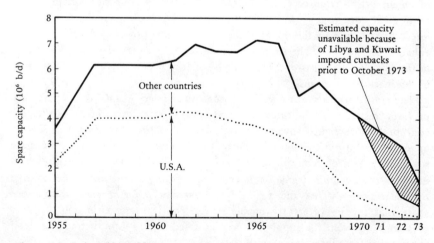

Figure 4.1 *Estimated world spare capacity (crude oil), excluding communist countries.* Source: Exxon

Table 4.1 Value of Net Oil Imports as Percentage of GDP

| | (1) GDP at purchaser prices | 1960 (2)* Oil imports, net value | (3) (2) as % of (1) | 1965 (1) | (2) | (3) | 1970[b] (1) | (2) | (3) | (1)[a] | 1974[b] (2) | (3) |
|---|---|---|---|---|---|---|---|---|---|---|---|---|
| Germany ($ million) | 72 036 | 550.2 | 0.76 | 112 500 | 1 044.4 | 0.93 | 187 694 | 2 024.5 | 1.08 | 384 530 | 11 243 | 2.9 |
| France (million French francs) | 301 400 | 2 463.2 | 0.81 | 461 000 | 4 571.0 | 0.99 | 147 311 | 1 593.4 | 1.08 | 266 100 | 9 717 | 3.7 |
| Italy (million lire) | 21 751 000 | 186 421 | 0.86 | 35 244 000 | 379 507 | 1.08 | 92 645 | 1 236 | 1.33 | 149 810 | 7 960 | 5.3 |
| Netherlands (million goulder) | 42 400 | 364.4 | 0.86 | 68 000 | 698.7 | 1.03 | 31 356 | 325.7 | 1.04 | 69 180 | 1 541 | 2.2 |
| BeLux (million francs) | 588 900 | 6 010.3 | 1.02 | 864 600 | 9 907.5 | 1.15 | 26 698 | 404.8 | 1.52 | 55 550 | 2 331 | 4.2 |
| Denmark (million kroner) | 41 098 | 1 012.0 | 2.46 | 69 300 | 1 458.8 | 2.11 | 15 598 | 364.3 | 2.34 | 30 400 | 1 481 | 4.9 |
| Ireland (£ million) | 642.9 | 13.845 | 2.15 | 975 | 20.366 | 2.09 | 3 888 | 82.9 | 2.13 | 6 730 | 461 | 6.8 |
| UK (£ million) | 25 403 | 378.9 | 1.49 | 34 900 | 500.3 | 1.43 | 119 182 | 1 797.7 | 1.51 | 188 990 | 8 984 | 4.8 |

* After deduction of exports of oil products
a At market prices
b All figures in columns (1) and (2) for 1970 and 1974 are in $
Source: UN Yearbook for International Trade Statistics; IMF International Financial Statistics; OECD Trade Statistics

weapon could be applied successfully in 1973 while it had failed on previous occasions: the problem of interruptions from the consumers' point of view had changed from one of logistics (re-routing principally available additional supplies from the USA, Venezuela and Iran) to one of resource scarcity.

(2) A continuation of supplies is indispensable. This is obvious in the case of oil: any shortfall over and above the amount which could be absorbed by cutting back on wasteful energy consumption and luxury use would affect the performance of the economy, reduce the standard of living, cause disintegrative strains in the consumer societies and political systems, and dangerously weaken the power of the importing countries.

(3) The losses as a consequence of trade interruptions must be unevenly distributed. A useful way of analysing this is to look at the net gains derived from the trade relationship for both exporters and importers. Net gains can be analysed in terms of both direct gains and indirect gains, the latter referring to intra-actor effects of the interaction in the trade relationship. Intra-actor effects will be dealt with in the next chapter; here, we shall focus on direct gains from the oil trade both for exporters and importers.

For the importers, direct net gains can be considered as the difference of the value of imported goods over the value of the same goods produced domestically with the resources allocated to pay for the imports. Clearly, the net gain for the European oil importers is very high: a full substitution of indigenous energy supply for imported oil is for most European Community countries virtually impossible because of lack of resources, and even a partial substitution would demand comparatively much higher costs, both in economic and social terms. As *Table 4.1* shows, the percentage of GDP allocated to pay for oil import bills in the European Community countries has been rising, but from a fairly low level. Compared with investment and production costs for indigenous energy sources,

**Table 4.2** OPEC Import Bills (*annual value of imports c.i.f.* $10^6$)

|      | Ecuador | Indonesia | Bahrain | Oman  | Qatar  | UAE   | Gabon | Trinidad and Tobago |
|------|---------|-----------|---------|-------|--------|-------|-------|---------------------|
| 1960 | n.a.    | 578       | n.a.    | n.a.  | n.a.   | n.a.  | n.a.  | n.a.                |
| 1965 | 274     | 718       | n.a.    | n.a.  | n.a.   | n.a.  | n.a.  | n.a.                |
| 1970 | 340     | 1 002     | (264)[a]| (29)  | (64)   | (267) | 264   | 510                 |
| 1971 | 317     | 1 103     | (319)   | (97)  | (108)  | (309) | 319   | 663                 |
| 1972 | 532     | 1 562     | 280     | 161   | 138    | 482   | 280   | 765                 |
| 1973 | 939     | 2 729     | 318     | 169   | 195    | 821   | 318   | 789                 |
| 1974 | 678     | 3 842     | 1 127   | 711   | 271    | 1 705 | 332   | 1 847               |
| 1975 | 943     | 4 709     | 1 198   | 668   | 413    | 2 669 | 469   | 1 471               |
| 1976 | 993     | 5 673     | 1 664   | 1 174 | 817    | 3 329 | 694   | 1 966               |
| 1977 | 1 508   | 6 230     | 2 029   | 1 176 | 1 289  | 4 583 | 844   | 1 778               |
| 1978 | 1 583   | 6 690     | 2 045   | n.a.  | n.a.   | 4 892 | n.a.  | 1 886               |

a Figures in parentheses relate to pre-independence imports
Source: IMF International Financial Statistics, May 1978 and Oct. 1979

these percentages constitute a very small figure, indicating the enormous net gain of the European countries. Consequently, the costs of the adjustment process towards lower imported oil supplies would in the long term be very high, and in the short term impossible, if the shortfall exceeds the limits a society and economy can absorb; the result would be the collapse of this industrialized country.

For the exporters, the gains present themselves in a somewhat different light. If we look at the importance of the oil sector, and its revenues, for the economies of these countries, then the gains appear to be very high[3]. At the same time, the high percentage shares of the oil sector in GNP, in total foreign exchange earnings, and in government revenues indicate a serious problem for some producers: oil revenues are so high that they cannot be absorbed by the domestic economies and their societies alone. This becomes clear when one compares the development of imports of oil producers with that of oil revenues and foreign exchange holdings (see *Tables 4.2 and 4.3*): the accumulation of foreign exchange as a result of a growing gap between import costs and export revenues are essentially excess earnings with marginal immediate usefulness. Even in the longer run, such excess earnings over and above what can be absorbed domestically in consumption and investment are only economically sound if the real purchasing power of the income per barrel invested externally, plus the interest derived from this investment, can reasonably be expected to exceed the future value of a barrel of oil produced at a later time. It has to be stressed that for producers with incomes exceeding their demand it is not enough to assume a fall in oil prices at some later stage in the future as a temporary phenomenon (for then the decision not to produce oil now but leave it in the ground would still be perfectly rational — the producer would simply wait until the price rises again); what would be necessary to motivate increased production to control investment capital is the expectation of a long-term

**Table 4.2** *continued*

| Iran | Iraq | Kuwait | Saudi Arabia | Algeria | Libya | Nigeria | Venezuela |
|---|---|---|---|---|---|---|---|
|  | 389 | 242 | 235 | (1 265) | 169 | 604 | 1 188 |
| 860 | 451 | 377 | 506 | 672 | 320 | 772 | 1 518 |
| 1 662 | 509 | 625 | 711 | 1 257 | 554 | 1 059 | 1 994 |
| 1 873 | 694 | 650 | 815 | 1 221 | 699 | 1 511 | 2 302 |
| 2 469 | 713 | 797 | 1 136 | 1 472 | 1 038 | 1 505 | 2 404 |
| 3 393 | 906 | 1 052 | 1 944 | 2 242 | 1 734 | 1 865 | 2 825 |
| 5 433 | 2 365 | 1 552 | 2 859 | 4 058 | 2 763 | 2 772 | 4 191 |
| 10 343 | 4 205 | 2 390 | 4 214 | 5 861 | 3 586 | 6 042 | 6 031 |
| 12 894 | 3 470 | 3 319 | 8 694 | 5 465 | 3 203 | 8 199 | 7 058 |
| 14 070 | 4 052 | 4 796 | 17 196 | 6 952 | 5 148 | 10 050 | 9 269 |
| 14 202 | n.a. | 4 613 | 22 852 | 8 530 | 6 144 | 11 913 | 11 643 |

**Table 4.3** *Oil Revenues ($10^6$) and Total Central Bank Reserves of Selected OPEC Countries ($10^6$)*

(A) Oil Revenues

| | Iran | Iraq | Kuwait | Saudi Arabia | Algeria | Libya | Nigeria | Venez-uela | Indo-nesia | Qatar | Abu Dhabi |
|---|---|---|---|---|---|---|---|---|---|---|---|
| 1960 | 385 | 266 | 465 | 355 | n.a. | — | — | 877 | n.a. | — | — |
| 1961 | 301 | 266 | 464 | 400 | n.a. | 3 | — | 938 | n.a. | — | — |
| 1962 | 334 | 267 | 526 | 451 | n.a. | 39 | — | 1 071 | n.a. | — | — |
| 1963 | 398 | 325 | 557 | 502 | n.a. | 109 | — | 1 106 | n.a. | 59 | 6 |
| 1964 | 470 | 353 | 655 | 561 | n.a. | 197 | — | 1 122 | n.a. | 65 | 12 |
| 1965 | 522 | 375 | 671 | 655 | n.a. | 371 | — | 1 135 | n.a. | 69 | 33 |
| 1966 | 593 | 394 | 707 | 777 | n.a. | 476 | — | 1 112 | n.a. | 92 | 100 |
| 1967 | 737 | 361 | 718 | 852 | n.a. | 631 | — | 1 254 | n.a. | 102 | 105 |
| 1968 | 817 | 476 | 766 | 966 | 262 | 952 | — | 1 253 | n.a. | 109 | 153 |
| 1969 | 938 | 484 | 812 | 1 008 | 299 | 1 132 | — | 1 289 | n.a. | 115 | 191 |
| 1970 | 1 136 | 521 | 895 | 1 200 | 325 | 1 295 | 441 | 1 406 | 185 | 122 | 233 |
| 1971 | 1 944 | 840 | 1 400 | 2 149 | 350 | 1 766 | 915 | 1 702 | 284 | 198 | 431 |
| 1972 | 2 380 | 575 | 1 657 | 3 107 | 700 | 1 598 | 1 174 | 1 948 | 429 | 255 | 551 |
| 1973 | 4 100 | 1 500 | 1 900 | 5 100 | 900 | 2 300 | 2 000 | 2 800 | 900 | 400 | 900 |
| 1974 | 17 500 | 5 700 | 7 000 | 22 574 | 3 700 | 6 000 | 8 900 | 8 700 | 3 300 | 1 600 | 5 536 |
| 1975 | 18 500 | 7 500 | 7 500 | 25 676 | 3 375 | 5 100 | 6 570 | 7 525 | 3 850 | 1 700 | 6 000 |
| 1976 | 22 000 | 8 500 | 8 500 | 33 500 | 4 500 | 7 500 | 8 500 | 8 500 | 4 500 | 2 000 | 7 000 |
| 1977 | 23 000 | 9 600 | 8 500 | 37 800 | 5 600 | 9 400 | 9 400 | 3 000 | 5 000 | 1 900 | 8 300 |

Sources: Comité Professionnel du Pétrole; Petroleum Economist, July 1978

(B) Total Financial Assets of OPEC Countries[a]

| | Iran | Iraq | Kuwait | Saudi Arabia | Algeria | Libya | Nigeria | Venez-uela | Indo-nesia |
|---|---|---|---|---|---|---|---|---|---|
| 1960 | 184 | 258 | 72 | 185 | n.a. | 82 | 434 | 609 | n.a. |
| 1965 | 236 | 234 | 124 | 726 | 184 | 832 | 246 | 843 | 21 |
| 1970 | 208 | 462 | 203 | 662 | 339 | 1 590 | 224 | 1 021 | 160 |
| 1971 | 621 | 600 | 288 | 1 444 | 506 | 2 666 | 432 | 1 572 | 187 |
| 1972 | 960 | 782 | 363 | 2 500 | 493 | 2 835 | 385 | 1 734 | 574 |
| 1973 | 1 237 | 1 553 | 501 | 3 876 | 1 134 | 2 127 | 592 | 2 420 | 807 |
| 1974 | 8 383 | 3 273 | 1 397 | 14 285 | 1 658 | 3 616 | 5 629 | 6 529 | 1 492 |
| 1975 | 9 020 | 2 829 | 1 717 | 24 186 | 1 402 | 2 277 | 6 019 | 9 190 | 608 |
| 1976 | 8 778 | 4 572 | 1 917 | 26 855 | 1 975 | 3 185 | 5 170 | 8 524 | 1 489 |
| 1977 | 12 263 | 6 436 | 2 989 | 30 027 | 1 918 | 4 600 | 4 249 | 8 212 | 2 515 |

a Official Central Bank assets only. Figures are for end of year
Source: IMF International Financial Statistics

decline in oil prices. In fact, however, the world has now entered a long period of rising energy prices[4]. Excess earnings, therefore, do not appear to be in the interest of the producers — but even if they were intent on exporting oil at a rate producing such excess earnings, they still could easily decide to cut back production for any amount of time: there simply exists a discrepancy between the importance of oil for the consumer countries, and of oil revenues for some producer countries. For some of the rich producers, a large share of the annual oil revenues has become, at best, of marginal utility.

Production rates, in any case, are only one of the two basic variables determining the level of oil revenues; to the extent producers can exert influence over oil prices, they can regulate their income by changing the latter. As we shall argue later, there is little reason for the producers not to use their influence over prices in strictly economic terms. Since about 1970, the producers have to a growing degree asserted their influence over oil prices, thereby further reducing the need to rely on greater export quantities[5]. In addition, control over prices also provided a further argument against producing oil in order to amass investment funds — a decision to do so would not only endanger the very control over the market which was the precondition for influence on price levels, but also demonstrate lack of faith in the ability of producers to continue to maintain their influence.

Therefore, economic self-interest advocates at the least the stabilization of present oil price levels and production quantities, if not a reduction of production levels, and further increases in oil prices. The net gain from the present trade relationship is therefore very unevenly distributed — at least in the case of some oil producers, such as Saudi Arabia, Abu Dhabi, Qatar, Kuwait and Libya, this gain is marginal above a fairly low level of indigenous revenue requirements. This analysis, of course, is strictly economic — it has to be noted that political considerations might influence each of the crucial variables such as revenue demand, price policy, and production levels, and modify them considerably.

CHAPTER FIVE

# The oil export sector in a developing country: The political economy of oil

As the analysis of trade patterns between Europe and her energy suppliers has shown, this trade is basically an exchange of oil against manufactured goods. We encounter here another fundamental imbalance between the two groups — one which is somewhat more difficult to establish than the previously analysed asymmetries. This imbalance is the result of the domestic effects of the trade in oil at both ends.

In part, the asymmetry stems from the peculiarities of the exchange of the raw material petroleum against manufactured goods, but it is also partly rooted in the differences in economic and social structures between the two groups. These differences in structure have developed since the industrial revolution, and favour industrialized countries so long as the interaction with developing countries takes place within a framework basically shaped by the former[6].

Crude oil exports involve a very limited amount of industrial activity; they represent almost 'pure' nature, to follow Galtung's distinction between 'natural' and 'cultural' products[7]. Manufactured goods, on the other hand, involve a high degree of industrial activity and organization. The industrial activity necessary to produce manufactured goods elicits a wide range of *spin-off effects*, impulses spreading across a society, and radiating throughout the whole economy. Manpower is trained, knowledge is produced, technologies are developed, new products are created, new forms of organization are designed, and these in turn help to improve the utilization of resources and generate a self-sustaining process of growth[8]. This is so much the case that 40—50% of total internal demand in an industrialized economy actually comes from other sectors of production, rather than from the final consumer[9]. Economic spin-off effects also, of course, radiate outwards into political, military and cultural areas, and are transmuted into power and influence in these spheres[10]. These far-reaching processes in the industrialized countries have to be compared on the side of the oil-exporting developing countries with a rather mixed bag of limited positive, and quite substantial negative, spin-off effects. This, at least, has to be the conclusion of an analysis of the evolution of oil-exporting developing countries over the two decades until 1973. Whether the drastic increase in oil revenues since 1973 makes a qualitative difference, and substantially changes the situation in favour of the suppliers, remains to be seen. Meanwhile, the evidence of the past is highly instructive.

**Table 5.1** *Refinery Capacity at Beginning of Selected Years, by Region* ($10^6$ t/a)

|  |  | Western Europe | Middle East | Africa | North America | Latin America | Far East | Socialist bloc | World total |
|---|---|---|---|---|---|---|---|---|---|
| **1951** | | | | | | | | | |
| (1) | capacity | 46.220 | 46.450 | 3.250 | 369.500 | 73.810 | 14.680 | 49.190 | 603.100 |
| (2) | % of world total | 7.8 | 7.7 | 0.5 | 61.3 | 12.2 | 2.4 | 8.2 | 100 |
| (3) | (% of world oil production) | (0.8) | (15.7) | (0.02) | (54.2) | (18.9) | (2.2) | (8.1) | (100) |
| (4) | (% of world oil consumption) | (11.1) | (1.9) | (1.2) | (65.0) | (7.2) | (4.3) | (9.4) | (100) |
| **1961** | | | | | | | | | |
| (1) | | 221.245 | 74.820 | 9.360 | 538.945 | 161.390 | 70.525 | 163.500 | 1 239.785 |
| (2) | | 17.8 | 6.0 | 0.8 | 43.5 | 13.0 | 5.7 | 13.2 | 100 |
| (3) | | (1.5) | (25.0) | (1.7) | (35.2) | (17.9) | (2.4) | (16.3) | (100) |
| (4) | | (19.7) | (2.3) | (1.2) | (46.9) | (7.8) | (7.6) | (14.6) | (100) |
| **1965** | | | | | | | | | |
| (1) | | 368.065 | 91.135 | 28.470 | 564.995 | 194.610 | 148.615 | 253.150 | 1 649.040 |
| (2) | | 22.3 | 5.5 | 1.7 | 34.3 | 11.8 | 9.0 | 15.4 | 100 |
| (3) | | (1.4) | (27.3) | (7.0) | (29.1) | (15.9) | (2.2) | (17.5) | (100) |
| (4) | | (24.2) | (2.0) | (1.2) | (41.4) | (6.7) | (9.2) | (15.1) | (100) |
| **1974** | | | | | | | | | |
| (1) | | 908.335 | 138.315 | 59.290 | 800.575 | 345.960 | 420.255 | 518.000 | 3 190.730 |
| (2) | | 28.5 | 4.3 | 1.9 | 25.1 | 10.8 | 13.2 | 16.2 | 100 |
| (3) | | (0.8) | (37.8) | (9.6) | (21.9) | (7.8) | (3.9) | (18.3) | (100) |
| (4) | | (25.5) | (2.5) | (1.7) | (31.8) | (6.5) | (14.8) | (17.1) | (100) |
| **1978** | | | | | | | | | |
| (1) | | 1 007.550 | 161.955 | 75.295 | 938.605 | 396.050 | 498.250 | 695.000 | 3 772.705 |
| (2) | | 26.7 | 4.3 | 2.0 | 24.9 | 10.5 | 13.2 | 18.4 | 100 |
| (3) | | (2.7) | (35.0) | (9.0) | (18.1) | (7.8) | (4.5) | (22.9) | (100) |
| (4) | | (23) | (3) | (2) | (32) | (7) | (14) | (19) | (100) |

*Source: Comité Professionnel du Pétrole*

To analyse the relative distribution of benefits derived from the oil industry, one must consider the distribution (a) of refinery capacity and (b) of total investment in the industry. Refining is the most obvious way to use crude oil as an instrument of industrialization. Refining adds value to exports and creates employment[11]. But even this stage of processing a natural product is now heavily concentrated in the industrialized countries, as *Table 5.1* shows. Moreover, it also demonstrates that this is the result of a historical shift of refinery capacity from the producing to consuming areas. A variety of reasons can be given for this development[12]:

(1) *Strategic reasons:* sources of crude oil can be more readily replaced by others than sources of products, since refineries have a certain flexibility in the input of crude they can process and the 'product mix' they deliver.
(2) *Economic reasons:* refining at the consumer end reduces the foreign exchange burden of oil imports, since the value added by refining is generated within the consuming area. Refining also creates employment. The by-products of refining are also useful raw materials for the petrochemical industries. Tanker freight costs for products are higher than for crude since the crude can be transported in greater bulk and so benefits by economies of scale. Moreover, heavy fuel oil produced in the refineries is difficult to pump through pipelines. Finally, lower raw material losses in the process of refining have reduced the incentive to refine near the source of production.

While there are thus perfectly good reasons for the concentration of refining in consumer areas, they reflect in the last analysis the ability of importing countries to make their interests prevail over those of the LDC oil exporters. The siting of refineries is subject not only to economic considerations, but to political pressures. These political pressures can easily overcome economic considerations. This is strongly suggested by the vast expansion of refining capacity now under way in the Persian/Arabian Gulf and other producer areas[13], as well as the past success of some producers in obliging oil companies to construct refining capacity in conjunction with higher production[14].

Investment patterns in the industry are another index of the relative distribution of benefits (see *Table 5.2*). While the low investment share of LDC oil producers (and particularly the Middle East) certainly reflects the high probability of success in exploration, the extraordinary productivity of wells, and lower wage levels, the investment share of the LDC producers in areas other than production points to the unbalanced allocation of gains from the industry's activity: it means that downstream operations have developed elsewhere. Again, we can put forward perfectly sound economic reasons why, say, chemical industries have been constructed in the consuming countries, but this does not alter the fundamental bias of the present situation. Nor can the pattern of investment be explained totally in terms of comparative economic advantage. If this were the only consideration, production and exploration would be concentrated much more heavily in the oil-exporting areas, since their reserves would

**Table 5.2** *Average Annual Investment of the Petroleum Industry, Selected Periods ($10^6$)*

|  | Production | Pipelines | Tanker | Refineries | Chemical industry | Marketing | Total[a] |
|---|---|---|---|---|---|---|---|
| **World** | | | | | | | |
| 1952–1955 | 4 621 | 390 | 578 | 1 200 | 64 | 746 | 7 738 |
| 1956–1960 | 5 994 | 507 | 1 076 | 1 395 | 282 | 1 245 | 10 740 |
| 1961–1965 | 5 300 | 569 | 1 050 | 1 577 | 697 | 1 879 | 11 625 |
| 1966–1970 | 6 319 | 892 | 1 744 | 3 083 | 1 444 | 2 761 | 17 330 |
| 1971–1975 | 13 807 | 2 423 | 6 268 | 6 204 | 1 840 | 2 612 | 34 014 |
| **USA** | | | | | | | |
| 1952–1955 | 3 656 | 281 | 58 | 658 | 38 | 333 | 5 094 |
| 1956–1960 | 4 047 | 200 | 89 | 567 | 150 | 450 | 5 640 |
| 1961–1965 | 3 635 | 268 | 50 | 397 | 352 | 545 | 5 755 |
| 1966–1970 | 4 132 | 362 | 63 | 875 | 680 | 1 240 | 7 905 |
| 1977–1975 | 7 514 | 1 240 | 155 | 1 375 | 740 | 920 | 12 258 |
| **Canada** | | | | | | | |
| 1952–1955 | 294 | 63 | — | 73 | 5 | 58 | 498 |
| 1956–1960 | 374 | 34 | 4 | 89 | 11 | 122 | 646 |
| 1961–1965 | 369 | 31 | — | 45 | 18 | 95 | 612 |
| 1966–1970 | 580 | 58 | — | 130 | 39 | 147 | 1 070 |
| 1971–1975 | 1 095 | 90 | — | 327 | 28 | 164 | 1 761 |
| **Venezuela** | | | | | | | |
| 1952–1955 | 258 | 15 | — | 29 | — | 5 | 308 |
| 1956–1960 | 536 | 53 | 6 | 57 | — | 13 | 672 |
| 1961–1965 | 165 | 5 | — | 11 | — | 13 | 198 |
| 1966–1970 | 164 | 11 | — | 37 | — | 7 | 236 |
| 1971–1975 | 226 | 6 | — | 24 | — | 2.5 | 264 |
| **Other Western Hemisphere** | | | | | | | |
| 1952–1955 | 119 | 10 | — | 75 | — | 36 | 245 |
| 1956–1960 | 334 | 74 | 3 | 126 | 6 | 91 | 647 |
| 1961–1965 | 345 | 32 | — | 132 | 52 | 100 | 690 |
| 1966–1970 | 308 | 99 | — | 287 | 120 | 95 | 1 016 |
| 1971–1975 | 741 | 475 | — | 790 | 72 | 130 | 2 174 |
| **Western Europe** | | | | | | | |
| 1952–1955 | 81 | 4 | — | 203 | 21 | 188 | 509 |
| 1956–1960 | 145 | 26 | 5 | 370 | 103 | 388 | 991 |
| 1961–1965 | 115 | 83 | — | 568 | 190 | 570 | 1 630 |
| 1966–1970 | 239 | 140 | — | 955 | 435 | 820 | 2 718 |
| 1971–1975 | 1 733 | 271 | — | 1 810 | 830 | 815 | 5 879 |
| **Africa** | | | | | | | |
| 1952–1955 | 31 | — | — | 9 | — | 39 | 80 |
| 1956–1960 | 240 | 49 | 1 | 15 | — | 83 | 407 |
| 1961–1965 | 317 | 79 | — | 74 | — | 82 | 578 |
| 1966–1970 | 425 | 94 | — | 65 | 5 | 67 | 712 |
| 1971–1975 | 755 | 105 | — | 173 | 17 | 95 | 1 168 |
| **Middle East** | | | | | | | |
| 1952–1955 | 99 | 16 | — | 56 | — | 15 | 221 |
| 1956–1960 | 233 | 59 | 15 | 76 | 1 | 20 | 390 |
| 1961–1965 | 255 | 39 | — | 50 | 6 | 24 | 401 |
| 1966–1970 | 280 | 105 | — | 138 | 25 | 25 | 614 |
| 1971–1975 | 773 | 161 | — | 300 | 37 | 25 | 1 379 |
| **Far East** | | | | | | | |
| 1952–1955 | 84 | 1 | — | 101 | — | 74 | 265 |
| 1956–1960 | 105 | 15 | 2 | 115 | 11 | 128 | 325 |
| 1961–1965 | 99 | 32 | — | 300 | 77 | 230 | 757 |
| 1966–1970 | 191 | 23 | — | 570 | 140 | 360 | 1 358 |
| 1971–1975 | 970 | 75 | — | 1 405 | 116 | 460 | 3 099 |

a Includes miscellaneous other investments
Source: *Chase Manhattan Bank, Capital Investment of the World Petroleum Industry, 1961 1972 and 1976*

offer the most rewarding possibilities for increasing future production. The diversification of sources of supply, and, more recently, the shift of oil company exploration activity into the industrialized areas cannot be fully explained by existing opportunities for discoveries — they reflect a political strategy of the companies to keep the oil-exporting countries in a weak bargaining position. The distribution of investment has, then, to be accepted as a result of the political economy of the oil industry and of producer–consumer relations. The unevenness of this distribution certainly constitutes a potential source of conflict. The chemical industry (one of the downstream industries of oil and the main area of consumer-concentrated investment apart from refineries) is a case in point. It was one of the leading growth industries in the EEC during the 1960s, its rate of expansion being consistently higher than that of industrial production as a whole[15]. What would happen if the oil producers continued to press for it to be established at home, as they have begun to do? It could but exacerbate the difficulties of European industries.

Another aspect of the uneven distribution of resources — though one difficult to quantify — is the concentration of organizational and technological knowhow in the developed countries. There the oil companies have their administrative headquarters which direct the worldwide management of the international oil market; there, also, are the R & D departments of affiliates producing oil (and energy) related technologies.

The relative distribution of benefits of the activities of the oil industry itself, then, sustains the argument of a severe bias against the LDC oil exporters.

Let us now turn to spin-off effects related indirectly to oil industry activities.

Theoretically, at least six factors related to the development of an oil-exporting industry could be expected to produce positive impulses for development. These are:

(1) *Forward linkages:* the demand for locally produced goods and services created by the oil industry in the country itself.
(2) *Technological linkages:* the training of manpower and the acquisition of technical and organizational know-how transferable to other sectors.
(3) *Employment* opportunities through the activities of the industry.
(4) *Demand* created by local wage and salary payments of the companies.
(5) *Infrastructural development* by the industry which could be used for other purposes.
(6) *Capital and foreign exchange availability*[16].

In fact a closer look at these factors and their importance for the development of the host country demonstrates that their impact, with the exception of the last factor, is quite limited[17].

*Forward linkages* play a minor role in most oil producer economies. Since the bulk of the services and the equipment required for the exploration and production stages are highly sophisticated, they must, generally, be imported[18]. The share of local suppliers and contractors in total expenditures for goods and

**Table 5.3** Breakdown of Operating and Investment Expenditure, Middle East ($10^6$)

| Country | 1948 | | | | | 1953 | | | | | 1958 | | | | |
|---|---|---|---|---|---|---|---|---|---|---|---|---|---|---|---|
| | A | B | C | D | E | A | B | C | D | E | A | B | C | D | E |
| Iran (AIOC/consortium only) | 56.0 | 18.8 | 3.5 | 85.2 | 46.7 | — | — | — | — | — | 47.1 | 16.0 | 6.6 | 53.8 | 35.6 |
| Iraq (IOC only) | 12.8 | 5.6 | 4.4 | 25.2 | 4.8 | 10.9 | 3.6 | 2.0 | 24.1 | 7.0 | 13.7 | 5.0 | 5.6 | 33.6 | 7.8 |
| Kuwait (KOC only) | 4.8 | 7.6 | 2.3 | 67.5 | 1.0 | 6.8 | 5.2 | 0.6 | 15.6 | 1.5 | 16.3 | 18.0 | 8.0 | 40.7 | 5.5 |
| Qatar (QPC only) | 1.2 | 1.6 | 2.0 | 8.4 | 0.9 | 2.2 | 0.8 | 0.8 | 14.0 | 0.3 | 4.8 | 2.2 | 0.8 | 9.0 | −0.3 |
| Saudi Arabia (Aramco only) | 23.5 | 4.1 | — | 80.0 | 0.9 | 56.1 | 10.3 | 0.5 | 71.5 | 2.4 | 70.0 | 11.6 | 5.2 | 65.7 | 8.4 |
| Total | 98.3 | 37.7 | 12.2 | 266.3 | 53.8 | 76.0 | 19.9 | 3.9 | 125.2 | 11.2 | 151.9 | 52.8 | 26.2 | 202.8 | 57.0 |

A = *wages and salaries*
B = *payments to local contractors*
C = *purchases from local suppliers*
D = *imports*
E = *other*
Source: C. Issawi and M. Yeganeh, *The Economics of Middle Eastern Oil*, Faber, London 1962, pp. 192–193

services required in the host country can be seen in *Table 5.3* for the period 1948 to 1958. It never exceeded 20% of total operating and investment costs, excluding wages and salaries. Data for three countries (Venezuela, Iran and Nigeria) in later years point to the same conclusion[19]. Moreover, a detailed study of the Nigerian oil economy has revealed that (even though the share of local suppliers and contractors is higher than in the Middle East during the 1950s) the vast majority of those local suppliers and contractors are actually foreign-owned (see *Table 5.4*). Their share in total locally purchased supplies and

**Table 5.4** *Selected Ancillary Firms' Receipts from the Nigerian Petroleum Industry 1965 (N£$10^6$)*[a]

|  | Total | Paid abroad | Paid locally to expatriate-owned supplier | Paid locally, to Nigerian supplier |
|---|---|---|---|---|
| Suppliers (goods purchases locally by oil industry) | 2.735 | — | 2.713 | 0.022 |
| Contractors | 20.540 | 10.079 | 9.588 | 0.873 |
| Total | 23.275 | 10.079 | 12.301 | 0.895 |

a Nigerian receipts are slightly undervalued, since not all Nigerian firms are included in the sample. The error is, however, marginal
Source: S.R. Pearson, Petroleum and the Nigerian Economy, Stanford University Press, Stanford, 1970, p. 88

services has been estimated at 83% (1965). Pearson concludes this analysis as follows:

> 'In short, the firms supplying and servicing the Nigerian oil industry are most likely to be expatriate-owned and managed, to be large relative to other firms operating in Nigeria, and with a couple of exceptions to have been working in Nigeria on a large scale prior to the emergence of crude oil production. Many of the most highly-specialized international corporations service and supply the petroleum industry in its operations all over the world.'[20]

Similarly, data from Iran and Venezuela point towards a large share of foreign-owned suppliers in total local supplies and services to the oil industry[21]. It appears that forward linkage effects (a) are small and (b) tend to integrate the developed sectors (mostly foreign-owned) of the company, thereby emphasizing the dualistic character of oil-exporters' economies, which will be discussed later.

*Technical linkages*, the provision of training for indigenous employees and sometimes also non-employees (scholarships etc.) as well as general education facilities established by the oil companies as part of their goodwill policy *vis-à-vis* the host countries[22], have also played only a limited role[23]. The (originally quite high) rate of horizontal mobility of indigenous employees of the oil industry has been falling as a consequence of high wage levels and substantial fringe benefits accorded by the industry[24] and of increased vertical mobility within it[25]. Pearson concludes his analysis about Nigeria with the following remark:

> 'On the whole, the establishment of the petroleum industry has resulted in a fairly significant upgrading of a very limited spectrum of Nigeria's human resources. If, however, the Nigerian economy reaches a point where further growth were constrained by lack of skilled manpower, it is highly doubtful that the petroleum industry in the course of its normal operations will be able to do much about breaking this bottleneck.'[26]

*Employment* in the oil industry has always been relatively limited, and has in recent years tended to fall. This reflects the highly capital-intensive nature of crude oil production. The significance of employment created by the oil industry clearly depends on the overall size of the labour force. In larger LDC oil exporters, the importance of the oil industry as an employer has only become substantial where operations were not confined to the production stage, and national oil companies have begun to play an important role in the economy. The indirect impact of the oil industry on the employment situation as a consequence of demand for locally produced goods and services also appears to be relatively restricted — in Nigeria, the number of Nigerian employees in ancillary industries totalled 14 051 (1965)[27]. Some data about employment in the oil industry in various countries are given in *Table 5.5*.

Since employment in the industry itself is generally limited, a similar conclusion can be drawn about the demand created by wages and salaries paid locally (it has to be noted, however, that such demand is relatively high since levels of wages and salaries in the oil industry are normally well above general levels)[28]. Some details of the exact amount of such payments can be found in *Table 5.3*. The impact on the local economy is, however, reduced to the extent that the earnings of expatriate employees are transferred abroad. Moreover, the impact on the local economy can be expected to be reduced also by the import-oriented demand from expatriate employees and even (as a consequence of 'display effects' by a Western lifestyle) from indigenous personnel[29].

The evaluation of *infrastructural development* through the arrival of an export-oriented oil industry in an LDC — roads, port facilities, hospitals etc. — again points to less than favourable conclusions. The construction of such facilities does of course create jobs and demand for local goods, such as building materials, but this effect is only temporary. Of course, the facilities can afterwards be put to more permanent uses not linked to petroleum production or export[30]. But normally they are so closely tailored to the needs of the industry as to be useless for other purposes[31]. The infrastructure of roads and ports is, moreover, very much a reflection of the export-orientation given to the whole economy by oil. By the same token, imports are also facilitated. If not corrected by specific measures, local production can become less competitive as a consequence of low import costs[32].

**Table 5.5** Direct Employment in the Oil Industry, Selected Years

| Country | 1948 | 1956 | 1960 | 1964 | 1969 |
|---|---|---|---|---|---|
| Bahrain | 6 078 | 8 785 | 7 684 | 6 200 | 4 312[a] |
| Iran | 50 393 | 55 234 | 54 030 | 41 926 | 41 960 |
| (+ employment by local contractors) | (13 603[b]) | (7 913) | (3 206) | (727) | n.a. |
| Iraq | 14 241 | 15 832 | 16 702 | 11 245 | 8 610 |
| Kuwait | 10 233 | 7 814 | 7 161 | 7 490[c] | 6 392 |
| (+ employment by local contractors) | (n.a.) | (n.a.) | (6 409[d]) | | |
| Saudi Arabia | 18 637 | 19 632 | 14 834 | 12 280 | n.a. |
| Qatar | n.a. | 4 500 | n.a. | n.a. | n.a. |
| Algeria | – | – | 8 500[e] | 11 000 | n.a. |
| Libya | – | 1 550 | 10 900 | 12 600[f] | 6 395 |
| Venezuela | – | 45 652[g] | n.a. | ca. 30 000[h] | ca. 21 000[i] |
| Nigeria | – | – | n.a. | 3 075 | 3 901[k] |

a 1971
b 1951
c Includes Neutral Zone
d 1958
e 1959
f 1963
g 1957
h 1965
i 1974
k 1967

Sources: C. Issawi and M. Yeganeh, The Economics of Middle East Oil, Faber, London 1962, p. 97; N. Gall, 'The challenge of Venezuelan oil', in Foreign Policy, Spring 1975, pp. 44–67 (63); The Middle East and North Africa, 1974–1975, Europa Publications, London 1974; United Nations, The Growth of World Industries 1938–1961, New York, 1965; W.G. Harris, 'The impact of the petroleum export industry on the pattern of Venezuelan economic development', in R.F. Mikesell (ed.), Foreign Investment in the Petroleum and Mineral Industries, Case Studies of Investor-Host Country Relations, Johns Hopkins University Press, Baltimore 1971, pp. 129–156 (132); K.S. Sayegh, Oil and Arab Regional Development, Praeger, New York 1968, p. 302

The last — and crucial — factor of potentially positive importance for the economic development of the host countries is the revenue derived from oil exports, the share of the host government in the rent derived from crude oil production. Since this rent can be considered a compensation for the exploitation of a finite resource, it should be used for economic development and diversification eventually capable of replacing the oil export sector by other sectors. This is now generally recognized by the LDC oil exporters, and they have all initiated development programmes designed to industrialize and so diversify their economies. Huge sums of government money are available to promote this. These LDC industrialization strategies normally rely on two pillars: the injection of capital, and the use of oil and gas as cheap sources of energy and raw material. The core of industrialization is to be formed by the nationalized or nationally controlled oil industry[33].

Whether the dramatically increased level of oil revenues since 1973 has indeed pushed the oil exporters beyond the critical threshold of self-sustained growth cannot be assessed with any confidence so far. All that can be said is that in the past the impact of oil revenues on economic development has been less beneficial than might be expected. While the huge sums available undoubtedly provide a unique opportunity, they also create strong temptations which, on past experience, are difficult to reject. This is not to suggest that they cannot be avoided. A developing petroleum economy with huge revenues is not bound to become a dramatic example of how not to develop. Nevertheless, the dangers are greater, and the difficulties more serious, than is generally appreciated.

The chosen path to development leads all oil producers to stress industrialization. This raises the question of where the markets for the new industries will be. There are two possible answers: domestic demand, or exports. In fact, both seem to be necessary, even for countries with potentially large domestic markets. There is ample evidence that industrialization does not imply a reduction of imports but rather international specialization. This means that in the long run industrial and agricultural exports will have to replace oil as the main foreign exchange earner. Even while oil exports continue, other exports should be built up to achieve economies of scale, to increase employment, to reduce dependence on one source of export earnings, and possibly also to allow some conservation of hydrocarbons, since there will be an element of future competitive advantage for energy-based industries[34].

Oil-producing countries with small populations have no choice but to seek export markets. Those with a larger population, however, can hope to exploit opportunities through imports. However, this must not be geared only to the limited markets provided by the upper and middle classes. As long as these classes grow the consumption flow can certainly stimulate a substantial expansion of demand over a long period. But the limits are bound to be reached earlier than in the case of demand expansion throughout society, i.e. by a policy which ensures sizeable 'trickle-down-effects'. If the development strategy fails to spread the benefits, limits on import substitution will slow down the initially rapid growth of industry[35]. Great attention should therefore be given to agriculture, which employs the bulk of the population in the large supplier countries (Nigeria, Iran, Iraq, Algeria). Comprehensive statistical studies have shown that the link between industrial and agricultural growth is extremely close, with the latter providing the crucial sustained stimulus for the former[36]. For the large suppliers agricultural prosperity appears to be a necessary condition of successful development.

The record seems to show that in spite of impressive industrial growth rates, Europe's energy suppliers face substantial difficulties on the way to industrialization. The contribution of oil revenues is often counter-balanced by the inability to correct the negative effects which accompany them. These tend to perpetuate dependence on oil exports as the engine of growth[37]. This implies a risk of failure to diversify away from oil exports before they begin to decline. The economic and social implications could be catastrophic, at home and abroad.

The impact of an export-oriented oil sector on a developing economy is profound and complex, causing widespread economic, but also social, political, cultural and psychological changes. This can be divided into three aspects: the *direct economic* impact of the oil sector, the general *social and cultural* consequences, and finally the effects of *government expenditures*.

That the positive economic impact of an oil export industry is quite limited has been pointed out already. However, two potentially negative effects have not yet been mentioned. First, there is the effect of high wages and salaries in the oil industry[38], on other industries and the service sector as a consequence of competition[39]. An interesting example of this is quoted by Wells[40]. When Petromin, the state-owned Saudi oil company, wanted to take over some 150 Aramco employees, it had to pay salaries 25–33% above the equivalent level of the civil service pay scale. The consequence of generally raised wage and salary levels is a potential reduction of industrial competitiveness[41]. The exchange rate will be unlikely to adjust to this, because of the overall trade surplus as a result of oil exports. Second, the arrival of an oil industry can emphasize already existing centrifugal tendencies and regional break-away movements[42].

The social and cultural impact of the oil industry cannot be dissociated from the general impact of Western civilization on developing societies. We are dealing here with the general phenomenon of 'social modernization'. The oil industry, however, has historically often been the 'wedge' by which Western civilization, technology and culture intrude[43]. It also seems to have exacerbated some of the effects of social change which Westernization would have produced anyway. Take some general symptoms of such change:

(1) Rapid urbanization[44] and economic concentration.
(2) Growing expectations and demands.
(3) Adoption of Western lifestyles and patterns of consumption.
(4) Dualistic economies with sharp dividing lines between modern and traditional sectors of the economy[45].
(5) Large-scale open or hidden unemployment or underemployment, together with a shortage of skilled labour[46].
(6) Marginalization of large groups both within the rural and the urban sectors (landless peasants, agricultural workers, slum dwellers)[47].

These can all be found in a particularly pronounced form in many oil-exporting countries, although the small oil exporters with insignificant populations obviously do not suffer severely from unemployment or the impoverishment of some social groups.

The role of government expenditures is clearly crucial for development: it is the oil revenues injected through governments into the economies which offer the only important stimulus for growth. Here, too, there are inherent tendencies working against a successful use of oil income for economic diversification and development.

First, the huge balance of trade surpluses of the oil exporters tend to overvalue their currencies. This reduces the international competitiveness of industries other than oil, and harms not only their export prospects, but even their chances of capturing their own domestic markets. The home market can be secured by tariff barriers[48], but exports are not so easily promoted.

Second, oil revenues are channelled into the economies of the exporters through government bureaucracies. This means (a) the expansion of the tertiary sector, and in particular administrative jobs, and (b) a severe test for the efficiency and competence of the civil service. Although most exporters now have a thin layer of highly competent and hard-working officials, their overall strength is clearly insufficient for the enormous task most countries face. Often, bureaucracy becomes a convenient form of social security, which reduces the quality of the civil service[49].

Third, the basic decision with which the national policy-makers are faced is the allocation of oil revenues for consumption (normally through the regular budget) and for investments (often in separate development budgets in addition to regular budget items). The high level of expectations which results from the 'bonanza' of oil revenues builds up strong pressures for consumption. To respond to or to anticipate these pressures, governments set up unproductive job-creation programmes[50], and elaborate social service systems. Groups and individuals close to the decision-makers or with power and influence, such as the armed forces, demand and receive their share of the oil riches, and small and vulnerable countries will also be under obligation to others, as the generosity of the Persian/Arabian Gulf sheikdoms *vis-à-vis* their poor Arab neighbours demonstrates. Corruption flourishes in this climate[51], and inequality increases since different social groups have different leverages on and access to the governing elite. Thus, the army, the industrial work force, the civil service and business will get the biggest share from the cake of oil revenues while the poorer and less important groups, e.g. the rural population, end up with little. In the absence of strong leadership which can impose, or create support for, a policy of investment for the future, rather than a policy of present consumption, spending patterns shift towards consumption. This may not be a problem for the small Gulf exporters, or even Libya, but it could imply lost opportunities for other suppliers.

Fourth, the revenues available after demands for consumption have been met (with the armed forces generally as the most privileged recipient) flow into investment either abroad or at home. But the productivity of investment differs, and often money has been spent on unproductive, or at best indirectly productive, projects such as large luxury airports and industrial 'white elephants'. Such investments heavily increase imports, and encourage a boom in the construction industry, but do little to foster a sound economic base[52]. For investments which do go into productive projects, the preference seems to lie with modern, capital-intensive technologies, such as petrochemical industries[53]. Around those industries, consumer-good industries develop — often oriented towards luxury products such as cars. Quite often, these industries are only assembly plants with a very high import dependence, which create only limited employment[54]. Prestige plays an important part in such investment decisions — indeed the whole concept of industrialization is strongly influenced by the desire to emulate the West. Ultimately, this type of industrialization is liable to create a vast gap between a rapidly growing modern sector and a traditional sector comprising the majority of the population, which falls further and further behind. This applies both to industry and agriculture. Such a dualism might be an acceptable evil for a limited period, as is argued by some of the development planners of those countries, but apart from the potentially explosive social and political tensions this implies, the economic soundness of the proposition still has to be tested.

To develop a coherent strategy for economic diversification is anything but easy and although none of the tendencies and barriers outlined above mean that oil is a curse rather than a blessing for a developing country, the difficulties should not be underestimated. To sum them up (see *Figure 5.1*) the problems of development of an oil-exporting country can be presented as follows:

(1) An oil-exporting developing country tends to evolve into a dualistic economy and society with a disproportionately large service sector, and an industrial sector chiefly oriented towards the production of consumption goods[55].

(2) Import requirements are very high, partly as a result of the elevated level of expectations and demands. To the extent that domestic industries are built up to substitute imports, they tend to form a consumption-oriented industrial structure[56]. It is doubtful whether such an incomplete, distorted industrial sector can become the base for self-sustained growth.

(3) The specific path of industrialization chosen generally reflects dominant patterns in industrialized countries — i.e. labour-saving, capital-intensive technologies. While this might be suitable for the industries and agricultural sectors of thinly populated exporters, it tends not to decrease unemployment in countries with large populations[57], at least for a fairly long period until the spin-off effects from the new industries become visible (if, indeed, these industries can be integrated into their economic environment). Unemployment, however, is still one of the most urgent problems in those countries.

(4) Politically, such systems tend to develop towards consumption-oriented pressure group politics of a peculiar kind, with demands based on very high expectations of the benefits which will be immediately available from oil wealth, and rewards accorded in line with a group's access to the state, and the threat it can pose to the ruling group by effective organization. The weakest groups and classes will secure a very small share of the wealth, although in some cases this wealth is great enough to satisfy all demands. The long-term political stability of such a system remains in doubt, although while the oil wealth lasts it may prove quite flexible. However, the decision-making process does not appear favourable to the effective and productive use of available resources.

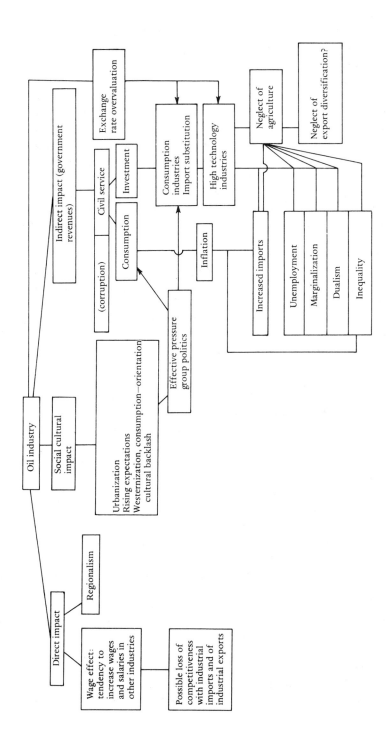

**Figure 5.1** *The impact of an export-orientated oil industry on a developing country (a model)*

This picture, of course, bears only a rough resemblance to the different realities which can be found in the various oil-exporting countries. In practice, it is necessary to distinguish between several types of exporters, the differences being due to (a) population size, (b) potential for economic diversification, (c) the political strategy of development, and (d) vulnerability to external and internal pressures. Closest to the model are Iran, Nigeria, Iraq and Algeria, all countries with a large population, a substantial potential for development, and a fairly low level of per capita income. Algeria has the most pronounced and elaborate development strategy based on an egalitarian aim; similarly, Iraq has opted for a 'socialist' path. Most aspects outlined above and in *Figure 5.1* can, however, be found in these societies. Algeria is the country most closely tied to Western Europe, one of the major centres of economic activity, and is therefore particularly vulnerable to economic developments there — a factor limiting her freedom of action in oil politics, and one possible explanation for the moderation of her 'hawkish' position on oil prices[58].

At the other end of the spectrum are countries like Kuwait and Abu Dhabi, with small populations, very limited possibilities of diversification outside hydrocarbon and hydrocarbon-based industries, and huge oil reserves. The only real possibility of diversifying sources of income for these countries appears to be investment abroad, with all the risks and dependencies this implies — again a factor limiting their freedom of action. Libya also comes close to this type, although it has a somewhat larger population, more scope for economic diversification, and also a very aggressive development strategy. Saudi Arabia also belongs to the group of desert economies with a somewhat broader population base, some scope for economic diversification (e.g. through other mineral deposits), and very large oil reserves which enable her to pursue a much more drawn-out strategy of diversification, which, however, implies a long-term dependence on oil markets. Singularly large oil revenues also give Saudi Arabia a large scope for investment abroad — implying the reinforcement of ties with, and dependence on, Western countries.

All these desert economies suffer from a general shortage of labour. The result tends to be the development of an indigenous ruling stratum composed of full citizens, with manual and skilled labour provided on a large scale by immigrants and guest workers with second-class citizenship. Egyptians in Libya, Yemenites in Saudi Arabia, Palestinians in Kuwait, and Pakistanis in other small sheikdoms add up to significant proportions of the population, posing a variety of social and political problems for these countries.

What conclusions for the future can be drawn from this analysis? First, it should be repeated that what is said above is based on experience before 1973. Although the inherent problems and tendencies working towards a perpetuation of dependence on oil exports will not disappear, there are some arguments for assuming that the massive increase in oil revenues since then has changed the situation. The collateral element of this influx of wealth has been a dramatic expansion of supplier power and influence. Arguably, the result has been a qualitative change in the opportunities open to an oil-exporting developing country. The following points can be made to support this:

(1) The massive influx of petro-revenues has fundamentally changed the prospects for successful economic development into self-sustained growth independent of oil-exports. Depending on continued OPEC control of the international oil market, and on sufficient hydrocarbon reserves, these countries might at least in some cases succeed by 'investment overkill' – such a lavish expense on new industries and agricultural projects that the desired result will be attained, albeit with a huge waste of resources.
(2) Some of the states with little prospect of economic diversification will now be able to undertake such large investments abroad that income from these investments will provide a sizeable share of national income. This development of 'rentier states' implies a very high degree of dependence on the countries where the investments are placed.
(3) Oil power and oil wealth has given the suppliers, for the first time, a considerable degree of influence in international affairs. This influence could be used to support development efforts, and to increase their chances of success. For example, new industries in supplier countries might not have to be competitive on a world scale if these countries can use their influence to shape bilateral and regional economic relations in such a way as to secure markets for the products of those industries. This might become particularly evident in relationships with other developing countries, but will also be a factor to be considered in relations with Europe.

On the other hand, arguments could be advanced which indicate limitations even in the new situation of oil exporters. These can be summed up as follows:

(1) The fundamental problem of the suppliers is their very heavy dependence on exports of a non-renewable resource; this implies a very serious economic security problem. In the short run, this problem has been solved by the suppliers taking control of the international oil market. In the medium term, it has to be tackled by reducing dependence on this one source of income, and in the long run an oil-export independent economy has to be created. The time gained by the oil resource then becomes of crucial importance. How much time and how much revenue will be needed to achieve national objectives? What is the time frame of capital creation through oil exports, and is this consistent with available hydrocarbon reserves?
(2) The problems of the highly populated suppliers are still very substantial, while their available reserves are limited (except, possibly, for Iraq). This implies that a very long road has to be covered in a relatively short time; economic and social development has to be telescoped. While it cannot be said that this will be impossible, there must remain serious doubts as to whether the social and political strains involved in such an acceleration of economic change have been taken into account in the optimistic forecasts.

Table 5.6  Dependence of Government Income on Oil Revenues, Selected Countries, 1960–1971

| | 1960 | 1961 | 1962 | 1963 | 1964 | 1965 | 1966 | 1967 | 1968 | 1969 | 1970 | 1971 |
|---|---|---|---|---|---|---|---|---|---|---|---|---|
| Venezuela | | | | | | | | | | | | |
| (1) Government receipts | 1 482.98 | 1 746.27 | 1 765.07 | 1 465.78 | 1 585.11 | 1 614.72 | 1 722.44 | 1 897.56 | 1 950.00 | 1 924.67 | 2 110.67 | 2 644.77 |
| (2) Oil receipts | 877 | 938 | 1 071 | 1 106 | 1 122 | 1 135 | 1 112 | 1 254 | 1 253 | 1 289 | 1 406 | 1 702 |
| (3) (2) as % of (1) | 59.1 | 53.7 | 60.7 | 75.5 | 70.8 | 70.3 | 64.6 | 64.3 | 65.1 | 67.0 | 66.6 | 64.4 |
| Iran | | | | | | | | | | | | |
| (1) | 598.98 | 657.25 | 912.46 | 1 176.38 | 1 330.77 | 1 498.34 | 1 269.97 | 1 456.48 | 1 723.27 | 1 667.98 | 1 872.28 | 2 225.77 |
| (2) | 285 | 301 | 334 | 398 | 470 | 522 | 593 | 737 | 817 | 938 | 1 136 | 1 944 |
| (3) | 47.4 | 45.8 | 36.6 | 33.8 | 35.3 | 34.8 | 46.7 | 42.8 | 47.4 | 56.2 | 60.8 | 87.3 |
| Iraq | | | | | | | | | | | | |
| (1) | 372.74 | 423.61 | 524.66 | 481.12 | 524.49 | 587.30 | 624.85 | 619.94 | 801.84 | 841.23 | 944.08 | 1 114.77 |
| (2) | 266 | 266 | 267 | 325 | 353 | 375 | 394 | 361 | 476 | 484 | 521 | 840 |
| (3) | 71.4 | 62.8 | 50.9 | 67.6 | 67.3 | 63.9 | 63.0 | 58.3 | 59.4 | 57.5 | 55.2 | 75.4 |

Sources: IMF International Financial Statistics; UN Statistical Yearbook; Petroleum Press Service

Table 5.7  Value of Oil Exports as Percentage of Total Exports (selected oil exporters and years)

| Exporter | 1958 | 1961 | 1962 | 1963 | 1964 | 1965 | 1966 | 1967 | 1968 | 1969 | 1970 | 1971 | 1972 | 1973 |
|---|---|---|---|---|---|---|---|---|---|---|---|---|---|---|
| Iran | 87.7 | 84.8 | 88.2 | 88.3 | 88.6 | 87.2 | 87.6 | 90.6 | 89.7 | 88.7 | 88.7 | 91.4 | 91.5 | 91.8 |
| Nigeria | 0.7 | 19.1 | 15.3 | 10.6 | 15.0 | 25.4 | 32.5 | 29.8 | 17.1 | 42.8 | 57.4 | 73.7 | 83.3 | 82.5 |
| Iraq | 91.7 | 94.4 | 90.5 | 92.9 | 94.0 | 93.3 | 92.0 | 91.9 | 92.6 | 93.1 | 93.7 | 95.8 | 93.3 | 95.2 |
| Venezuela | 91.1 | 92.1 | 92.7 | 92.6 | 93.4 | 93.0 | 92.4 | 92.3 | 92.8 | 91.3 | 90.2 | 91.5 | 90.7 | 92.1 |
| Indonesia | n.a. | n.a. | n.a. | 38.6 | 36.8 | 38.4 | 30.0 | 36.0 | 38.9 | 44.9 | 34.8 | 45.3 | 51.4 | 46.6 |
| Libya | — | 51.0 | 93.1 | 97.5 | 97.7 | 98.4 | 98.7 | 99.0 | 99.6 | 99.7 | 99.6 | 99.7 | 99.5 | 97.1 |
| Algeria | n.a. | n.a. | n.a. | n.a. | n.a. | n.a. | n.a. | 69.4 | 63.6 | 67.6 | 66.0 | 70.6 | 79.1 | 75.7 |
| Kuwait | n.a. | n.a. | n.a. | n.a. | n.a. | n.a. | n.a. | 96.9 | 96.7 | 95.8 | 95.6 | 96.1 | 94.9 | 93.8 |

Source: IMF International Financial Statistics

(3) While the expansion of power and influence of the suppliers is undeniable, this has to be qualified: their power rests almost exclusively on their market control over oil, and the wealth derived from it. If we used a complex indicator of power comprising a variety of variables, the suppliers would have moved up the hierarchy of the international system quite considerably. In fact, however, it might be more realistic to look at power as a multi-dimensional capability — and if one accepts that, then suppliers have made much less progress: there are many dimensions of power where they have hardly any importance and where they are consequently vulnerable and dependent. Oil power has to be seen within the context of its limits and this also applies to its ability to shape the environment.

(4) Past experience makes one wonder whether the exporters will really manage rapidly to reduce their dependence on exports. The 1960s certainly do not offer any evidence of this. As *Tables 5.6* and *5.7* show, none of the exporters listed had managed between 1961 and 1969 to reduce significantly the share of oil exports in total exports, and of oil revenues in total government revenues. (It should be noted that during this period oil prices were fairly constant, although government revenues increased substantially. Nevertheless, the implication is that other export sectors of the economy did not take over some of the role which oil played in those economies.)

CHAPTER SIX

# Co-operation and conflict

What conclusions can be drawn from the previous chapters for the future evolution of Europe's relations with her main energy suppliers? The various structural elements of mutual dependence, the asymmetries cutting both ways, should (broadly speaking) work towards co-operation between the two groups. This pressure towards co-operation results from the vulnerability of both sides, from the need to avoid developments ultimately damaging to both sides, and from differences in positions between the two groups with regard to power and status in the international system.

The interaction between the two groups has already reached a level which renders both sides vulnerable to adverse developments on the other side, and this trend will intensify in the future. The suppliers no doubt have the capacity to damage European economies and societies very seriously, either by drastically increasing energy prices, or by cutting exports. Yet this damage would not be confined to Europe. It would spread through the international system, affect the world economy and in particular other developing countries. It would, therefore, not only threaten the large investments of the suppliers in the West and in non-oil developing countries; the political ramifications of such a crisis could include local and regional changes in the balance of power between East and West, thereby potentially threatening vital political and security interests of the suppliers themselves. Even economically, the suppliers could not insulate themselves from a severe recession in the world economy. This would spill over into their economies, affecting exports other than hydrocarbons, domestic rates of inflation, employment (for example, a possible reflux of emigrants in the case of Algeria) and ultimately demand for hydrocarbon exports.

Thus, both sides face substantial costs if co-operation breaks down, while the rewards of successful co-operation are positive. At any given time for the foreseeable future, the longer-term interests of both sides advocate mutually constructive approaches.

The mutual vulnerability of both sides is really an expression of the different composition of their respective powers to shape the environment. Europe's strength is multidimensional: economic, social, cultural and military elements combine in a broad-based tissue of power. Its main vulnerability relates to military and energy security, as well as to economic security in a broader sense, given the somewhat greater dependence on international trade which characterizes the European Community in comparison with North America.

Apart from hydrocarbons, however, this difference can be exaggerated. The dependence of the European Community on trade with other regions was less than 9% of GNP in 1975, while that of the USA was around 7%. The much higher rates that pertain in most European countries, taken individually, are due to the intensity of international trade in Europe itself. The suppliers, on the other hand, have only one source of strength — their hydrocarbon resources. To diversify their sources of power and strength, of wealth and influence, is precisely the objective of the development strategies of these countries. To what extent, and when, these objectives will be reached is difficult to foresee, and the obstacles are substantial. During the period of transition, and quite possibly in many cases even afterwards, the suppliers will in many ways have to draw on the resources of the consumers. These resources will not always easily be made available. This implies not only huge expenditures on European goods, but also a constant process of bargaining over concessions. A strategy of confrontation by OPEC would diminish, perhaps even destroy, the ability and willingness of Europe and the other industrial consumers to provide these resources on which the exporters' aspirations depend. Confrontation would increase the risks of excessive vulnerability for both sides, and therefore be considered by the suppliers only if attempts to secure a co-operative solution fail. Until late 1978 co-operation in this sense consisted of a voluntary moderation of price increases and the maintenance of high production levels by the suppliers as a group (leaving aside for the moment the mechanisms within OPEC producing those results), while the consumers refrained from attempts to break the suppliers' control over markets (which could only come from a military adventure without much chance of success, or from engineering a split within OPEC).

The diversification of sources of strength for the suppliers, pursued by investment in consumer countries, by the building up of new industries (which will need markets, not least in Europe) and by the development of their economies and societies on the basis of resources and know-how acquired from the West, makes co-operation a likely prospect for the future.

The picture, however, needs also to take into consideration some structural elements of conflict, as well as the possibility of various specific political and economic configurations which at one point or other will create strong conflicts between the two groups (as well as within them). Thus, a more accurate way of envisaging the future will be a series of conflicts and crises evolving above an underlying pattern working towards co-operation.

The first element of potential conflict is the short-term leverage available to the suppliers through Europe's strong dependence on energy imports. This leverage can be mobilized, as it was in 1973/4, in the form of export cutbacks and selective embargoes, but could also take more subtle forms of 'oil diplomacy'. Against this leverage, available at any moment, Europe has no response other than the consumers' common defence mechanisms discussed below (in Part 3). Counter-embargoes or counter-cartels are unlikely to succeed. This leverage could be used in connection with another Israeli—Arab crisis, it could be used in the context of open conflict between states in the Persian/Arabian Gulf, or in issues relating to the international energy market itself, such as conflicts over price levels (see Chapter 3).

The second area of potential conflict is more structural in character — it relates to the price of energy. Here, interests of the suppliers and of Europe will continue to be at least partially incompatible. The question of appropriate price levels will, therefore, continue to rank high on the agenda of issues between suppliers and importers. Again, this difference of interest will require a permanent effort of management.

The third area of potential differences in interest concerns the investment of supplier countries in the West. The continuing large current balance surplus of the suppliers as a group poses difficult problems of recycling, which have by no means been fully solved. The terms of OPEC investment, its distribution and best allocation to turn it into productive assets will continue to be a subject of negotiations between the two groups.

The efforts to industrialize the OPEC countries have begun to pose a fourth series of issues between Europe and its energy suppliers. As the latter succeed in building up new export industries, they will have to search for markets in the West, and in particular in Europe. Thus, problems of overcapacity will arise in the future — first with regard to refining capacity, and later possibly in petro-chemical industries, steel, and others[59]. The suppliers have begun to demand what in effect will turn out to be a shift of industrial capacity from Europe to the supplier countries. A whole complex set of differences in interest will, therefore, develop in the relationship between the two groups, and again this will require persistent bargaining and mutual adjustment.

A fifth structural element of conflict could result from domestic instability and revolutionary change in key producer countries. The previous chapter has shown the vulnerabilities and weaknesses of oil-exporting developing countries: managing such economies is extremely difficult and runs the risk of major failures. Most, if not all, OPEC countries seem to insist on economic policies which not only have doubtful chances of succeeding, but which also produce massive social and political strains. Yet there is obviously no alternative to social and economic change in a situation characterized by intensive interaction with the rest of the world.

In Iran, a mixture of disastrous planning errors not unrelated to massive and endemic corruption, and of the persistent neglect of the creation of political support in favour of repression and isolation from the mood of the people, widened the chasm between traditional institutions and values and the caricature of a modern society presented by the regime to such an extent that the social fabric broke. While the roots of opposition had only little to do with a radical refusal of any kind of modernization and social change, the upheaval also produced a severe backlash against the kind of transformation strategy pursued by the Shah, and against those countries which had supported this strategy and profited from it — principally the USA. Some of the backlash took on extreme, irrational and xenophobic forms.

Undoubtedly, the situation in Iran was marked by a number of specific factors. Yet it also is a pointer to the danger of a broader conflict between supplier countries and industrialized importers: that of periods of violent reaction against the process of transformation and the imposition of models of development which are not easy to reconcile with traditional identities of the supplier countries. The result could be regimes with a great deal of hostility towards any kind of close co-operation with the Western world, or simply the absence of any national authority in a meaningful sense for prolonged periods. The first would seem less dangerous since the necessities of interdependence would soon assert themselves; producer societies cannot function without oil exports. Whatever type of regime develops, it would have to bow to the logic of co-operation with the importers (although the *terms* of co-operation could be thrown open to re-definition, as happened with Libya in 1970, or again with Iran in 1979. This in itself could have extremely serious implications). The second development could be even more serious: it could not only affect oil supplies for prolonged periods in spite of serious implications for the producer country itself, but domestic instability in a key supplier could spill over into the region — either as a result of the 'demonstration effect', or through new and delicate regional political configurations. The regional politics under such circumstances could become a powder-keg, with governments hardly equipped to handle them cautiously. On the contrary, there might be foreign brinkmanship to compensate for lack of domestic support. Such regional instability could also result from a change of regime in a key supplier in the Gulf. In either case, it is unnecessary to indicate the risks of international confrontations resulting from such an evolution: Iran's revolution has provided ample illustration.

Possibly the most serious and important structural element of conflict, however, relates to the time factor in the evolution of interdependence. The result of a mismatch in the timing of energy policies of suppliers and importers could be very serious indeed, as the following scenarios indicate:

(1) For example, some suppliers might run out of export capacity, or at least be confronted with declining availabilities, before their objectives of economic diversification have been achieved. These countries are likely to press very hard for higher prices, without much consideration of the consequences of price rises for the world economy, and their longer-term self-interest — these effects might simply be conveniently misjudged. In addition, there might be a temptation to acquire new sources of exports through open or clandestine military action based on territorial claims (in offshore areas?). In any case, such a situation could easily bring countries with declining production (e.g. Iran in the later half on the 1980s) into conflict with other, energy-rich countries (Iraq, Saudi Arabia) whose relative position is likely to be upgraded considerably *vis-à-vis* the former as a consequence of shifting patterns of production. This scenario, which could be pursued further, offers unpleasant prospects of either open confrontation among supplier countries, or a serious crisis in supplier—consumer relations.

To indicate the potential explosiveness of such a scenario, it is sufficient to point out that Saudi Arabia's capacity to export will continue to be very large in the future, that Iraq's position will strengthen very substantially to a point where it could become the second most important OPEC exporter with a substantial lead over Iran, while the latter's position will decline.

(2) Another scenario has export demand increasing further to very high levels, while the incentives for the suppliers, and in particular for the countries with large potential but small absorption capacity for revenues, will diminish, indeed be replaced by a preponderance of disincentives (50% of oil revenues invested abroad over ten years could produce sufficient income to reduce dependence on hydrocarbon exports rather dramatically). Conservation of energy resources for future domestic demand, and for utilization in indigenous industries, would then become economically even more easily feasible at a time when general concern about the finiteness of hydrocarbon resources will increase. Besides, the suppliers might be faced with export demands so high that the depletion of their reserves, the impact of enormous oil revenues on their economies, and the problems of finding investment outlets abroad will compel them to limit their willingness to supply the quantities required by importers. At the same time, these countries are likely to be extremely sensitive to economic and political developments in the West, because of their investments there, because of their own political objectives, and because of their concern about shifts in regional and global East—West balances. Such a situation would confront the suppliers with a very difficult dilemma: they would have to limit their exports. This would even be in the interest of the importers, if they are seen to be unable to switch sufficiently to other energy sources before hydrocarbon reserves are seriously depleted, since in these circumstances the adjustment crisis would only be postponed to a later time, when the quantitative dimensions of the problem would become even more serious. Nonetheless, such a limitation would risk doing serious damage to the world economy. This scenario would correspond politically to the so-called 'second energy crisis', which has been projected for a point towards the end of this century. In fact, it is unlikely to be a single crisis, but rather a permanent dilemma, which could easily find expression in a series of crises. Such crises could arise over Saudi Arabian reluctance to increase installed capacity, over preventive price increases to reduce demand, and over production or export ceilings.

(3) Finally, there is the scenario of the suppliers' loss of control over the international energy market. Historically, events in this market which heavily affected rates of production, the price of oil, and therefore also government revenues and economic development, were largely outside the producers' power to influence developments[60]. Even now, the rates of production are not only set by demand for revenues in the supplier countries, but often as much by oscillations of demand for oil in the international market, by oil

company decisions, and by the so-called 'international responsibilities'[61] of the exporters: they are expected as a group to meet the demand of the world oil importers. This dependence of suppliers on the international market has also been an important element in their oil policies (see Part 3)[62]. As long as they depend on the international market, only their ability to control prices, and even to some extent demand for oil (should there be a real possibility of oil actually losing markets to other sources of energy) gives the suppliers some guarantee that they will be able to meet their economic objectives. This situation exists at present, and can be expected to continue for some time, as we shall see in Part 3. The reason for this lies in the strategic importance of oil to the consumer countries — an importance which far exceeds the importance of the continuation and even expansion of oil exports to the suppliers as a group, and to some suppliers individually. But this relationship could change over time. Should the consumers as a group manage to reduce their dependence on hydrocarbon exports to generate income and wealth, the suppliers would lose this strategic advantage. Such a development could result from excessive price increases over a short period. Oil prices could 'overshoot' the equilibrium point needed to reduce demand and encourage alternative energy production sufficiently to bring oil import demand in line with producers' preferred oil export revenues. The evolution over time of these two curves of dependence on the international trade in hydrocarbons obviously differs from country to country in both groups — a further factor complicating the picture.

To sum up: the future evolution of the relationship between Europe and its suppliers is likely to be marked by an underlying need for co-operation, which offers some hope for a successful management of discontinuities and crises bound to arise over a variety of issues. The element of stability lies in the mutual dependence of both groups, of their vulnerability to adverse developments on the other side, and their common incentive to co-operate, as well as the shared risks in a failure to do so. Nevertheless, such successful crisis management can by no means be taken for granted, and one crucial element of its success will lie in the establishment of the necessary instruments and institutions of crisis prevention, as well as crisis management, well before an actual contingency (see Part 4). If habits and mechanisms of co-operation are developed, if an atmosphere conducive to constructive problem solution evolves, then this should also help to defuse situations where all advance preparation will be insufficient, where narrow interests replace broader and longer-term calculations, where decision-makers will be under intense pressure from domestic constituencies and demands, and where political considerations unrelated to the world energy situation spill over into the international energy system. Such crises could arise over regional conflicts such as the Israeli—Arab confrontation, the various latent conflicts in the Persian/Arabian Gulf, or even out of domestic instability in one key supplier country such as Saudi Arabia. Even with regard to the critical risk to consumer—producer relations from destabilization or change of regime in a central OPEC country — be it as a result of the 'time trap' of unsuccessful diversification, or of domestic revolution against the process of transformation — international co-operation might be useful in containing the damage, moderating policy fluctuations, and devising measures to alleviate the consequences.

# Notes to Part two

1. I follow and expand ideas put forward by A.O. Hirschmann, *National Power and the Structure of Foreign Trade*, University of California Press, Berkeley 1945
2. See for example C. Issawi, *Oil, the Middle East, and the World*, New York 1972 (Washington Papers No. 4), pp. 38–39
3. See next chapter
4. See J.-M. Chevalier, *Le nouvel enjue pétrolier*, Calmann-Levy, Paris 1973
5. See below, Chapter 9 (Energy Policy Objectives III : Price Formation and Evolution)
6. H.W. Singer, 'The distribution of gains revisited', in H.W. Singer, *The Strategy of International Development : Essays in the Economics of Backwardness*, Macmillan, London 1975, pp. 58–66. A vast literature exists on this subject; useful bibliographies can be found in D. Senghaas,(ed.), *Imperialismus und strukturelle Gewalt, Analysen über abhängige Reproduktion*, Suhrkamp, Frankfurt 1972, and D. Senghaas,(ed.), *Peripherer Kapitalismus, Analysen über Abhängigkeit und Unterentwicklung*, Suhrkamp, Frankfurt 1974
7. J. Galtung, 'Eine strukturelle Theorie des Imperialismus', in D. Senghaas,(ed.), *Imperialismus und strukturelle Gewalt, Analysen über abhängige Reproduktion*, Suhrkamp, Frankfurt 1972, pp. 29–104 (43)
8. H.W. Singer, 'The distribution of gains between investing and borrowing countries', in H.W. Singer, *The Strategy of International Development : Essays in the Economics of Backwardness*, Macmillan, London 1975, pp. 43–57
9. H.B. Chenery and T. Watanabe, 'International comparisons of the structure of production', *Economoetrica* 1958, p.504
10. A detailed analysis can be found in J. Galtung 'Eine strukturelle Theorie des Imperialismus', in D. Senghaas, (ed.), *Imperialismus and strukturelle Gerwalt Analysen über abhängige Reproduktion*, Suhrkamp, Frankfurt 1972, pp. 44–47, 75–80
11. Refining is about 3.5 times as labour-intensive as crude oil production (C. Issawi and M. Yeganeh, *The Economics of Middle East Oil*, Faber, London 1962, p.81
12. C. Issawi and M. Yeganeh, *The Economics of Midle East Oil*, Faber, London 1962, pp.81–83
13. *Petroleum Economist*, Sept. 1974, pp.242–343; April 1975, p.182
14. Saudi Arabia's concession to ARAMCO stipulated the construction of refineries as soon as the production of certain fields would exceed 75 000 b/d. A later concession for a Japanese group in the Neutral Zone specified the construction of a refinery for 30% crude oil of the group. C. Issawi and M. Yeganeh, *The Economics of Middle East Oil*, Faber, London 1962, pp.79,170
15. The growth of chemical industrial production compares as follows to the growth of total industrial production (average annual rate, 1958–1967, in %)

|  | Industrial production (excluding construction) | Chemical production |
|---|---|---|
| UK | 3.1 | 6.6 |
| Germany | 5.1 | 11.0 |
| France | 5.1 | 9.9 |
| Italy | 8.9 | 13.4 |
| USA | 6.0 | 8.3 |

(B.G. Reuben and M.L. Burstall, *The Chemical Economy, A Guide to the Technology and Economics of the Chemical Industry*, Longman, London 1973, p.112
16. See P. Bairoch, *The Development of the Third World Since 1900*, Methuen, London 1975, p.58; C. Issawi and M. Yeganeh, *The Economics of Middle East Oil*, Faber, London 1962, Ch. 7; S.R. Pearson, *Petroleum and the Nigerian Economy*, Stanford University Press, Stanford 1970, Ch. 4; S.R. Pearson and J. Cownie, (eds.), *Commodity Exports and African Economic Development*, Lexington Books, Toronto 1974, Ch. 1
17. K.S. Sayegh, *Oil and Arab Regional Development*, Praeger, New York 1968, p.83; R.F. Mikesell, (ed.), *Foreign Investment on Investor–Host Relations*, Johns Hopkins University Press, Baltimore 1971; see also C. Morse, 'Potentials and hazards of direct international investment in raw materials', in M. Clawson, (ed.), *Natural Resources and International Development*, Johns Hopkins University Press, Baltimore 1964, pp.367–414

18 P. Bairoch, *The Development of the Third World Since 1900*, Methuen, London 1975, p.59; R.F. Mikesell, (ed.), *Foreign Investment in the Petroleum and Mineral Industries: Case Studies on Investor—Host Relations*, Johns Hopkins University Press, Baltimore 1971; S.R. Pearson, *Petroleum and the Nigerian Economy*, Stanford University Press, Stanford 1970, pp. 87—89; R.K. Meyer, and S.R. Pearson, 'Contributions of petroleum to Nigerian economic development', in S.R. Pearson and J. Cownie, (eds.), *Commodity Exports and African Economic Development*, Lexington Books, Toronto 1974, pp.155—178 (171)

19 W.G. Harris, 'The impact of the petroleum export industry on the pattern of Venezuelan economic development', in R.F. Mikesell, (ed.), *Foreign Investment in the Petroleum and Mineral Industries: Case Studies, on Investor—Host Relations*, Johns Hopkins University Press, Baltimore 1971, pp.129—156 (141); W.H. Bartsch, 'The impact of the oil industry on the economy of Iran', in R.F. Mikesell, (ed.), *Foreign Investment in the Petroleum and Mineral Industries: Case Studies on Investor—Host Relations*, Johns Hopkins University Press, Baltimore 1971, pp.237—263 (257); Venezuela has succeeded best in creating backward linkages: the country produces pipes, oil well tubings, metal cans, engine oils, cable, paper, tyres, chemicals, paint and other products for the oil industry. In the case of pipelines, the stage of plastic coating (until recently done in the USA) is now also handled locally.

20 S.R. Pearson, *Petroleum and the Nigerian Economy*, Stanford University Press, Stanford, 1970, p. 90

21 This seems to be implied by the large share of locally-purchased, but foreign-produced goods required by the oil industry in Venezuela and Iran. For details, see W.G. Harris, 'The impact of the petroleum export industry on the pattern of Venezuelan economic development', in R.F. Mikesell, (ed.), *Foreign Investment in the Petroleum and Mineral Industries: Case Studies on Investor—Host Relations*, Johns Hopkins University Press, Baltimore 1971, p.141; W.H. Bartsch, 'The impact of the oil industry on the economy of Iran', in R.F. Mikesell, (ed.), *Foreign Investment in the Petroleum and Mineral Industries: Case Studies on Investor—Host Relations*, Johns Hopkins University Press, Baltimore 1971, p.257

22 For details of this policy, see C. Issawi and M. Yeganeh, *The Economics of Middle East Oil*, Faber, London 1962, p.151; K.S. Sayegh, *Oil and Arab Regional Development*, Praeger, New York 1968, pp.88—92; L. Mosley, *Power Play, The Tumultuous World of Middle East Oil*, Penguin, London 1974, p.327

23 S.R. Pearson, *Petroleum and the Nigerian Economy*, Stanford University Press, Stanford, 1970, p. 85; K.S. Sayegh, *Oil and Arab Regional Development*, Praeger, New York 1968, pp. 92—92; a slightly different and more positive picture emerges in the case of Iran where the National Iranian Oil Company (NIOC) retains the monopoly for training indigenous oil industry employees (W.H. Bartsch, 'The impact of the oil industry on the economy of Iran', in R.F. Mikesell, (ed.), *Foreign Investment in the Petroleum and Mineral Industries: Case Studies on Investor—Host Relations*, Johns Hopkins University Press, Baltimore, 1971, p.255. See also C. Issawi and M. Yeganeh, *The Economics of Middle East Oil*, Faber, London 1962, p. 151; P.E. Church, 'Labour relations in mineral and petroleum resource development', in R.F. Mikesell, (ed.), *Foreign Investment in the Petroleum and Mineral Industries: Case Studies on Investor—Host Relations*, Johns Hopkins University Press, Baltimore, 1971, pp.81—98 (94—95)

24 K.S. Sayegh, *Oil and Arab Regional Development*, Praeger, New York 1968, pp.85—86

25 K.S. Sayegh, *Oil and Arab Regional Development*, Praeger, New York 1968, pp.93—94; C. Issawi and M. Yeganeh, *The Economics of Middle East Oil*, Faber, London 1962, p. 151

26 S.R. Pearson, *Petroleum and the Nigerian Economy*, Stanford University Press, Stanford, 1970, p. 85

27 S.R. Pearson, *Petroleum and the Nigerian Economy*, Stanford University Press, Stanford, 1970, p. 98

28 See for details P.E. Church, 'Labour relations in mineral and petroleum resource development', in R.F.

Mikesell, (ed.), *Foreign Investment in the Petroleum and Mineral Industries: Case Studies on Investor—Host Relations*, Johns Hopkins University Press, Baltimore 1971, p.83

29 P. Bairoch, *The Development of the Third World Since 1900*, Methuen, London 1975, p.60

30 A good example for positive side-effects of oil-related investments is the Iranian-Soviet gas pipeline which supplies about half of Teheran's industrial energy requirements, as well as domestic and urban gas requirements of areas along the pipeline (W.H. Bartsch, 'The impact of the oil industry on the economy of Iran', in R.F. Mikesell, (ed.), *Foreign Investment in the Petroleum and Mineral Industries: Case Studies on Investor—Host Relations*, Johns Hopkins University Press, Baltimore 1971, p. 259

31 S.R. Pearson, *Petroleum and the Nigerian Economy*, Stanford University Press, Stanford, 1970, p. 101

32 P. Bairoch, *The Development of the Third World Since 1900*, Methuen, London 1975, p. 59

33 J.-M. Chevalier, *Le nouvel enjeu pétrolier*, Calmann-Levy, Paris 1973. pp.129—142; Nicolas Sarkis, 'Pétrole et développement économique dans les pays arabes', in *Etudes Internationales*, Dec. 1974, pp.562—574. The most elaborate strategy in this direction is probably the Algerian path to development — see, for example J.-M. Chevalier, *Le nouvel enjeu pétrolier*, Calmann-Levy, Paris 1973, pp. 179—180; G. Destanne de Bernis, 'Les problèmes pétroliers algériens', in *Etudes Internationales*, Dec. 1974, pp.575—609

34 P. Bairoch, *The Development of the Third World Since 1900*, Methuen, London 1975, p.61

35 P. Bairoch, *The Development of the Third World Since 1900*, Methuen, London 1975, p.70; see also W.H. Bartsch, 'The impact of the oil industry on the economy of Iran', in R.F. Mikesell, (ed.), *Foreign Investment in the Petroleum and Mineral Industries: Case Studies on Investor—Host Relations*, Johns Hopkins University Press, Baltimore 1971, p.263

36 P. Bairoch, *The Development of the Third World Since 1900*, Methuen, London 1975, p.91. This evidence is derived from (a) an analysis of historical development of the economies of the industrialized countries, (b) an overall statistical assessment of gross agricultural and gross industrial production which shows strong correlations, and (c) a case-by-case analysis of 40 LDCs which showed the same strong correlation for all countries with a fairly developed manufacturing sector (i.e. advanced import substitution); the correlation refers to the response of the industrial sector to developments in the agricultural sector in the preceding year.

37 The following paragraphs draw mainly on arguments developed in R.F. Mikesell, (ed.), *Foreign Investment in the Petroleum and Mineral Industries: Case Studies on Investor—Host Relations*, Johns Hopkins University Press, Baltimore 1971. H.W. Singer, *The Strategy of International Development: Essays in the Economics of Backwardness*, Macmillan, London 1975; D. Seers, 'Some reflections on the lessons for Scotland of the experience of other oil producers' (unpublished paper, Sussex University, 1975)

38 For wage levels and fringe benefits, see C. Issawi and M. Yeganeh, *The Economics of Middle East Oil*, Faber, London 1962, pp.97—98; K.S. Sayegh, *Oil and Arab Regional Development*, Praeger, New York 1968, pp. 94—96, p.301; P.E. Church, 'Labour relations in mineral and petroleum resource development', in R.F. Mikesell, (ed.), *Foreign Investment in the Petroleum and Mineral Industries: Case Studies on Investor—Host Relations*, Johns Hopkins University Press, Baltimore 1971, p.830. This pull-effect does not imply total equalization of wage levels throughout the industrial sectors

39 P.E. Church, 'Labour relations in mineral and petroleum resource development', in R.F. Mikesell, (ed.), *Foreign Investment in the Petroleum and Mineral Industries: Case Studies on Investor—Host Relations*, Johns Hopkins University Press, Baltimore 1971, pp.91—92; W.G. Harris, 'The impact of the petroleum export industry on the pattern of Venezuelan economic development', in R.F. Mikesell, (ed.), *Foreign Investment in the Petroleum and Mineral Industries: Case Studies on Investor—Host Relations*, Johns Hopkins University Press, Baltimore 1971, p.144; R.F.

Mikesell, 'Major problem areas', in R.F. Mikesell, (ed.), *Foreign Investment in the Petroleum and Mineral Industries: Case Studies on Investor–Host Relations*, Johns Hopkins University Press, Baltimore 1971, pp.423–440 (430)

40  D.A. Wells, 'ARAMCO: the evolution of an oil concession', in R.F. Mikesell, (ed.), *Foreign Investment in the Petroleum and Mineral Industries: Case Studies on Investor–Host Relations*, Johns Hopkins University Press, Baltimore 1971, pp.216–236 (234)

41  R.F. Mikesell, 'Major problem areas', in R.F. Mikesell, (ed.), *Foreign Investment in the Petroleum and Mineral Industries: Case Studies on Investor–Host Relations*, Johns Hopkins University Press, Baltimore 1971, p.430; D. Seers, *Some reflections on the lessons for Scotland of the experience of other oil producers*, (unpublished paper, Sussex University, 1975), p.4

42  Scotland is not the only example which springs to mind: the secession of Biafra and the consequent Nigerian civil war were certainly encouraged by the existence of oil in Biafra (see S.R. Pearson, *Petroleum and the Nigerian Economy*, Stanford University Press, Stanford, 1970, Ch. 9)

43  D.A. Wells, 'ARAMCO: the evolution of an oil concession', in R.F. Mikesell, (ed.), *Foreign Investment in the Petroleum and Mineral Industries: Case Studies on Investor–Host Relations*, Johns Hopkins University Press, Baltimore 1971, p.234

44  Within 25 years (from 1936 to 1961), Venezuela turned from a predominantly rural (65%) into a predominantly urban (68%) society.

45  For the concept of dualism, see H.W. Singer, 'Dualism revisited', in H.W. Singer, *The Strategy of International Development: Essays in the Economics of Backwardness*, Macmillan, London 1975

46  See, for example, W.H. Bartsch, 'The impact of the oil industry on the economy of Iran', in R.F. Mikesell, (ed.), *Foreign Investment in the Petroleum and Mineral Industries: Case Studies on Investor–Host Relations*, Johns Hopkins University Press, Baltimore 1971, p.253; W.G. Harris, 'The impact of the petroleum export industry on the pattern of Venezuelan economic development', in R.F. Mikesell, (ed.), *Foreign Investment in the Petroleum and Mineral Industries: Case Studies on Investor–Host Relations*, Johns Hopkins University Press, Baltimore 1971, p.134

47  In 1965, Venezuela's bottom 20% of the population received 3% of total national income, while the top 5% received 26.5%. The Gini-Index for land (90.9 in 1956), and for income distribution (48.7 in 1963) marked Venezuela as one of the countries with very high inequality for its overall level of GDP. K. Ruddle and K. Burrows, (eds.), *Statistical Abstract of Latin America 1972*, University of California Press, Los Angeles 1974, p.26; C.L. Taylor and M.C. Hudson, *World Handbook of Political and Social Indicators*, 2nd ed., Yale University Press, New Haven 1972

48  R.F. Mikesell, 'Major problem areas', in R.F. Mikesell, (ed.), *Foreign Investment in the Petroleum and Mineral Industries: Case Studies on Investor–Host Relations*, Johns Hopkins University Press, Baltimore 1971, p.430; D. Seers, *Some reflections on the lessons for Scotland of the experience of other oil producers*, (unpublished paper, Sussex University, 1975) p.3; W.G. Harris,'The impact of the petroleum export industry on the pattern of Venezuelan economic development', in R.F. Mikesell, (ed.), *Foreign Investment in the Petroleum and Mineral Industries: Case Studies on Investor–Host Relations*, Johns Hopkins University Press, Baltimore 1971, p.150

49  E.T. Penrose, 'Oil and the state in Arabia', in E.T. Penrose, *The Growth of Firms, Middle East Oil and other Essays*, Frank Cass, London 1971, pp.281–295 (282). In 1953, about 10% of the total population (sic!) of Kuwait worked in the civil service; one third of them were illiterate. Another example is the Saudi recruitment policy, which effectively encourages school dropouts by providing well-paid government jobs

virtually without formal educational requirements. (R. Knauerhase, 'Saudi Arabia's economy at the beginning of the 1970s', *Middle East Journal*, Spring 1974, pp. 126–140 (128))
50 See, for example, 'Venezuelan measures to alleviate unemployment', *The Economist*, 27 Dec. 1975 (Survey on Venezuela)
51 See for recent corruption affairs, and attempts to fight them, *Le Monde*, 1–2, 25 Feb, 1976, and *Newsweek*, 8 March 1976 (Iran); *The Financial Times*, 30 March 1976 (Abu Dhabi)
52 K.S. Sayegh, *Oil and Arab Regional Development*, Praeger, New York 1968, pp.138–148 (e.g. Kuwait)
53 See *The Financial Times*, 15 April 1976: also S. Bluff, 'Mixed prospects for a Middle East petrochemical industry', *Middle East Economic Digest*, 5 April 1974
54 See W.G. Harris 'The impact of the petroleum export industry on the pattern of Venezuelan economic development', in R.F. Mikesell, (ed.), *Foreign Investment in the Petroleum and Mineral Industries: Case Studies on Investor–Host Relations*, Johns Hopkins University Press, Baltimore 1971, p.134: in Venezuela, employment growth in industry did not keep up with the rate of population growth, and the share of the labour force in manufacturing stagnated in spite of the expansion of this sector
55 W.G. Harris, 'The impact of the petroleum export industry on the pattern of Venezuelan economic development', in R.F. Mikesell, (ed.), *Foreign Investment in the Petroleum and Mineral Industries: Case Studies on Investor–Host Relations*, Johns Hopkins University Press, Baltimore 1971, contrasts the structure of the Venezuelan economy with a hypothetical 'normal' structure of output (based on historical experience and statistical analysis, and using per capita GNP and population size as main variables), which clearly indicates distortions:

|  | Normal output, % distribution (1950) | Venezuelan output (1950) (%) | Normal output (1964) (%) | Venezuelan[a] output (1964) (%) |
|---|---|---|---|---|
| Agriculture | 19.2 | 11.7 | 13.1 | 10.1 |
| Mining/Quarrying | 1.9 | 0.2 | 1.6 | 1.6 |
| Manufacturing | 21.1 | 12.2 | 30.1 | 16.4 |
| Construction | 6.1 | 9.5 | 6.0 | 6.6 |
| Transport/Communication | 8.8 | 7.7 | 9.2 | 5.5 |
| Services | 42.9 | 58.7 | 40.0 | 59.8 |
|  | 100.00 | 100.00 | 100.00 | 100.00 |

[a] excluding petroleum

56 W.G. Harris, 'The impact of the petroleum export industry on the pattern of Venezuelan economic development', in R.F. Mikesell, (ed.), *Foreign Investment in the Petroleum and Mineral Industries: Case Studies on Investor–Host Relations*, Johns Hopkins University Press, Baltimore 1971, p.136; Harris contrasts normal industrial output by sector with Venezuelan output (again excluding petroleum):

|  | Normal output, industrial sector (%) (1964) | Venezuelan output (%) (1964) |
|---|---|---|
| Investment goods | 38.9 | 20.8 |
| Intermediate goods | 21.8 | 30.0 |
| Consumer goods | 39.3 | 49.2 |
|  | 100.00 | 100.00 |

See also D.A. Wells, 'ARAMCO: the evolution of an oil concession', in R.F. Mikesell, (ed.), *Foreign Investment in the Petroleum and Mineral Industries: Case Studies on Investor—Host Relations*, Johns Hopkins University Press, Baltimore 1971 pp.234—235

57 See W.G. Harris, 'The impact of the petroleum export industry on the pattern of Venezuelan economic development', in R.F. Mikesell, (ed.), *Foreign Investment in the Petroleum and Mineral Industries: Case Studies on Investor—Host Relations*, Johns Hopkins University Press, Baltimore 1971, p.136 (Venezuela). A similar picture is presented by Iran: agricultural output between 1956 and 1966 grew at an average annual rate of 3.3%, while the average growth of the labour force in this sector fell by 0.5% p.a.; over the same period, industrial production increased by 12.0% annually, but the labour force fell by 4.6% only; for the service sector the figures were 8.0% and 1.4% respectively. Overall, this meant that higher output was not sufficiently translated into higher employment; since the agricultural sector still provided the bulk of employment facilities, this meant a substantial increase in unemployment over the period 1956—1966. (United Nations Commission for Asia and the Far East, *Economic Survey of Asia and the Far East* 1973 pp. 18—19

58 *The Financial Times*, 28 May 1976
59 L. Turner and J. Bedore, 'Saudi and Iranian petrochemicals and oil refining: trade warfare in the 1980s?', in *International Affairs*, Oct. 1977, pp.572—586
60 See C. Issawi and M. Yeganeh, *The Economics of Middle East Oil*, Faber, London 1962, p.25. For the impact of stagnating petroleum exports on the Venezuelan economy between 1957 and 1966 (when in spite of a 40% increase in retained value from the oil sector over the period 1960—1966 annual real growth for 1958—1966 was by far the lowest in South America, at about 1% p.a.) see W.G. Harris, 'The impact of the petroleum export industry on the pattern of Venezuelan economic development', in R.F. Mikesell, (ed.), *Foreign Investment in the Petroleum and Mineral Industries: Case Studies on Investor—Host Relations*, Johns Hopkins University Press, Baltimore 1971, pp.129, 132
61 Y. Sayigh, 'Arab oil politics: self-interest and responsibility', *Journal of Palestine Studies* No. 15, Spring 1975, pp.59—73
62 See Y. Sayigh, 'Arab oil politics: self-interest and responsibility', *Journal of Palestine Studies* No. 15, Spring 1975, and H. Madelin, *Pétrole et politique en Méditerranée Occidentale*, A. Colin, Paris 1973, p. 81

PART THREE

# Objectives for a European energy policy

Our conclusion so far has been that Europe will continue to depend substantially on energy imports and that this dependence will rest on a small group of developing countries, themselves dependent on continued energy exports to achieve their social and economic objectives. This situation basically favours co-operation between Europe and her suppliers. The question which arises concerns the objectives to be pursued — what guidelines should a European energy policy follow? Answers will be discussed in this part, Part 3, which will focus on the energy self-interest of the Community. Within this fairly narrow approach, three overriding objectives will be identified: security of energy supplies; stability of supplies; and the price of imported energy.

Narrow self-interest alone does not provide a sufficient guide for long-term energy policies. Energy is already, and will increasingly become, a global concern, and the difficulties confronted by the international system with regard to energy will far exceed the problem of assuring Europe of sufficient and safe supplies of energy in the long run. As will be argued later in this study, energy now requires some form of international management. A more equitable and prosperous world will call for a more even-handed international energy system. In particular, oil-importing LDCs, whose energy requirements can be expected to increase very rapidly during the rest of the century and beyond, need special attention: there is a great danger that these countries will be pushed into excessive dependence on hydrocarbons — from their point of view, the most convenient and flexible source of energy. If this demand for hydrocarbons manifests itself in the international markets fully, supply—demand inbalances will be impossible to avoid.

Obviously, future demand of oil-importing LDCs for oil and natural gas has to be taken into consideration — and this is a strong moral argument for constraining, as effectively as possible, hydrocarbon consumption in developed countries. However, at the same time efforts will be needed to develop alternative energy strategies for those LDCs — strategies based on indigenous sources of supply, and on alternative sources, in particular 'soft' technologies. This will have to be done by the LDCs themselves, but assistance both with technology and capital from the developed countries will be necessary; and Europe will have to play an important role in this.

Nevertheless, Europe will also have its own interests to pursue, and the formulation of policies designed to protect these interests is indispensable. The following chapters take a closer look at the international energy policy interests of the Communities. These interests will, in any case, be injected into the broader framework of international energy politics, where they will have to be reconciled with the interests and objectives of other participants. At the same time, European objectives in security, commerce and politics, and traditional alliances and links with the rest of the world, must also be harmonized with a strategy pursuing security and stability of energy supplies; this will inevitably tend to deform the 'ideal' strategy adequate to secure these basic objectives. Contradictory domestic pressures and influences will further tend to distort the path. Nevertheless, a European international energy policy will be necessary; and it could already be considered successful if it manages gradually to reconcile the various contradictory pressures in a zig-zag course towards the achievement of security and stability of supplies, together with some influence on price levels in the international energy markets.

CHAPTER SEVEN

# Security of supply

The Arab—Israeli conflict of 1973 provided an opportunity for a group of oil-exporting countries to exploit their strong position as a major supplier of European energy in the pursuit of objectives which had nothing to do with energy issues. Such pressure could be repeated and to assess the capacity to exert pressure through supply interruptions we first must determine the *size* of a critical shortfall. Clearly, minor shortfalls in external supplies could be absorbed without too much strain by emergency schemes directing energy consumption into the most vital sectors and reducing it in others. Taking the guidelines of the International Energy Agency (IEA) as a yardstick, this manageable deficit would appear to be about 10% of total energy consumption. The threshold, therefore, must be somewhat higher. Rather arbitrarily, this chapter sets two thresholds of 15 and 20% of total energy consumption, the former indicating serious political vulnerability and the latter major political dependence. In addition, problems of shifting available energy supplies into different sectors of consumption necessitate inclusion of a third, more tentative, threshold of a 25% dependence on imports for one particular fuel. This is based on the fact that such a shortfall in the case of oil and natural gas could have a serious impact on areas of consumption which do not allow replacement by other fuels[1]. Taking these thresholds as a rough guide, we can now look at the present energy situation of the Community as summarized in *Table 7.1*. What emerges clearly is that only oil currently poses a problem of major political dependence. In all other cases, the share of imports in total supplies is too small to be significant. *Table 7.2* shows two scenarios of European energy dependence in 1985, illustrative of each end of a spectrum of possible energy positions. The threshold levels show how the picture has changed: while coal import dependence is still insignificant, natural gas imports by 1985 will have reached a level of political significance. Moreover, indigenous supplies might well decline after 1985, exacerbating political vulnerability thereafter. Overall import dependence remains much beyond the critical thresholds of 15 and 20% of total energy demand. Dependence on oil will continue to be the main area of concern (nuclear imports do not enter this picture, as import dependence is not considered problematical in terms of politically motivated supply interruptions, given the specific characteristics of reactor fuel, i.e. long burning times and small quantities needed).

Our analysis, therefore, focuses on oil and natural gas. Three motives could

Table 7.1 European Community Energy Position, 1974 and 1977 ($10^6$ toe)

| | Indigenous production | | Imports net | | Consumption | | Imports as % of consumption | | Imports as % of total energy consumption | |
|---|---|---|---|---|---|---|---|---|---|---|
| | 1974 | 1977 | 1974 | 1977 | 1974 | 1977 | 1974 | 1977 | 1974 | 1977 |
| Solid fuels | 182.2 | 198.8 | 27.2 | 29.2 | 218.3 | 204.6 | 12.4 | 14.3 | 2.9 | 3.1 |
| Oil | 13.0 | 48.1 | 598.9 | 487.2 | 584.3 | 537.6 | 102.5 | 90.6 | 64.5 | 52.4 |
| Natural gas | 130.2 | 142.3 | 5.7 | 17.0 | 134.6 | 157.7 | 4.2 | 10.8 | 0.6 | 1.8 |
| Nuclear | 16.4 | 27.0 | — | — | 16.4 | 27.0 | — | — | — | — |
| Other | 28.7 | 36.5 | 0.8 | 1.4 | 29.5 | 37.9 | 2.8 | 3.4 | 0.1 | 0.2 |
| Total | 370.6 | 433.5 | 632.6 | 534.7 | 927.9 | 929.8 | 68.2 | 57.5 | 68.2 | 57.5 |

Note: % figures may not add up because of rounding
Source: OECD

Table 7.2 Europe's Energy Position in 1985 ($10^6$ toe): Two Scenarios

| | | Aggregate energy consumption | Solid fuel consumption | Liquid fuel consumption | Natural gas consumption | Coal imports | | Crude imports | | | Natural gas imports | | |
|---|---|---|---|---|---|---|---|---|---|---|---|---|---|
| | | | | | | Socialist | Other | Arab | OPEC | Socialist | Other | Arab | OPEC | Socialist | Other |
| Scenario I | | 1 400 | 230 | 830 | 270 | 20 | 30 | 480 | 610 | 20 | 50 | 40 | 70 | 30 | 20 |
| Scenario II | | 1 300 | 240 | 700 | 260 | 20 | 30 | 350 | 480 | 20 | 50 | 40 | 60 | 20 | 20 |
| *Thresholds* | | | | | | | | | | | | | | | |
| 15% shortfall of total energy | I | 210 | — | — | — | | | | | | | | | | |
| | II | 195 | — | — | — | | | | | | | | | | |
| 20% shortfall of total energy | I | 280 | — | — | — | | | | | | | | | | |
| | II | 260 | — | — | — | | | | | | | | | | |
| 25% shortfall in individual fuel sectors | I | — | 57.5 | 207.5 | 67.5 | | | | | | | | | | |
| | II | — | 60.0 | 175 | 65 | | | | | | | | | | |

155

bring a sufficiently large group of Europe's suppliers together: objectives relating to the international energy system itself (such as prices); objectives carried into the energy system but not originating from it (Israeli–Arab or Persian/Arabian Gulf conflicts); and the general relationship between industrialized and developing countries. Given these motives it is clear that either the Arab producers, or OPEC, or parts of both, might be willing to co-operate in supply interruptions to the industrialized consumers. A fourth objective related to the East–West conflict can also be included here. In the two scenarios given in *Table 7.2*, the communist bloc accounts for only 4–5% of the total energy supplies of the Community. Although this is clearly below the threshold of vulnerability, allowance must be made for the fact that Iran's natural gas exports to the Nine are partly controlled by the USSR, and this could increase Europe's import dependence on supplies arriving from the communist countries to close to 25%. Soviet interference with natural gas supplies to the Community is unlikely, for it would constitute a direct confrontation with the West. Much will depend on the future development of East–West relations, but retaliation in trade, for example grain, could make the costs of such actions rather high for the Soviet Union, depending on the circumstances.

The cumulative leverage of the Arab producers through oil *and* gas supplies is clearly great – varying between 30 and 40% in the two different scenarios, depending on the evolution of trade patterns, but only oil exceeds the critical thresholds alone. In the case of OPEC, the Community's dependence is even higher, but again natural gas would only pose a problem in scenario I; cumulative dependence will be between 32% and 45% for the two scenarios.

The next factor to be considered is the level of exports to the Community. In 1985, to reach the 20% overall threshold, the producers will have to be able to cut back exports to the Community by $5.2 \times 10^6$–$5.6 \times 10^6$ b/d. Assuming a total Arab production of between $20 \times 10^6$ and $30 \times 10^6$ b/d, and OPEC production of $30 \times 10^6$–$40 \times 10^6$ b/d, this would appear to be easily within the means of the producers, or even only some of them. The costs for them depend on (a) the capacity of a producer to absorb a shortfall in revenues and (b) the impact of supply reductions on oil income. Some countries at least would have considerable room for a shortfall in revenues in a situation likely to arouse nationalist sentiments, which would lead to strong support for governments whose measures demand some sacrifices. Besides, oil prices might again accelerate in a situation of artificial scarcity, as they did in 1973/4 and 1979, compensating producers for production cutbacks.

The most important criterion for defining potential candidates for an action group thus appears to be the export level to Europe. This, obviously, cannot be predicted precisely, but it is plausible to assume that the critical threshold could be reached by not only all Arab producers, OPEC or even all producers, but also by a smaller group of suppliers. Another variable is the amount of oil available from other sources of supply for Europe, notably stock levels and standby capacity outside the producer action group. Stocks are only a temporary buffer against the oil producer's power. They can postpone but not avoid the crisis. While they might help to prolong the crisis and allow consumers to execute counter-measures, they could not fundamentally reduce the leverage of the producers. Standby capacity, on the other hand, could change the distribution of power, although such readily available capacity in substantial quantities would most likely be concentrated in the producer countries pursuing a conservation policy. While some standby capacity may be available within the Community (natural gas from the Netherlands, possibly oil from Britain), the bulk would presumably be located in the Middle East, particularly in countries like Saudi Arabia, Kuwait, Abu Dhabi and Libya. The attitude of such producers, in particular Saudi Arabia, which has the largest production capacity and resources, will therefore be vital for the success or failure of a producer action group and Saudi Arabia's active opposition would almost certainly cause it to fail.

However, even the potential to achieve the critical cutback thresholds might not be enough to give producers the leverage they want. A further condition is the impact on the international distribution system. There is a crucial distinction between oil and gas in that the international distribution system in oil is highly flexible, while the equivalent for natural gas is, and will remain, extremely rigid. Gas is transported over long distances either by pipeline or by special carriers. In 1985 the EEC will receive about $40 \times 10^9$ m$^3$ by pipeline from suppliers other than Norway (see *Table 2.6*) and this amount might be even higher. Moreover, since economic factors favour pipeline transport, this sector can be expected to expand more rapidly in the future than trade in LNG carriers. The rigidity of deliveries by pipeline needs no elaboration; it is obviously very easy to cut supplies and very difficult to replace them by others.

The much smaller size of an LNG gas network in comparison with oil works in the same direction: to service a $9.5 \times 10^9$ m$^3$/a ($8.17 \times 10^6$ toe/a) contract between Algeria and Germany, Switzerland and Austria, only two 125 000 m$^3$ LNG carriers will be required[2]. The total number of carriers employed by 1985 will be very small, in comparison with the tanker fleet servicing the international oil market. One recent estimate is that by 1985 some 113 carriers with a capacity of 125 000 m$^3$, and more, each, and some 21 carriers of smaller capacity, will be needed[3]. With such a limited number of tankers, any attempt to reallocate existing supplies will be extremely difficult, even with no other barriers to overcome. While there might be some surplus capacity available, this could be more than balanced by a growing share of producer countries in the world LNG tanker fleet (summarized in *Table 7.3*)[4]. Similarly, the number of gasification and liquefaction plants for natural gas will be small, again emphasizing the rigidity of the system. The implications of this are that if Arab, or OPEC, producers want to exert political pressure they might well favour a shift away from oil to natural gas. The lack of flexibility in the distribution system of gas allows political pressure to be directed fairly precisely at one particular target. Interference with natural gas supplies could be attractive within the context of a general 'energy war' against the European Community.

**Table 7.3** *World LNG Tanker Fleet, Existing and on Order (early 1978)*

|  | Existing fleet | | On order | |
| --- | --- | --- | --- | --- |
|  | Number of carriers | Total capacity (m³) | Number of carriers | Total capacity (m³) |
| Shell | 7 | 528 800 | – | – |
| Exxon | 3 | 123 000 | – | – |
| France | 5 | 365 081 | 1 | 129 500 |
| UK | 5 | 352 000 | – | – |
| Norway | 3 | 184 188 | 1 | 128 600 |
| Belgium | – | – | 1 | 131 580 |
| Spain | 1 | 40 000 | – | – |
| USA | 8 | 1 006 800 | 14 | 1 760 000 |
| Algeria | 3 | 294 609 | 3 | 379 500 |
| AMPTC[a] | 9 | 837 634 | 9 | 1 173 200 |
| Total | 44 | 3 732 112 | 29 | 3 752 380 |
|  |  | (OPEC share 30.3%) |  | (OPEC share 41.3%) |

[a] Arab Maritime Petroleum Transport Company (involving several Arab states)
Source: H.P. Drewry, Shipping Consultants, *The Seaborne Transportation of LNG, 1977–1985*, Drewry, London 1977

On the other hand, there are certain cost factors which would seem to favour restraint by the producers. The concentration of future natural gas suppliers to the Community will be much higher than for oil: the three biggest, apart from Norway, will be the Soviet Union, Iran and Algeria. The case of the Soviet Union has already been considered briefly. For Iran, the triangular character of its trade contributes a special factor. Any interference with supplies to Europe would have to be either agreed or initiated by Moscow; to do otherwise would antagonize the Soviet government — a situation Iran would certainly try to avoid. Besides, the revolution has created considerable uncertainty about the expansion of Iranian natural gas exports to the EEC. As for Algeria, income from oil and natural gas sustains her development, and the impact of severely reduced revenues would be felt directly and strongly, particularly since Algeria is closely tied to Europe economically and therefore sensitive to sanctions, such as the removal of preferential access for her products[5]. It should also be noted that the large reserves of natural gas needed in consumer countries for technical reasons will provide a considerable measure of protection against supply interruptions, at least in the short run. Such stocks of natural gas can be established cheaply in underground reservoirs, or in depleted natural gas fields.

For oil, the International Energy Programme Emergency Allocation Scheme, operated under the auspices of the International Energy Agency (IEA), represents a formalized solidarity plan of key consumers. In the event of a cut in oil supplies this scheme aims to share available supplies among all member states, provided the shortfall to the whole group, or any one of its members, can be expected to exceed 7% of total oil supplies of the group, or country, affected.

Security of supply 159

The procedure demands a mixture of restraints on consumption, withdrawal from stocks ('drawdown'), and reallocations. The following hypothetical example will serve to illustrate the functioning of the scheme. Let us assume that the Community is subjected to a selective embargo by a group of exporters in the next decade: to reach the critical threshold, exporters cut back by 20% of the Community's energy consumption of, say $30 \times 10^6$ b/d, creating a shortfall of $6 \times 10^6$ b/d. Let us assume that France takes part in the sharing, and further that total IEA oil consumption will be $50 \times 10^6$ b/d, of which 60% is imported, and that European Community oil consumption would be $16 \times 10^6$ b/d, with production running at $2.4 \times 10^6$ b/d. The various steps of the scheme can be described as in *Table 7.4*.

Table 7.4[a]

|  |  | European Community ($10^6$ b/d) | Rest IEA ($10^6$ b/d) | Total IEA ($10^6$ b/d) |
|---|---|---|---|---|
| (1) | *Initial situation:* | | | |
|  | Demand for oil | 16 | 34 | 50 |
|  | Imports | 13.6 | 16.4 | 30 |
|  | Production | 2.4 | 17.6 | 20 |
| (2) | *Emergency reserves:* (members have to maintain 90 days' import equivalent in stocks). | | | |
|  |  | 1.224 | 1.476 | 2.700 |
| (3) | *Actual supplies:* | | | |
|  | Availability after embargo | 10 | 34 | 44 |
| (4) | *Demand restraint:* (10% of consumption, since shortfall exceeds 12% of consumption of group). | | | |
|  | Consumption of oil after restraint | 14.40 | 30.60 | 45 |
| (5) | *Oil allocation:* | | | |

(a) Emergency obligation to draw down stocks (calculated as the product of the country's share of the group's total reserves, and the group shortfall, i.e. the difference between permissible consumption and available supplies):

European Community    $\dfrac{1 \times 45.3}{100}$    $= 0.453 \times 10^6$ b/d

Rest IEA    $\dfrac{1 \times 54.7}{100}$    $= 0.547 \times 10^6$ b/d

(b) Supply rights (difference between permissible consumption and emergency drawdown obligation).

For the European Community: $14.4 - 0.453 = 13.947 \times 10^6$ b/d
For the rest IEA: $\qquad\qquad\qquad\qquad\qquad 30.053 \times 10^6$ b/d

(c) Allocations (difference between supply rights and actual supplies).

For the European Community: $13.947 - 10 = 3.947 \times 10^6$ b/d
For the rest IEA: $30.053 - 34 = -3.947 \times 10^6$ b/d

[a] The table is based on the explanation in the OECD *Observer*, Dec. 1975. A mathematical expression will be found in *Table 7.5*

This would mean that the European Community would receive about $4 \times 10^6$ b/d from other IEA members. Note that stocks in this example would last for over seven years. In practice, the duration of stocks would largely depend on their character — commercial stocks could hardly be depleted very much, since this would probably affect the smooth functioning of the industry.

The selective embargo of the exporters against Europe would thus become a general confrontation between the producers and the consumers organized in the IEA, much like the supply crisis 1973/4, when an informal emergency allocation scheme was operated by the oil companies. This would mean that cutbacks would have to be much more severe to reach the critical thresholds. Based on actual data for 1975, E.N. Krapels has calculated the impact of a $9 \times 10^6$ b/d shortfall on the IEA members[6]. This constitutes about one-third of total OPEC exports, i.e. a very serious reduction. The result is reproduced in Table 7.5. It shows clearly that the effect of such a cutback would be severe: oil supply rights fall to between 63 and 80% of normal demand levels — which indicates that countries with a large production are somewhat favoured: they have to draw less heavily on stocks. The 10% shortfall in oil consumption has, of course, a different impact on different countries, depending on their particular 'mix' of energy production and consumption. The burden on stocks would be much heavier in this example than in the previous one: apart from the Netherlands (a special case because of its large re-exports), where the figure is particularly low, stocks would last for between 200 and 600 days, according to the country. This indicates that proper emergency stocks are important in establishing the credibility of the IEA scheme.

Such a level of supply cutbacks ($9 \times 10^6$ b/d) would pose serious problems for the suppliers, too. It would correspond to a 50% reduction of Arab exports — a level where there is at least a serious risk of revenues losses. To put this figure in perspective, one should remember that in 1973 Arab cutbacks reached a maximum level of 25% of pre-crisis levels. Nevertheless, such a cutback of 50% would be possible, depending on the situation. But it would be an irrational policy of self-inflicted damage: the uncertainties and risks of that kind of sustained economic pressure on the world economy would be very great.

Overall, then, these examples demonstrate that the vulnerability of all consumers is in principle significantly reduced by the IEA Emergency Scheme. This is not only a question of quantities — it is also related to the impact of a general confrontation between producers and consumers. As the damage will spread through the whole system, the weakest countries, rather than specific targets of the embargo, will suffer most. Weakness in this connection has two connotations: first a domestic structure of energy consumption which relies heavily on oil and cannot absorb shortfalls easily (as in Japan, where oil accounts for the overwhelming share of total energy consumption, and is used highly efficiently, leaving little room to absorb shortfalls); and second the level of development. Historically, early periods of industrialization are accompanied by a more rapid

| Member | Con-sumption ($10^3$ b/d) | Production ($10^3$ b/d) | Imports ($10^3$ b/d) | Imports as % of group total | Supply right ($10^3$ b/d) | Required stock drawdown ($10^3$ b/d) | Supply right as % of con-sumption | Draw down as % of con-sump-tion | Average stock level ($10^3$ b) | Days' imports in stock | Draw-down main-(days)[b] |
|---|---|---|---|---|---|---|---|---|---|---|---|
| Austria | 215 | 40 | 175 | 0.8 | 147 | 46 | 68 | 21 | 8 729 | 50 | 189 |
| Belgium/Luxembourg | 545 | — | 545 | 2.6 | 343 | 148 | 63 | 27 | 39 907 | 73 | 270 |
| Canada[c] | 1 735 | 1 735 | — | 4.0 | 1 333 | 228 | 76 | 14 | 135 133 | 159 | 592 |
| Denmark | 310 | — | 310 | 1.5 | 193 | 86 | 63 | 27 | 35 831 | 115 | 416 |
| Ireland | 105 | — | 105 | 0.5 | 66 | 29 | 63 | 27 | 7 494 | 71 | 258 |
| Italy | 1 925 | 20 | 1 905 | 9.2 | 1 209 | 524 | 63 | 27 | 162 263 | 84 | 309 |
| Japan | 4 905 | 10 | 4 895 | 23.6 | 3 070 | 1 345 | 63 | 27 | 324 000 | 66 | 240 |
| Netherlands | 710 | — | 710 | 3.4 | 445 | 194 | 63 | 27 | 18 729 | 26 | 96 |
| Spain | 880 | — | 880 | 4.2 | 553 | 239 | 63 | 27 | 50 584 | 57 | 211 |
| Sweden | 530 | — | 530 | 2.6 | 329 | 148 | 63 | 27 | 39 362 | 74 | 265 |
| Switzerland | 260 | — | 260 | 1.3 | 160 | 74 | 63 | 27 | 28 509 | 110 | 385 |
| Turkey | 275 | 60 | 215 | 1.0 | 190 | 57 | 69 | 21 | n.a. | n.a. | n.a. |
| Britain | 1 875 | 25 | 1 850 | 8.9 | 1 180 | 507 | 63 | 27 | 149 462 | 80 | 295 |
| USA[c] | 15 845 | 9 995 | 5 850 | 28.2 | 12 654 | 1 607 | 80 | 10 | 922 366 | 158 | 574 |
| W. Germany | 2 665 | 115 | 2 550 | 12.3 | 1 698 | 701 | 64 | 26 | 168 990 | 62 | 241 |

[a] The IEP allocation formula can be expressed as follows:

$$\text{Supply right} = cC_x - P_x - \frac{I_x}{\Sigma I}(c\Sigma C - \Sigma P - \Sigma I_{r}),$$

where $c$ is the level of oil consumption allowable under the IEP agreement (the value will be 0.93 or 0.9, depending upon the severity of the oil supply reduction); $C$ is consumption of oil; $x$ refers to an individual member's consumption, production or imports; $\Sigma C$, $\Sigma P$ and $\Sigma I$ refer to the total consumption production and imports of the IEA as a group; $P$ is production of oil; $I$ is imports of oil; and $\Sigma I_r$ is imports remaining to the group during the crisis. Required drawdown is established by the expression $(I_x/\Sigma I)(c\Sigma C - \Sigma P - \Sigma I_r)$.

The hypothetical import reduction is $9 \times 10^6$ b/d. The values of the $\Sigma$ variables are:

$\Sigma I = 21 \times 10^6$ b/d  $\quad \Sigma P = 12 \times 10^6$ b/d
$\Sigma I_r = 12 \times 10^6$ b/d  $\quad \Sigma C = 33 \times 10^6$ b/d

[b] This calculation is based on the assumption that all the stocks could be used in an emergency. This is not in fact the case, as the IEP agreement itself implies. Article 15 stipulates that when 50 per cent of the emergency reserve is exhausted, the Governing Board must meet to consider whether additional actions are required, including further decreases in permissible consumption.

[c] Canada and the United States are special cases: their oil markets are 'incompletely integrated', in the parlance of the IEP agreement (i.e., oil is produced in a distinct region of the country). Article 17 of the agreement stipulates that the allocation programme can be activated if a major region in such a market suffers a loss of oil imports exceeding 7 per cent of consumption in the previous four quarters.

Source of consumption and production data: BP Statistical Review of the World Oil Industry, 1975.

Source: E.N. Krapels, Oil and Security, Problems and Prospects of Importing Countries, IISS, London, 1977 (IISS Adelphi Paper, No. 136)

growth in the consumption of energy than in total production[7]. Thus, developing countries would suffer most from any interference with oil supplies, as indeed the aftermath of the 1973/4 supply crisis demonstrated.

The countries best protected will be IEA members, because the oil companies which manage the allocation system will probably give them priority, both as important customers and from fear of internal political repercussions, although the scheme explicitly states that this is to be avoided. However, a crisis could stretch the capacity of the international distribution system to its limits and one could reasonably expect that non-members of the IEA, in particular the LDCs, would not be so well protected by the general system of rationing.

Thus, the leverage of the oil producers is reduced and the risks of unintended repercussions in friendly countries, such as the LDCs, are increased. Furthermore, economic consequences of a shortfall for all major industrialized countries would be much more serious and the impact much more direct than in the case of a selective embargo, where the recession in one area would take time to work its way through the system. These considerations point to one conclusion: the risks of severe pressure through oil cutbacks rise to very high levels, while limited cutbacks lose credibility as a political threat. All-out confrontation still cannot be excluded, but appears to be a last resort, likely to cause chaos for consumers and producers alike.

Can the Emergency Allocation Programme therefore be considered a credible deterrent? There could be three reasons for failure: a lack of political will to activate the scheme, an inability to operate it once it has been activated, and political disagreement during its application.

Let us look first at the decision-making process which activates the Emergency Allocation Programme. The secretariat of the IEA monitors supply positions and if it finds either a reduction of oil supplies by 7% or more of the total oil consumption of the IEA member states, or a reduction of 12% for the whole group, or a reduction for one or more IEA members by at least 7%, it immediately reports to the Management Committee and the Governing Board. The Management Committee has to meet within 48 hours of notification and report to the Governing Board within the next 48 hours. If the Governing Board decides to activate the Emergency Allocation Programme, it has to be implemented within 15 days by all members countries[8]. This gives a perfectly adequate reaction time, since supply interferences need about the same time to work their way through the system to the consumer. However, there is, strictly speaking, no automatic reaction. Since political decisions are part of the process, it is possible that there will be a lack of political will to activate the scheme.

The voting procedure in the Governing Board is based on a complicated system of weighted votes (see *Table 7.6*). The special majority, necessary to prevent activation of the scheme, required 60% of total combined voting weights and 36 general voting weights in the case of a shortfall of over 7% or 12% respectively for the whole group, or 42 general voting weights in the case of one or more members being affected by a shortfall of 7% or more. If more than one condition to activate

Table 7.6

|  | General voting weights | Oil consumption voting weights | Combined voting weights |
|---|---|---|---|
| Austria | 3 | 1 | 4 |
| Belgium | 3 | 2 | 5 |
| Canada | 3 | 5 | 8 |
| Denmark | 3 | 1 | 4 |
| Germany | 3 | 8 | 11 |
| Ireland | 3 | 0 | 3 |
| Italy | 3 | 6 | 9 |
| Japan | 3 | 15 | 18 |
| Luxembourg | 3 | 0 | 3 |
| Netherlands | 3 | 2 | 5 |
| Spain | 3 | 2 | 5 |
| Sweden | 3 | 2 | 5 |
| Switzerland | 3 | 1 | 4 |
| Turkey | 3 | 1 | 4 |
| UK | 3 | 6 | 9 |
| United States | 3 | 48 | 51 |
|  | 48 | 100 | 148 |

the scheme is met, then the vote has to be taken for each condition separately and, if applicable, for each concerned country separately[9]. In effect, this means that nothing short of three-quarters of IEA members voting together could prevent the activation of the scheme if the shortfall exceeds 7% of the group's oil consumption, and 14 countries have to co-operate to prevent it if only some countries are affected. Moreover, in the former case, the requirement of 60% of the total combined votes gives the major consumers an effective power of veto over any decision to block the activation of the scheme: if the four countries voting in effect for activation are the USA, Japan, Germany, and the UK or Italy, the scheme has to be activated in spite of the fact that 12 countries vote against it. If the Community were to be subjected to a selective embargo, then the scheme would have to be implemented; on the other hand, if the USA was the target it is theoretically conceivable, though highly unlikely, that the application of the scheme could be prevented.

While the political nature of the procedures somewhat reduces the deterrent effect of the scheme, this is small enough to make the programme virtually automatic. Discarding this possibility, one must turn to the second conceivable cause of failure — the operation of the programme. Some of the problems which can be discussed under this heading relate to the structure of the scheme itself. As noted, it calls for drawing on stocks, which in the case of a prolonged crisis would run out. This might take a long time — in the second scenario given above, for example, the stock drawdown obligation (i.e. the volume of oil countries will have to provide from their emergency stocks under the agreement) would in effect result in a depletion after 540 days. Nevertheless a time limit exists and the IEA programme at a certain point of stock depletion allows for the Governing Board to take subsidiary measures, such as a reduction of consumption. Both the IEA and the European Community now have stockpile regulations which are compulsory for the member states of the two organizations — in the case of the IEA, the present level of stockpiles required is 70 days of total oil *imports* (a target to be raised to 90 days in 1980); the Communities expects its member states to have strategic reserves equivalent to 90 days of total oil *consumption*. However, management of stocks differs widely between the member countries, and actual levels of stockpile availability are in any case below theoretical levels, since a substantial quantity of stocks is required to guarantee the smooth functioning of the oil distribution system. Thus, actual protection from stockpiles is less than it first appears, and differences in vulnerability between European countries are substantial, depending on different definitions of stocks, different stockpiling arrangements, different trading positions, etc[10]. Nevertheless, present levels of strategic reserves appear sufficient to deal with the type of politically motivated crisis which appears plausible. (Stockpile levels in some countries are given in *Table 7.7*).

**Table 7.7** *Average Stocks in Several Industrialized Countries (millions of barrels)*

|  | 1973 | 1974 | 1975 | 1976 |
|---|---|---|---|---|
| (A) Average stocks of crude oil and principal finished products in selected countries[a] | | | | |
| USA | 661.0 | 710.0 | 761.0 | 764.0 |
| Germany | 128.8 | 146.3 | 153.6 | 164.4 |
| France | n.a. | 217.3 | 222.3 | 197.3 |
| Italy | 126.4 | 149.4 | 136.8 | 122.1 |
| Netherlands | 58.9 | 73.0 | 72.4 | 66.3 |
| Japan | 203.5 | 254.0 | 266.5 | 272.4 |
| (B) Average stocks in days of previous year's domestic sales of principal finished products | | | | |
| USA | 56 | 56 | 63 | 64 |
| Germany | 59 | 63 | 79 | 80 |
| France | n.a. | 125 | 138 | 113 |
| Italy | 94 | 105 | 93 | 85 |
| Netherlands | 176 | 228 | 279 | 258 |
| Japan | 67 | 75 | 82 | 90 |
| (C) Average stocks in days of previous year's net oil imports[b] | | | | |
| USA | 154 | 123 | 136 | 137 |
| Germany | 50 | 52 | 60 | 68 |
| France | n.a. | 88 | 92 | 99 |
| Italy | 71 | 77 | 69 | 71 |
| Netherlands | 108 | 129 | 138 | 159 |
| Japan | 49 | 52 | 55 | 61 |

a Average month-end stock levels for all countries except Italy, for which only end-of-quarter data are available after 1973
b Net imports, that is, total imports less exports (including international marine bunkers)
Source: E.N. Krapels, *Emergency Oil Reserve Programs in Selected Importing Nations*, Royal Institute of International Affairs, London 1977 (mimeographed paper)

A second problem in the operation of the IEA programme is the obligation to constrain oil consumption. The impact of such a constraint would be quite different, since the position of oil in the total energy balance of the IEA member states varies. Countries such as Italy or Japan, where oil imports supply 70% or more of total energy demand, will find a reduction of oil consumption much more difficult to deal with than other countries with a more diversified energy demand pattern. The impact of the level of oil demand constraint prescribed by the IEA could pose serious risks for the most vulnerable countries; however, this could be eased by additional petroleum stocks beyond the levels committed to the IEA scheme. Such additional reserves could be utilized in the case of a crisis to reduce the burden of adjustment. However, only Japan now seems to consider seriously such additional stocks beyond the 90-day target set for 1980. In the case of a prolonged crisis, this problem of quite different damage from similar shortfalls could arise and create political tensions within the IEA.

Finally, there is the task of actually managing the distribution of oil under crisis conditions among members of the IEA. As indicated above, the 'Agreement on an International Energy Programme' provides for an international advisory body of oil companies to implement the scheme. The oil companies have operated emergency supply allocation schemes in the past: in 1956 and 1967 the Middle East Emergency Committee, set up by 15 companies, dealt effectively with the supply crisis caused by the closure of the Suez Canal[11]. In 1973, the companies again managed to share out available supplies fairly effectively by reallocating and re-routing tankers, without actually breaking the conditions imposed by the producers[12]. This indicates that the oil companies are still in a position to deal with the logistics involved in the implementation of the emergency scheme. Two new factors, however, have become relevant.

The first concerns the spare capacity in the world tanker fleet. The causes of this reach back to the oil boom in the early 1970s, when worldwide consumption was growing 7% annually, and very large crude carriers (VLCCs — tankers of more than 200 000 dead weight tonnes (dwt) were in great demand. Coincidentally, an upsurge in banking liquidity injected speculative finance into tanker building and this led to new orders which would have far exceeded demand even without the 1973/4 price increases[13]. Net additions to the world tanker fleet in 1974 came to 212 ships of a total of $42.6 \times 10^6$ dwt, or 16.8% of the world total, and in 1975 a further net addition of $29.6 \times 10^6$ dwt (or another 11.8% on the 1974 total) produced a growth rate which was historically needed only at times of serious upheaval, such as the Suez Canal closure[14]. In a world of fourfold oil price increases, spare capacity was further aggravated by the re-opening of the Suez Canal which allowed tankers with a capacity of 50 000 dwt to pass with full cargo. This meant that the time needed for a return voyage was cut significantly for a substantial share of the oil delivered from the Persian/Arab Gulf to Western Europe and North America. A 45 000 dwt tanker en route from Mina al-Amadi to Rotterdam can gain 26.3 days (at 15 knots) per round voyage and average 8.65 voyages per year, instead of 5.24 by other routes[15].

Objectives for a European energy policy

The consequences of this combination of factors were dramatic. At the end of 1977, 219 tankers with a total capacity of 30.8 × 10⁶ dwt were laid up, among them 51 VLCCs[16]. *Table 7.8* summarizes the development of this surplus capacity. In addition, a reduction of average cruising speed provided a substantial hidden surplus of about 62 × 10⁶ dwt[Note 17]. Spot charter rates in 1975 plummeted to levels around 17.5 in the World Scale (the yardstick used to express the level of tanker charter rates) which meant that tanker operators

**Table 7.8** *Developments in the World Tanker Market*

|  | 1972 | 1973 | 1974 | 1975 | 1976 | 1977 |
|---|---|---|---|---|---|---|
| **I The oil fleet (10⁶ dwt)** | | | | | | |
| (a) Operational tanker fleet, including combined carriers, year end | | 238.4 | 265.7 | 292.7 | 329.6 | 377.5 |
| (b) Inactive tankers, year end | | 1.4 | 5.3 | 43.4 | 37.0 | 32.1 |
| (c) Net additions to world tanker fleet | | 25.9 | 42.6 | 35.5 | 31.3 | 11.7 |
| **II Inactive tankers (10⁶ dwt) (number)** | | | | | | |
| June | | | | | | |
| Number | 176 | 51 | 52 | 436 | 490 | 340 |
| 10⁶ dwt | 6.1 | 1.68 | 2.07 | 32.98 | 44.22 | 35.25 |
| November | | | | | | |
| Number | 101 | 54 | 97 | 482 | 389 | 338 |
| 10⁶ dwt | 2.67 | 1.69 | 3.64 | 37.9 | 32.70 | 36.74 |
| **III Charter rates, single voyage dirty (based on Worldscale)** | | | | | | |
| January | | | 176 | 78 | 81 | 88 |
| February | | | 176 | 71 | 83.5 | 99 |
| March | | | 171 | 65 | 77 | 83 |
| April | | | 176 | 66 | 77 | 79.5 |
| May | | | 151 | 69 | 84 | 77.5 |
| June | | | 155 | 73 | 87 | 75 |
| July | | | 119 | 84 | 84 | 68 |
| August | | | 110 | 77 | 84 | 59 |
| September | | 357 | 118 | 80 | 83 | 60 |
| October | | 400 | 129 | 79 | 86.5 | 63.5 |
| November | | 247 | 114 | 74 | 91 | 64 |
| December | | 172 | 109 | 83 | 107.5 | 78 |

Sources: H.P. Drewry, *Shipping Statistics and Economics*, various issues; *Petroleum Economist*

made heavy losses — total costs including recovery of capital outlays for a new tanker need a level of World Scale 75 to break even[18]. The extent to which the market has been destroyed is demonstrated by the fact that a Japanese VLCC was reportedly chartered for six months on a negative freight rate of 21 cents per ton a month[19]. The dangers of bankruptcy among builders have led to the establishment of the International Maritime Industry Forum to decide measures to deal with the tanker surplus. In addition to slow-steaming, already widely practised, their suggestions included[20]:

(1) Cancellations of new orders: this would be the most effective measure and eventually lead to some relaxation of the surplus as old tankers were not replaced. Given the fact that a further $51 \times 10^6$ dwt were scheduled for delivery in 1976 and 1977, and $20 \times 10^6$ dwt for 1978 or later, the urgency of cancellations became obvious. Equally obvious, however, was the impact cancellations had on the shipbuilding industry.
(2) Conversion of up to 20% of tanker capacity to ballast-water use: this would be an effective measure, but one which increases costs for the owner without any hope of recovering those costs.
(3) A programme of accelerated scrapping: this is already increasing, with newer and larger ships being scrapped[21]. The disadvantage is that it is often the older tankers which make the biggest profits should demand increase, since capital costs are low.
(4) Using VLCCs as floating storage tanks: this would not provide any reduction in insurance costs, unlike laying up. Main mooring spaces are also generally far away from consuming areas.

Five years after the crisis erupted, most of these ideas had been implemented. Yet, the problem persisted: the world tanker market was still characterized by very substantial overcapacity. Tanker owners received substantial aid from governments and tried to organize a cartel to reconstitute a minimum level of profitability through limitation of competition on charter tariffs[22]. Estimates about the future tanker requirements, and thus about market prospects, are still hardly encouraging: shipping consultants predicted that even with severe measures to reduce the surplus, and with widespread slow-steaming, a substantial surplus of VLCCs and ULCCs (ultra-large crude carriers) would continue into the early 1980s[23]. The conclusions to be drawn are the following:

(1) A substantial surplus of tanker capacity will continue to exist.
(2) The international tanker business will become economically unattractive. This could result in increasing concentration in the industry, since second-hand tankers are readily and cheaply available. As the oil companies decrease their investment in tanker fleets, the concentration would presumably be in the oil-producer countries and/or with consumer governments.

This brings us to the second new aspect to be considered – the appearance of the oil producers as a force in the international tanker market. In early 1978, OPEC countries owned only a small share of the world tanker fleet (3.3%), with Arab producers together accounting for about $9 \times 10^6$ dwt, and Iran for about $1 \times 10^6$ dwt[Note 24]. Almost all producers, however, have placed large orders with ship-yards and during the coming years Arab and Iranian tanker fleets will grow substantially. Present size and orders for the OPEC tanker fleets are summarized in *Table 7.7*. Beyond present orders, most producers have ambitious plans to build up their tanker capacity. Iran at one point expressed the desire to build a fleet of $7 \times 10^6$ dwt by 1978[25]; Saudi Arabia wanted to reserve about 50% of her total oil exports for her own fleet by 1980[26], and a senior executive of the Kuwait Oil Tanker Company has been quoted as saying that the Arabs should aim to export 60% of their oil production in 1980 in their own tankers[27].

Some of these objectives were unrealistic, but a significant build-up of Arab and Iranian tanker fleets can be expected. Given the distressed state of the tanker market, there exist ample opportunities to buy almost new or second-hand tankers at very favourable prices[28]. In 1975 for example, the Saudis bought the 102 504 dwt *Berge Sigval* for $\$6.3 \times 10^6$, the Kuwait Oil Tanker Company purchased a Japanese VLCC, built in 1973, for $\$18 \times 10^6$, and Algeria acquired a new tanker on completion from its previous owner at a 40% discount[29]. Control over tankers by acquiring long- or short-term charter contracts is also a possibility, and indeed several producers have done this. The Kuwait National Petroleum Company has at present 18 carriers under long-term charter and intends, eventually, to buy its fleet as the contracts run out[30]. One of several joint ventures between Saudi Arabian interests and foreign companies (SARAMCO, with 30% shares of Mobil and 15% of Fairfield Maxwell) is also known to have chartered two VLCCs[31].

The quoted study of H.P. Drewry, the shipping consultants, concludes that present and forecast OPEC fleets could by 1980 transport a maximum of 5% of OPEC crude oil exports, assuming full employment of their fleet. To build a fleet in line with their ambitious targets, OPEC would need a total fleet of $128 \times 10^6$ dwt for the transport of 40% of its crude, or $160 \times 10^6$ dwt to reach the 50% target. In the first case, OPEC would own 33% of the world tanker fleet, in the second about 43%. The cost of acquiring such a fleet would be between $13.1 (40% of crude on own tankers) and $17.1 \times 10^9$ (50% of crude on own tankers), assuming a mixture of new purchases (from the original contractor), second-hand purchases, and new tanker orders[32]. Such an investment seems entirely feasible but not very likely in the near future: some OPEC countries have already experienced the impact of the gloomy world tanker market — they, like other operators, have operated at losses. In the longer run, however, such a downstream diversification in the tanker business at the scale indicated seems entirely likely.

How would the oil producers having a substantial share in the world tanker fleet affect the viability of the IEA Emergency Allocation Programme? Could the combination of control over a large part of world production plus a share of the world tanker fleet destroy the flexibility of the international distribution system? To attempt an answer we need to look first at the size of the producer tanker fleet on the assumption of a 50% share of total exports in the case of the Persian/Arab Gulf and North African exporters. (Nigeria is unlikely to invest heavily in a tanker fleet until 1985.) A rough calculation based on the scenario given in *Table 7.2* indicates that OPEC would need a tanker fleet of 2704 (scenario I) or 1539 (scenario II) T-2 equivalents*, or $43.3 \times 10^6$ and $24.6 \times 10^6$ and $24.6 \times 10^6$ dwt respectively, to service 50% of the Community's imports. For the Arab tanker fleet to service 50% of the Community's imports would require $35.5 \times 10^6$ dwt or $20.1 \times 10^6$ dwt for scenarios I and II respectively, or between 2218 and 1256 T-2 equivalents**. Those figures are far higher than the present share of the oil producers (under 5%) and they would still cover only about one-third of their total exports. But assuming that the oil producers actually acquire a 50% share in the tanker fleets servicing their exports, would this mean that the flexibility of the distribution system could be destroyed?

The additional risk (or leverage, depending on one's viewpoint) to be considered in connection with the world tanker market is the *selective embargo* — the precise targetting of the oil weapon against one consumer. Were this to become possible, it would constitute a qualitative change in oil power. LDC oil importers might find themselves in this situation. They often receive all their imports from one or two suppliers, and they are inclined to conclude direct supply arrangements with the national oil company of the producer government, possibly at favourable conditions. They might, therefore, be the first to accept the principle of deliveries on producer tankers. In contrast, even a 50% share of OPEC in the world tanker fleet might not suffice to invalidate the IEA scheme. If a *selective embargo* is directed against one small or medium-sized consumer country and all tankers of the producer country (and possibly all other tankers delivering oil from the producer group to the target country) are withdrawn, then neither the shortfall in supplies nor the necessary logistic rearrangements would create serious problems, provided substantial surplus tanker capacity remains available. The problem becomes more complex if a selective embargo is decreed against one of the major oil consumers. A simplified model, which bears some resemblance to reality, allows several conclusions to be drawn.

Assume an IEA with only three major members: the Community, Japan, and the United States. Each region would, according to this model, import about $9 \times 10^6$ b/d in, say, 1985. Now, one of the three is made the target of a selective embargo, including an equivalent cutback in production of $9 \times 10^6$ b/d (this would be about 25–30% of OPEC exports) and there are strictly enforced orders to the producers' fleet to withhold supplies. The same orders are also given to all tankers normally supplying the target with OPEC oil, but not directly controlled by it. Finally, all re-routing of supplies to the embargoed area is effectively

---

* A T-2 is a tanker with 16 000 dwt, 330 days per year in operation, 18.2% port time, 92.4% effective deadweight (100% would be equivalent to full utilization of the capacity of 16 000 t for 50% of time at sea), and a speed of 15 knots = 360 nautical miles/d$^{32}$.

**Calculated on the basis of T-2 equivalents, with details as given above, and assuming a distance of 17 853 nautical miles for the return voyage — Persian/Arab Gulf, Cape of Good Hope/Western Europe/Suez Canal/Gulf. This distance actually refers to the type of voyage from Mina al-Amadi to Rotterdam and back. $100 \times 10^6$ t/a are assumed to come from North Africa, and $70 \times 10^6$ t via the Suez Canal from the Persian/Arab Gulf. Tanker requirements for these routes are calculated using the tanker factors given in H. Lubell, *Middle East Oil Crisis and Western Europe's Energy Supplies*, Johns Hopkins University Press, Baltimore 1963, p. 58.

blocked — in other words, all other exports from OPEC are handled as before both by OPEC and non-OPEC tankers; no changes in routes occur.

This model can be quantified as in *Table 7.9*.

Table 7.9

|  | USA ($10^6$ b/d) | European Community ($10^6$ b/d) | Japan ($10^6$ b/d) |
|---|---|---|---|
| Oil consumption | 22 | 12 | 9 |
| Oil imports | 9 | 9 | 9 |
| Oil production | 13 | 3 | — |
| Share of emergency stocks | (33%) | (33%) | (33%) |
| Permissible consumption | 19.8 | 10.8 | 8.1 |
| Shortfall |  |  |  |
| Case A | 9 | — | — |
| Case B | — | 9 | — |
| Case C | — | — | 9 |
| Emergency drawdown obligation | 1.5 | 1.5 | 1.5 |
| Supply rights | 18.3 | 9.3 | 6.6 |
| Reallocations$^a$ |  |  |  |
| Case A | +5.25 | −2.75 | −2.45 |
| Case B | −3.75 | +6.25 | −2.45 |
| Case C | −3.75 | −2.75 | +6.55 |

*a* Figures do not add up because of rounding

To ship 3.75 × $10^6$ b/d from the US East Coast to Europe, some 1705 T-2 equivalents, or about 27 × $10^6$ dwt tanker capacity, would be needed*. No precise data for the Japan—Europe route are available, but assuming a tanker factor of 1000 b/d transport capacity per T-2 on this route, a further 2450 T-2 equivalents, or 39.2 × $10^6$ dwt, would be required to ship 2.45 × $10^6$ b/d from Japan to Europe — a total of 66.2 × $10^6$ dwt. Against this, we have to discount the tonnage set free by the total embargo, the equivalent needed to ship 50% of pre-embargo exports to the target area. If the target was the United States (Case A), then tanker capacity released would be 2432 + 389 + 167 T-2 equivalents, or a total of 39.5 × $10^6$ dwt, assuming that 3.3 × $10^6$ b/d of the total was normally shipped via the Cape under cargo, and through the Suez Canal on the way back**, a further 0.7 × $10^6$ b/d via the Suez Canal in both directions, and 0.5 × $10^6$ b/d from North Africa. In an embargo on Europe (Case B), tanker capacity set free would be (on the same assumptions) 2988 T-2 equivalents or 47.8 × $10^6$ dwt. For Case C (an embargo on Japan) 2165 T-2 equivalents, or 34.6 × $10^6$ dwt would be set free***. Calculating for Cases B and C (data for the voyage from Japan to the USA are not available, but are assumed to amount

* Based on a tanker factor of 2600 b/d per T-2 (H. Lubell, *Middle East Oil Crisis and Western Europe's Energy Supplies*, Johns Hopkins University Press, Baltimore 1963, p. 58).
** The distance for one round trip is 20 723 nautical miles (Mina al-Amadi to Baltimore); a T-2 equivalent could, therefore, make 4.7 trips per year, or transport 67 840 t/a.
*** Distance for round trip 13 500 nautical miles, 7.2 trips per year[33].

Security of supply    171

Table 7.10

|  | Tankers required ($10^6$ dwt) | Tankers released ($10^6$ dwt) |
|---|---|---|
| Case A | 39.6 | 47.8 |
| Case B | 66.2 | 39.5 |
| Case C | 74.0 | 34.6 |

to a tanker factor of 2000 b/d per T-2 equivalent) the results would be as in *Table 7.10*.

If the USA were subject to an embargo, the number of tankers released would, in our model, actually be higher than capacity required, while in the other cases additional capacity of 26.7 and 39.4 $\times 10^6$ dwt would be needed. On the basis of present forecasts, sufficient surplus capacity should be available to meet those requirements, though in an embargo against Japan, the spare capacity required is substantial — an indication that the international distribution system could fail in such a contingency.

Apart from the fact that these calculations are somewhat simplified and arbitrary, it has to be stressed that the assumptions made postulate the worst conceivable crisis: joint action by all OPEC countries with no re-routing except for those tankers freed from servicing the embargoed areas. The only crisis more serious than a selective embargo would be one of total confrontation between all producers and all industrialized consumers. In this case, the capacity of the international distribution system would be most unlikely to cope, assuming the producers control a 50% share of the tanker fleet carrying their exports.

Another weakness of the IEA is the possibility of divisions within the organization, under the pressure of prolonged confrontation. The Emergency Allocation Scheme, concentrating on oil only, implies that a shortfall in oil consumption will have a different impact in terms of total energy consumption on different importing countries. This fact could create serious political problems in a long crisis, for tensions might develop within the IEA which could destroy consumer solidarity and the emergency allocation provisions. The alternative approach, tempting for countries hit by the emergency scheme but not themselves a target of the embargo, would be bilateral arrangements with the oil exporters. This reveals another weakness of the Emergency Allocation Scheme. While it provides additional security for all members in the case of an embargo, it also multiplies the risk of confrontation with the oil exporters. The most likely target of exporter dissatisfaction and political pressure is the USA, and if their Middle East policy led to confrontation in the international oil market Europe and Japan would be drawn in[34]. Against this, it could be argued that (a) Europe and Japan are likely to find themselves in such a position in any case, since it is unlikely that any future embargo would succeed in working selectively (as the shortfall can be spread throughout the system) and (b) the oil exporters might decide (as they did in 1973/4) to direct some of their pressure against Europe and Japan simply as a means of exerting additional, indirect, leverage on the USA.

The greatest risk to the IEA emergency allocation system may well be prolonged confrontation[35]. The viability of the scheme would then be tested for its political rather than its technical and managerial ingredients. Frictions could arise in such circumstances and, as in 1973/4, attempts might be made to trade off economic security against military guarantees. In a crisis, the link between different political issues could pose very serious problems. However, these considerations apply to the stage beyond the initial confrontation. Initially, the Emergency Allocation Scheme could provide a useful way to handle embargoes and cutbacks and thus constitute an important deterrent against confrontation in the oil market. This 'economic NATO', then, does not constitute a political answer to energy dependence, but it does attend to some of its immediate security implications. As in the case of East—West military security, a mechanism of defence and deterrence is necessary, but not sufficient: a political structure is needed to manage a relationship with such a vast potential for mutual destruction. However, unlike the East—West conflict, the relation between oil producers and consumers seems amenable to compromise, and hence co-operation should be (as it has been) more important than the ideological differences of the opposed social systems of East and West.

CHAPTER EIGHT

# Long-term availability of supply

The discussion in the previous chapter gave reasons for some moderate optimism: the danger of serious shortages as a consequence of supply interruptions seems limited and manageable. The risk of all-out confrontation between producers and consumers is small, and adequate arrangements exist to allow political crisis management to deal with politically motivated supply interruptions. Security of supply, however, raises another aspect of international energy: the balance between import demand and export availability in the longer term.

To analyse this problem, some perspective is needed on the likely future evolution of supply and demand on the global level. *Table 8.1* presents some

**Table 8.1** *Possible Evolution of World Oil Supply—Demand Balance* ($10^6$ b/d)

|  | 1974 | 1985 | 2000 |
|---|---|---|---|
| Import demand |  |  |  |
| North America | 5.7[a] | 12.1 | 10.4–15.8 |
| Western Europe | 14.2 | 13.3 | 12.5–16.5 |
| Japan | 5.2 | 6.9 | 7.9–15.2 |
| Non-OECD, non-socialist, other | 4.8 | 3.0 | 9.0–11.2 |
| Subtotal, excluding bunkers | 29.9 | 36.0 | 45.2–53.3 |
| Exports |  |  |  |
| Socialist bloc | 0.9 | –1.0 | – |
| OPEC | 28.9 | 32–34 | 34.5–38.7 |
| Gap/prospective shortage | – | 4–12 | 15.2–20.0 |

a *The figures might not add up because of rounding; estimates for 1985 and 2000 are based on differing scenarios. This means that shortage estimates are not directly derived from the maxima and minima of import demand (2000), or from the government estimates (1985)*
Sources: *1974: World Energy Outlook, OECD, Paris 1977; 1985: Energy Policies and Programmes of IEA Countries, 1977 Review, IEA, Paris 1978; 2000: WAES, Energy: Global Prospects 1985–2000, McGraw-Hill, New York 1977*

recent forecasts, and the figures reveal two different sets of problems with regard to the future availability of supplies. In the longer run, world demand for oil might meet physical constraints — according to present forecasts, production *capacity* in 2000 will be insufficient to meet requirements. With all due caution in interpreting such forecasts, it seems safe to conclude that the phasing-out of hydrocarbons as the dominant source of energy has begun.

A second, and probably more important constraint exists through political and economic choices and decisions taken by the oil exporters. The availability of internationally traded hydrocarbons will depend on the energy policies of the exporters. To some extent the two constraints are interdependent: a failure to find any large new reserves would probably exacerbate fears about 'running out' of petroleum, and thus accentuate conservation policies: on the other hand, the prospects of massive additional supplies of oil entering the international market might dampen enthusiasm for export ceilings. Over the long run, the discomforting conclusion must be that the decline of oil production will tend to be reinforced by political and economic choices, thus exacerbating the adjustment problems for the importers.

But the political factors also act independently of resource positions. It has already been shown that at present serious imbalances exist between exporters and importers. Some exporters today produce at levels which exceed their demand for oil revenues – not to mention the fact that their production levels work as a brake on world price levels, thus keeping down their income per barrel. Additional exports beyond present levels will, therefore, be of very limited use for these exporters, but the incremental barrels per day might be of great significance for the importers. In essence, the political problem with international oil stems from an uneven distribution of benefits from the trade relationship. The gap between supply and demand is today largely bridged by political considerations of Saudi Arabia's present regime.

But the political framework is subject to changes, and given the volatility of the Middle East, and the strains caused by rapid processes of modernization, the present configuration could change very quickly, as indeed it began to do in 1979\*. A long-term supply–demand analysis should not, therefore, be based on reserve and theoretical capacity assessment (this neglects the importance of the political factor) but on the level of production advocated by a strategy to optimize benefits from exports, and on the conceivable directions of exporters' energy policies.

Economic benefits for exporters fall into two categories: the direct benefits from oil and gas industries, and the indirect ones of government income from oil and gas sales. The direct benefits of energy exports seem only tenuously linked to rates of production. Employment in oil and gas production is small and would not vary much with higher or lower output levels. Oil and gas exporters might move into downstream activities – as many of them intend to do in the case of refining (see *Table 8.2*). But this contributes little to European security of supplies, for while refining adds value to oil production, and hence introduces a stabilizing element to the export of oil in the form of products, it reduces rather than increases the stability of crude exports, and also brings producers and consumers into conflict over the issue of refinery siting. This could be avoided if producers acquired refineries in consumer countries, a move which would increase interdependence by creating hostage investments in the

---

\* This will be taken up in the Postscript.

consumer areas. Until now, however, producers have generally been reluctant to pursue such a strategy, although some have invested downstream in refining in consumer countries – Iran in South Africa, India and Greece, and Libya in Yugoslavia, for example[36]. Only Iran, however, has considered taking a share in a major Western refinery and distribution company, the US Ashland Oil Company – an arrangement which never materialized[37].

Similarly, direct producer involvement in *marketing* in consumer countries contributes only marginally to the longer-term stability of supplies. It is in the

**Table 8.2** *OPEC Share in World Refining Capacity, 1940–1980 (primary distillation capacity, $10^6$ t/a)*

| | 1940 | 1950 | 1960 | 1970 | 1977 | Capacity under construction or proposed | Possible capacity early 1980s |
|---|---|---|---|---|---|---|---|
| Western Europe | 21.1 | 42.4 | 196.2 | 722.7 | 1 083.4 | 104.2 | 1 187.6 |
| Africa | – | – | 1.4 | 29.5 | 63.5 | 95.2 | 158.7 |
| Middle East[a] | 19.0 | 47.0 | 75.2 | 142.0 | 186.3 | 119.9 | 306.2 |
| Far East[b] | 13.5 | 13.7 | 67.8 | 255.7 | 523.6 | 76.3 | 599.9 |
| North America[c] | 234.8 | 354.3 | 579.9 | 687.9 | 986.2 | 111.7 | 1 097.9 |
| Caribbean South America | 37.2 | 61.1 | 122.5 | 237.9 | 357.6 | 42.9 | 400.5 |
| Total Free World | 325.6 | 518.5 | 1 043.0 | 2 075.7 | 3 200.6 | 550.2 | 3 750.8 |
| of which: | | | | | | | |
| Algeria | – | – | – | 2.3 | 6.2 | 27.1 | 33.3 |
| Ecuador | 0.1 | 0.3 | 0.4 | 1.8 | 2.2 | – | 2.2 |
| Gabon | – | – | – | 0.8 | 2.0 | – | 2.0 |
| Indonesia | 8.4 | 10.0 | 13.5 | 13.2 | 26.7 | – | 26.7 |
| Iran | 16.4 | 24.8 | 24.4 | 34.6 | 47.0 | 17.3 | 64.3 |
| Iraq | 0.2 | 0.4 | 2.8 | 4.1 | 9.9 | 18.5 | 28.4 |
| Kuwait | – | 1.2 | 10.9 | 24.1 | 35.2 | – | 35.2 |
| Libya | – | – | – | 0.5 | 3.8 | 41.0 | 44.8 |
| Nigeria | – | – | – | 2.0 | 3.0 | 10.0 | 13.0 |
| Qatar | – | – | – | 0.3 | 0.3 | 8.3 | 8.6 |
| Saudi Arabia | – | 6.9 | 9.3 | 22.9 | 31.2 | 44.2 | 75.4 |
| UAE | – | – | – | – | 0.8 | 11.5 | 12.3 |
| Venezuela | 2.7 | 12.5 | 43.7 | 65.3 | 74.1 | – | 74.1 |
| Total OPEC | 27.8 | 56.1 | 105.0 | 171.9 | 242.4 | 177.9 | 420.3 |
| OPEC's % share of free-world capacity | 8.5 | 10.8 | 10.1 | 8.3 | 7.6 | 32.3 | 11.2 |
| OPEC's % share of free-world capacity excluding North America | 30.6 | 34.2 | 22.7 | 12.4 | 10.9 | 40.6 | 15.8 |

a Including Egypt
b Including Australasia
c Including Mexico
Source: 'OPEC Oil Report', *Petroleum Economist*, London, Dec. 1977

interest of producers to establish direct access to markets for their exports, and several producers have done this: Iran has negotiated for a share in ENI's European and African distribution network, and Kuwaiti interests have set up a joint venture with several European distributors in the Netherlands, Italy and France[38]. But there is still a long way to go before even only a substantial part of OPEC oil will be distributed in Europe and elsewhere through its own channels.

A further direct benefit of oil production concerns associated natural gas, which is a by-product of crude oil production. Already in some countries, associated natural gas is a vital energy source, and will become increasingly so in other countries. In Kuwait, natural gas utilization has reached a high level and oil production could not be allowed to fall much below $1.5 \times 10^6$ b/d. In effect, then, the use of natural gas for domestic purposes draws a bottom line for oil production, with obvious implications for exports[39]. But this also works the other way round: if natural gas cannot be used productively for domestic uses or exports, it has to be flared (i.e. burned uselessly). This is a substantial waste of a valuable energy resource, and it has become increasingly clear that producer government policies will aim at eliminating such waste wherever possible. The refusal to accept large-scale flaring of associated gas could therefore become an important limitation on future oil production levels.

On balance, it seems as if direct benefits offer little prospect for supply stability beyond present production levels — in fact the last element could even be seen as a downward, rather than an upward pressure. We have to look to indirect benefits to fill the gap.

Government oil revenues are the product of quantities sold and prevailing price levels. From the consumer's viewpoint one way to stabilize export levels would be to reduce the price of oil. The problem with this policy (which was tried by the US government in 1974/5) is that it presupposes consumer control over oil prices. However, this control does not exist.

From the producers' viewpoint, the alternative ways to increase revenues are either to push up prices or to increase production. Price increases would have to be handled carefully: producers must consider their own long-term security objectives: oil might price itself out of the market[40]. Until 1978, however, producers had substantial room for manoeuvre. Costs of most alternatives to oil were certainly above the then prevailing oil price levels[41]. The economic self-interest of the producers favours oil conservation and higher prices, and OPEC is clearly moving in this direction. As an example, we may quote an OPEC consultant advocating a conservation policy comprising the following steps[42]:

(1) Allocation of resources for the requirements of consumption and investment.
(2) Adequate resources for defence capability.
(3) Allocation for resources in non-oil Arab countries.
(4) Formation of a contingency reserve of high liquidity.

(5) Synchronization of depletion rates with financial requirements as indicated by points (1) to (4), 'in a manner which betrays neither conspicuous consumption nor conspicuous investment'.

Apart from important sins against the rider in quotation marks, the policies of the OPEC governments seem to correspond fairly closely to the first four points of the list. The synchronization of depletion rates with financial requirements, however, has not yet taken place in Saudi Arabia and some other low absorbers. Yet indications are that the evolution will be in this direction: the economic self-interest of producers favours a reduction in export rates if and when they produce more revenue than required, or if and when prices can be raised. Only if demand for revenue exceeds export earnings, and prices cannot be increased sufficiently, would an increase in production become necessary. This gap is obviously a highly dynamic one, depending on a wide range of factors. One important new element is income derived from investments abroad. For Saudi Arabia alone this was estimated at about $\$4.9 \times 10^9$ in 1977 (about 20% of total oil revenues of the same year), and is expected to reach about $\$10 \times 10^9$ in 1981[43]. In other words: Saudi Arabia might soon be able to live off its investment income.

Economic self-interest is only one element in a whole range influencing export policies. Tactics to gain markets at the expense of other producers, so as to deprive them (or him) of oil income, and to induce a relative loss of military, economic and political strength could also influence these decisions — as well as tactical considerations in the opposite direction. Thus, Saudi Arabia, in the past, seems to have adjusted production downward to help other OPEC countries, but also upward to put pressure on them, depending on the particular political strategy pursued in a wider context.

The main element of synchronization between import demand and export requirements on a global level is today undoubtedly the fear that Saudi Arabia and other producers have about the impact of supply shortfalls and the consequent price increases on the world economy and the Western political system. These producer countries have a high stake in both: their investments and their export markets depend on the well-being of Western capitalism, and their survival as regimes depends on the capability and willingness of the West to keep them in power against external and internal threats. The close relationship between Washington and Riyad also provides a powerful channel of communication for such awareness of the producers' international responsibilities. Saudi Arabia, for one, certainly recognizes these responsibilities[44]. Nevertheless, this preparedness to elevate production levels (and restrain prices) in line with Western expectations cannot be relied upon. Already, there are powerful members of the Saudi elite advocating conservation policies[45]. The events in Iran in 1978/79 must give additional support to those arguments. Moreover, the dilemma of the low absorbers might well be growing. If demand for OPEC oil continues to grow the disincentives to increasing oil production will become more pronounced — on the one hand because absorption problems will be exacerbated (and with them the effects on social stability), on the other hand because pressure on reserves will grow. Higher prices are one line of defence against such a movement — but they would only increase oil income further, and would risk damaging the world economy if they occurred too rapidly and too steeply. The strain of the dilemma between national objectives and interests and international responsibilities will force a change. For example Saudi Arabia could resolve its own dilemma by setting an ultimatum to consumers to constrain import demand, by freezing production levels, or by a slow decline in output.

On balance, then, the analysis of factors influencing revenues indicates that in the longer-term conservationist attitudes will become more pronounced in producer countries with excess earnings. As early as 1976, Kuwait set a production ceiling of $2 \times 10^6$ b/d despite her comfortable reserve position[46], and a number of other countries, including Saudi Arabia (with an overall ceiling of $8.5 \times 10^6$ b/d annual average production, and specific ratios between light and heavy crude liftings) have followed that example. Overall emphasis in the composition of revenues will shift from quantity exported (as during the 1950s and 1960s) to prices per unit (as since 1971). These conclusions are fully borne out by the price increases and further reductions in production announced in 1979.

What are the factors influencing demand for oil revenues? As Part II has shown, three different sets of variables can be distinguished here:

(1) Oil has a lifecycle. Sooner or later, oil exports will have to be replaced by other sources of income. To avoid becoming a 'rentier state' relying on foreign investment income (with all its implications for national independence), producers will have to build up a diversified economy capable of self-sustained growth without oil income.
(2) Many oil exporters still share most aspects of underdevelopment with the rest of the Third World — low per capita income, unemployment, and great inequality. To overcome these, governments have to pursue a strategy of economic and social development, and even conservative regimes have been unable to resist pressure in this direction.
(3) The repercussions of an oil-exporting sector on the social system creates changes in attitudes and growing expectations. Generally, this can be expected to translate into demands for consumption and imports; but as the developments in Iran have shown, modernization policies can also produce profound counterreactions, with — at the extreme — a complete rejection of Western models and values. The implications of such an effect would appear to be introversion and pressure to cut links with the outside world.

All three variables influence import demand, mostly in the upward direction, and to a large extent it will be the future level of imports which will determine oil revenue requirements. Thus, import needs of OPEC countries have a significant bearing on Europe's future energy supply security.

The level of imports during the next decade will depend not only on the pull-effects outlined above, but also on the constraints of the producers' absorptive capacity. These constraints can be temporary or permanent; the main temporary ones, now already somewhat alleviated in most OPEC countries, are lack of infrastructure (port capacity, roads and railways, power networks, etc.), and shortages of skilled personnel (scarce skilled manpower to operate plants, of administrators to supervise and execute development projects, etc.). Permanent constraints are the lack of resources: raw materials, land, water, people. These criteria will be used in the following assessment of Europe's main energy suppliers.

## Algeria

Algeria has substantial human and natural resources, and a desperate need for capital. On the one hand, there is a population of about 18 million, with one of the highest birth rates in the world. On the other hand, per capita income in 1976 was only about $700 and unemployment is widespread: about 20% of the urban, and up to 60% of the rural population, are estimated to be without work or underemployed[47]. Agriculture — the traditional mainstay of the Algerian economy — employs about 57% of the workforce, but produces only 8% of GNP — an indication of widespread poverty in the countryside[48]. Inequalities of income distribution and wealth have been a major, and increasing, problem for the country: in 1972, 25% of agricultural land was owned by 3% of the farmers, and the agricultural reform since has largely been unsuccessful in redressing this situation[49].

Apart from moderate oil and large natural gas reserves, Algeria also possesses a variety of other minerals. Particularly important are iron ore, and phosphate zinc, where newly discovered deposits will allow a substantial expansion of production over the coming years[50]. Algeria's mineral wealth has played a key role in the regime's development strategy. A state-controlled heavy industrial sector clusters around the nationalized production of oil, gas and other minerals, providing the base for the transformation of industry and agriculture. At present, Algeria has a sizeable industrial sector with iron and steel, petrochemicals, and vehicle and farming machinery plants[51].

The main constraint on Algeria's progress in the past has been a shortage of capital, and even the windfall of increased oil and natural gas prices has not removed these limits. The external debt stood at $6.6 $\times 10^9$ at the end of 1976, and is expected to rise to $21 $\times 10^9$ in 1985; debt service took about 8% of total export earnings in 1975, and will probably claim 20% in 1980[52]. However, the main influx of capital for investment will continue to come from domestic savings — Algeria has one of the highest investment rates in the world (about 40% of the GDP)[53]. These figures indicate that Algeria's regime is taking quite a gamble: it is willing to accept a very high level of foreign debt, and a shift of resources from consumption to investment. Oil and gas play a crucial role in this: they are expected to finance most of the development outlays during the rest of the century, and they provide the main security for foreign creditors. Whether the policy of consumption deferral and austerity can be maintained in the future, seems at least uncertain: in recent years, growing opposition and social unrest has been noted, and the regime was forced to concede a 30% rise in the minimum wage as of January 1978, as a result of widespread strikes in protest against unemployment and inflation.

Some of the characteristic deficiencies and problems of an oil-exporting developing economy are thus clearly evident in the case of Algeria, too. In comparison to other oil exporters, however, the regime seems to have achieved a high degree of stability and acceptance for its firm leadership: the process of discussion of the new National Charter in 1976 was kept remarkably open, and vigorous criticism was encouraged. Bureaucracy and corruption still pose major problems, and conservative opposition against too radical a transformation had to be placated by a stronger emphasis on Islam. A greater willingness to concede demands by the population for a higher standard of living have also been in evidence over the last years. The most serious obstacle for future development has been the consistent weakness of agriculture, and the objective of achieving self-sufficiency by 1980 seems difficult to reach. Another doubt must be the push for capital-intensive, highly sophisticated export industries, which the regime is still emphasizing heavily.

The conclusions which can be drawn from this are, first, that Algeria will have to seek maximum production and export levels for oil and natural gas. The development plan for oil and gas until the year 2005 foresees a total income over the period of $250 \times 10^9$, coming increasingly from natural gas exports[53 a]. Large quantities of gas have already been committed for exports — in fact, there is some opposition from Algerian officials who consider the volume of exports too high[54]. Second, the Algerian strategy of development will bring the country into increasingly close contact with the European Communities. This will broaden the range of conflictual issues (migrant labour, new industrial markets) but also might turn Algeria into a moderating influence among Arab exporters: the country will be extremely vulnerable to economic recessions in Europe. The urgent need for high hydrocarbon revenues will, however, also mean that Algeria will press for higher oil and gas prices[55].

## Iran

Iran has an estimated population of about 35 million, and per capita income of approximately $2200 in mid-1977[56]. Like Algeria, Iran has substantial potential resources, quite apart from large oil and natural gas reserves. Only half of the total land under cultivation (about 19 million hectares ($19 \times 10^6$ ha) out of a total surface of $165 \times 10^6$ ha) are presently irrigated[57]. The country possesses rich mineral resources including iron ore, coal, copper, chrome, lead, zinc, and uranium. The majority of the population (about 55%) still live on the land which employs only 34% of the active work force[58].

Even more than Algeria, however, the mobilization of Iran's potential wealth has been plagued by severe problems, not least those caused by planning and administrative failures. At the same time, the present situation still requires urgent efforts — the relatively high overall per capita income conceals very large inequalities between town and countryside (where per capita income is thought to be about one-fifth of that in the towns)[59]. Official statistics show that during the regime of the Shah, 10% of the population accounted for about 40% of total consumption, while the poorest 30% had a share of only 8%[60]. And while the situation of the rural population might have improved in absolute terms since 1960, they have certainly fallen behind in relation to the urban groups.

The answer to this challenge of underdevelopment was supposed to come from a massive push for development: the Shah himself ordered an increase of the 1973–1978 development plan by 90% after the oil price rises in 1973/4. Growth rates in 1974/5 and the following year were extremely high — 41% and 42% respectively[61]. Yet by 1978 it had become clear that the Five Year Plan, 1973–1978, had been an utter failure. This could be concluded not only from the severe slow-down in the economy (a mere 2.8% growth in GNP was registered during 1977/8), but even more clearly from the socio-political unrest, which was shaking the very foundations of the system in late 1978. Although the social distortions caused by the massive push for development could not be foreseen in their full implications, the structural constraints facing the Plan were clear enough to counsel caution. They included:

(1) A shortage of qualified manpower: during the Fifth Plan, about 700 000 vacancies for skilled personnel were expected. At the end of the Plan period, there were still 93 000 jobs to be filled and this figure excludes the 100 000 plus foreign workers in Iran.
(2) Severe infrastructural shortcomings. Overcrowded ports, railway and road links, as well as major breakdowns in power supply, had by 1977 severely compromised growth in a number of sectors.
(3) An underqualified, corrupt and sycophant bureaucracy. The small layer of well-qualified planners and administrators was not backed up by good middle level personnel, and the autocratic and occasionally erratic rule of the Shah did not allow for serious criticism. Thus, the temptation for government officials to enrich themselves was enormous, and not only condoned, but supported by the regime[62].

Most important of all, however, was probably the type of political regime the Shah had built. The revolts in 1978 have revealed to what extent the system depended on iron-fisted suppression by Savak, the secret police, and how limited support the regime enjoyed even among those who benefitted from the system. At the same time, the push for development since 1973 showed that such a policy simply could not work. In spite of appearances, the political system of Iran seemed, in fact, rather 'soft': the Shah was forced to placate potentially powerful social groups — such as the industrial workers — with large concessions, and this in turn destroyed some of the inherent assumptions in the Plan. As the Plan unfolded, its failures and the ever more vociferous and powerful pressures for material benefits from strategically placed groups threw the government strategy more and more into disarray, and brought it to disintegrate into a set of measures without any clear sense of direction. Finally, the breakdown of economic policy spilled over into a general crisis of the regime. The symptoms of economic failure can briefly be described as follows (see Chapter 5):

(1) Inflation accelerated to at least 30% in 1977, before it began to fall to a level of 15—18%. This led to an escalation of project costs, exacerbated social cleavages, and to an artificial boom in housing and land.

(2) Agriculture did very badly. The country, which in 1968 still had achieved a small agricultural surplus for export, was forced to import foodstuff worth $1.6 \times 10^9$ (1975/6). Two factors accounted for this development — the very steep increase of demand for food caused by population growth and, mainly, by improved diet, and inadequate government policies, which have been unable to deal effectively with the two major structural constraints of Iranian agriculture — lack of manpower, and a shortage of land and qualified labour. The agricultural reform of the 1960s had led to the establishment of farmers co-operatives, but there was a desperate lack of managers capable of running them. Thus, government control has often been confined to book-keeping, and farmers have been able to receive large revolving credits even if they did not increase production. The attempts to develop a large agro-business sector have also been a dramatic failure: only one of seven major enterprises set up in Khuzestan did not go bankrupt. The result of all this has been that agricultural output made only modest (if any) progress during the Fifth Plan[63].

(3) Industry by-and-large performed more successfully, but the strategy of export-orientation turned out to be utterly unrealistic. Rapidly growing domestic demand soaked up all domestic production, as well as increased quantities of imports; besides, Iran was fast losing its competitiveness in third markets, and even at home, *vis-à-vis* foreign producers. This has been at least partly the result of the privileged position of the industrial proletariat, which exploited the sellers' market for skilled and semi-skilled labour, often backing up their demands with strikes. Wage rises of 25% were the rule, and employers also had to offer large fringe benefits. The policy of buying off the politically dissatisfied and alienated workers is also reflected in the scheme for workers' participation in shares which burdened Iranian industry with additional costs, which could only partly be recovered by productivity increases and higher sales as a consequence of higher purchasing power of workers — they often bought foreign products. Several industrial sectors

have been doing particularly badly — among them the steel industry, which is saddled with a huge plant of old-fashioned design bought from the USSR, and textiles. One example may serve to underline the difference between targets and reality: the cotton textile production in 1977 was planned to reach $960 \times 10^6$ m — in fact, it achieved less than half of this figure, or $450 \times 10^6$ m. This was 25% less than production levels in 1974. The reasons for this particular failure are instructive — they include power cuts, high cotton prices, and shortage of labour[64].

Although this is a particularly crass example, it has become clear that Iran can hardly hope to build up industrial exports to levels sufficient to replace oil as the principal source of foreign exchange. In fact, non-oil exports (traditional ones such as carpets, as well as manufacturing exports) have declined rather than increased; they were down in 1977 by 14% on the level of the previous year.

(4) Minerals have — apart from hydrocarbons — so far played only a minor role in the economy, and only coal and iron is produced in significant quantities. The much flaunted copper resources of Cheshmeh have been developed and production has begun, but the mine is thought to operate at a loss, at least at price levels on the world copper markets that prevailed in late 1978.

(5) Dependence on oil has not declined. The economic setback in 1977/8 was at least partly caused by a fall in oil revenues, and if any reminder was needed about the strategic role of the oil export industry, it was provided by the strike of oil workers which sent Iran's export earnings and credit rating tumbling.

Even more serious, however, have been the results of the social distortions caused by the rush for development. The inevitable opening of the gap between expectations and fulfilment, and the excesses of the scramble to get rich quick, have led not only to deep dissatisfaction with the regime, but, among a significant part of the population, also to a radical opposition to the whole process of modernization and Westernization. This grassroot discontent was channelled effectively by the Shi'a religious leaders into a fundamental challenge to the Shah and his regime. Among the victims of the ambitious programmes launched by the regime since 1973 in the attempt to placate opposition was the military expansion programme — hardly an unworthy candidate for such pruning down: originally, expenditure on defence over the five-year period was to be $\$19 \times 10^9$, as compared to a total state development expenditure of $\$40 \times 10^9$ [Note 65]. The plans included the purchase of several AWACs — airborne electronic surveillance and combat centres, which the European NATO countries originally found too expensive for their objectives.

At the time of writing (1979), the future political development of Iran is difficult to foresee. The revolution has probably not yet run its course, and the Islamic regime looks increasingly shaky. Thus, major uncertainties surround the future oil policy of the country. Given the fact, however, that Iran's non-oil exports account for only 1% of total exports[66], and that Iran is heavily dependent on imports to satisfy the needs of its population, it seems likely that future oil and gas exports will be higher than during 1979, provided the technical problems of neglect of oil installations can be overcome. Yet the present level of oil reserves is insufficient to support much increase in output, and much of future hydrocarbon export earnings will have to come from gas exports: oil demands in the country itself will certainly expand further, and it seems likely that production will peak somewhere in the 1980s, causing a fall in oil available for export. Estimates for Iran's exports in 1990 range between 1.5 and $3 \times 10^6$ b/d. Thus, in the longer run, Iran's importance as energy exporter will lie more with natural gas than with oil.

## Iraq

Iraq is traditionally an agricultural country — and the fact that it once fed about twice its present population of 12 million inhabitants[67] is an indicator of the potential of this sector. Much of the fertile land is threatened, however, by high water tables, which cause the roots of plants to rot, and the salination of the soil[68]. The water resources to combat salination are available, but difficult to develop: this requires the establishment of an extensive drainage and irrigation system.

Apart from agricultural resources and a sizeable population, Iraq also possesses rich mineral reserves. Oil and gas still seem to be inadequately explored, and the potential of the country is considered vast. In addition, there are deposits of iron ore, chromite, copper, lead, zinc, limestone, gypsum and salt. Phosphate and sulphur deposits are under development for export and domestic demand[69].

The relatively modest per capita income of $1260 (1976)[70] indicates that this potential is still far from realized: during the 1950s and 1960s, Iraq's economy stagnated. The reasons for this were largely political — perennial conflict with the concessionaire oil companies over terms of exploitation of oil resources, political instability (the monarchy was overthrown in 1958, resulting in a series of military-dominated governments) which ended only when the Ba'th—Army combination, which is still in power, took over in 1968, and the plurality of Iraq's society with its different power centres resulting in ethnical conflict (the independence struggle of the Kurds) or tension (Shi'ite majority versus Sunnite elite). In recent years, the conflict with Syria also developed into a major brake on the economy, for at stake were, among other issues, the distribution of the Euphrates water resources and the export routes for Iraqi oil. Thus, the country is in need of a large development effort, and the present regime has clearly staked its future on the success of such a programme.

The most promising sector for investment is undoubtedly agriculture, and after some initial neglect the regime has now come round strongly to support this idea: about one-third of the $34 $\times$ 10$^9$ Development Plan, 1976—1980, has been allocated to agriculture. The objective is to increase food production, by increasing the cultivation of land by 25%, in order to have an export surplus[71]. That such an effort is needed has been underlined by the fact that Iraq has had to import food over the past years — to the extent of $600 $\times$ 10$^6$ in 1974 and 1975[72].

In the past, Iraq has traditionally pursued an oil export policy of maximizing revenue: it did not cut back production in 1973/4, and achieved significant production increases in 1974 and 1975, indicating some price cutting at a time when other OPEC countries had to reduce output as a result of slack demand. Production capacity is to increase to 3.5 $\times$ 10$^6$ b/d in 1980[73]. Yet in 1976 and 1977 Iraq encountered a major shortfall on projected oil revenues, necessitating a drastic reassessment of its development strategy[74]. The brakes have been taken off somewhat since, but the push for development is now evolving much more cautiously than before. Reasons for this setback are the by now familiar ones of inflation (estimated at 30% in 1976), shortage of skilled labour, escalation of project costs, unexpected financial constraints as a result of lower than expected oil export revenues, infrastructural shortcomings, a decline in agricultural output, and severe mismanagement by an overcentralized and bureaucratic planning machinery. Thus, the doubling of development budgets between 1973/4 and 1974/5, and the large concessions to the population through wage rises, tax cuts, and a reduction of water and electricity prices, had to give way to austerity budgets in 1976 and 1977[75]. Nevertheless, the regime seems firmly installed through a mixture of brutal repression of any challenge to its power (e.g. by the strong local Communist Party, which came repeatedly under attack from the ruling Ba'thists, and by the Kurdish minority) and support for the basic policy of the rulers.

The country, thus, needs a period of consolidation, and the willingness of the regime to resolve old differences with conservative states in the region, and in late 1978 even with Syria, indicated the desire of the regime to concentrate on internal development. Events in 1979 (the bloody repression of opponents to President Saddam Hussein and the renewed rupture with Syria) underlined, however, the continuing precariousness of Iraq's political stability. The regime thus had to make economic progress in order to stifle discontent. Oil exports are a vital element in this connection, and Iraq can be expected to continue its push for high export volumes and prices. As a result of painful experience, pragmatism is clearly gaining ground in Iraq, and this should help the country to realize some of its potential over the coming years. If this analysis is correct, then Iraq's importance in the Gulf and the Middle East as a whole is bound to grow; and this assumption is corroborated by the large oil potential still to develop in this underexplored but very promising oil-bearing region. Estimates about Iraq's future production capacity for 1985 range around 5 $\times$ 10$^6$ b/d; for 1990 it is estimated that it will be even higher.

## Nigeria

Nigeria is the most populous country of the group of exporters under consideration — and it is, together with Iran, the only non-Arab state. With an average income of about $280, and 80—100 million people to feed, Nigeria is economically in a different situation from the other countries considered so far; at the same time there is an enormous potential for development. This potential concerns not only the oil sector, although there, too, the limits of expansion have still not yet been reached. The country also possesses rich deposits of columbite (Nigeria supplies 95% of world demand), tin, lead, zinc, iron ore, wolfram, casserite and limestone[76], as well as extremely fertile agricultural land. Seventy-five per cent of the population lives on the land, which accounts for 65% of national income. The large population base of Nigeria also provides a practically unlimited domestic market, once its potential has been realized.

However, Nigeria is not only a country with enormous potential. It also has vast problems, many of them not unrelated to the oil sector. The quadrupling of oil prices in 1973/4 seemed for the first time to remove all financial constraints on economic planning, and under the regime of General Gowon, a Five Year Plan was launched: it budgeted expenditure at $45 \times 10^9$, as opposed to the $3.5 \times 10^9$ of the previous five-year period[77]. This plan was described — with the benefit of some hindsight — by a normally level-headed observer as 'a departure from economic sanity'[78]. Four years later the Nigerian economy underwent a serious contraction of its previous expansion: growth rates fell from 10.9% in 1976/7, and 9.9% in 1977/8, to below 5% for 1978/9[78a]. The balance of payments had slipped back into deficit as early as 1976, and the government not only had to resort more and more to borrowing from abroad and domestically, but also had to make drastic cuts in public expenditure — the financial constraints had reappeared too rapidly. Rampant inflation — conservatively estimated at 30% in 1978 — a steep decline in agricultural production, and above all a fall in oil income as a result of slack demand, competition from new production areas in the North Sea, and a careless pricing policy, were the main factors behind the economic recession[79].

Nigeria, as much as the more dramatic example of Iran, provides in the five-year experience since 1973 many lessons about the mixed blessing of an oil export economy. Even before 1973, oil had been a factor in the outbreak of regional tensions which deteriorated into civil war (the Biafra secession 1967—1970); although not directly responsible for this war — the ethnic differences and regional diversities are a perennial burden for Nigeria —, oil had worked as a catalyst in the secession. The exaggerated hopes in 1974 pushed expectations upward, and resulted in the frantic attempts to achieve quantum leaps in development on all fronts at the same time. To satisfy expectations, the regime of Gowon conceded lavish pay increases in the public sector, which were then imitated by private industries, and thus contributed to an already high rate of inflation caused by bottlenecks, increasing import prices, and the rapidly opening gap between supply and demand. The infrastructure was stretched beyond its capacity: in August 1975, 306 ships were waiting to unload their cargoes outside the port of Lagos. The government had ordered cement for five years, although port capacity was capable of handling only one-fortieth of the $20 \times 10^6$ t ordered[80]. This provides at the same time an illustration of the waste of resources and the spread of corruption, which went hand-in-hand with the boom. Manpower requirements were also severely neglected: the Plan demanded a work force of 450 000 alone for the public sector construction projects — double the available number[80a].

The oil sector continued in its central role in the economy; in fact it expanded further. Industry made only slow progress (production rose only 42.5% between 1972 and 1977, according to optimistic estimates[80b]. But the major problem area was agriculture, which went into strong decline. Once an exporter of palm oil, Nigeria now had to import the products of one of its major traditional export sectors. And this picture applies across the whole range of agricultural products. No wonder therefore that food imports have soared. The reasons for this decline are complex; not all relate to the oil sector (such as the inefficient land tenure structure). But the effect of the overwhelming oil export industry has clearly been felt — through the migration from rural into urban areas and the negative change in the rural age distribution, and through the exchange rate which made non-oil exports uncompetitive, and favoured food imports displacing domestic production. On the other hand, the positive effects of oil investments have not sufficiently trickled down into the countryside, whereas inflation has: the result is that the position of the farmer *vis-à-vis* the towns has severely deteriorated. Even in absolute terms, many farmers' incomes now appear lower than in 1973[80c].

Thus, one of the results of development since 1973 has been an increase in inequality — not only between the countryside and the town, but also within urban areas. At the same time, corruption and purges have severely affected the quality of the civil service. The dangerous aberrations and distortions of the development process in Nigeria since 1973 have led to increased social tensions, and have contributed to the fall of General Gowon, possibly also the assassination of his successor, General Muhammed. Whether the return to civilian rule in late 1979, to a mixture of an American-style presidential with a British-style parliamentary regime, will provide the necessary political stability and sense of direction, remains to be seen.

Clearly, government planners and politicians have learned some lessons from the experience of 1973–1978, and a new sense of austerity and realism now prevails. But even the revised targets, and the formulation of objectives for the next plan, imply heavy reliance on earnings from oil. Thus, all indications are that Nigeria will continue to have an interest in maintaining exports of hydrocarbons at a high level. However, in the short and medium term there is little prospect for production increases beyond the ceiling imposed by the government for oilfield conservation purposes (about $2.3 \times 10^6$ b/d; production capacity is slightly higher): exploration and development activity has been slack for several years[81]. In the longer run, the potential of Nigeria's oilfields could be significant enough to allow a marginal increase, if markets could be found. In any case, Nigeria seems likely to be a reliable source of oil imports, if not a very large one.

## Libya

Libya is the first country in the group widely labelled as 'low absorbers'. With a population of about 2.6 million, a per capita income of $4400 (1976)[82], and a cultivated area of only 1% of the total surface, the resources of Libya are clearly limited. This is accentuated by the absence of important natural resources apart from oil and natural gas, although an iron ore deposit estimated at $700 \times 10^6$ t is now under development[83]. Libya's agriculture employs about 30% of the active population (183 000 in 1973)[84]; industry is rudimentary and almost exclusively confined to food processing and oil-related plants[85].

The small population base has accentuated the habitual problem of skilled labour shortage in oil-exporting LDCs, and Libya has to rely to a growing extent on foreign workers: their numbers increased between 1973 and 1975 by an annual rate of about 60% to over 220 000, accounting for 32.9% of the total workforce. During the present Five Year Plan (1976–1980), this share is supposed to increase to 41.3%[85a]. Egyptians and Tunisians make up the bulk of foreign labour. One of the factors exacerbating the labour shortage is the large claim of the army on the Libyan workforce.

Since 1969, Libya has been under the control of a revolutionary military regime with ambitious development plans. Since 1973, agriculture has been given priority, with the Three Year Plan, 1973–1975, allocating over 20% of the total expenditure to this sector, and the present 1976–1980 Plan over 17% (about $4 \times 10^9$)[86]. Investment is being directed into very capital-intensive projects involving expensive irrigation in an attempt to expand the amount of land under cultivation (e.g. the famous Kufrah Oasis Project). Intensive farming of this sort, based on deep water reservoirs established with oil-drilling equipment, faces costly transportation of products to markets, and problems of manpower and expertise, as well as climatic and soil disadvantages. The two exemplary Kufrah projects (one for sheep breeding, and one for agricultural production and settlement) have been extremely expensive exercises with capital costs for the former of LD $30 \times 10^6$, and annual running costs of LD $10 \times 10^6$; whether these projects will ever be profitable remains doubtful[87].

The strategic objective for the Libyan development planners is, however, not so much to assure profitability. For them, agricultural development is designed to reduce dependence on food imports (for political reasons considered dangerous), to create alternative sources of revenues apart from oil, and to stop the migration to the towns. In these objectives, the regime has been to some extent successful: according to official statistics, the Three Year Plan, 1973–1975, managed to double wheat and barley production, to increase output of fruit and vegetables by two-thirds, and to raise the number of sheep and poultry by 48% and 544% respectively[87a]. Between 1976 and 1980, wheat production is supposed to double again, while demand is expected to go up by 20%.

The rural exodus, which created massive problems in the early 1970s, now seems to have been stopped, and even reversed somewhat. But the ultimate objective of lessening Libya's dependence on oil exports is still very far removed. This is even more in evidence in the industrial sector, which the regime is trying to develop: the prestige projects of a large steel works and a petrochemical complex, as well as the textile industry, will inevitably have to look for export markets, given the tight constraints on the evolution of domestic demand with a population of less than three million. The profitability of these industrial white elephants must, however, be considered highly doubtful: high wage levels, insufficient economies of scale, and manpower shortages will all contribute to make the success of these projects very difficult.

What are the conclusions of this analysis for Libya's future oil export policy? First, a continuation of the present emphasis on rapid development seems likely: the regime appears firmly installed and well-rooted in public support, and although more caution might become evident in future project planning, the absorption capacity of Libya will continue to be quite substantial – indicating the superficiality of the term 'low absorber': the type of regime in control in those countries can make for substantial differences. Nevertheless, Libya's oil revenues can be expected to be superior to its domestic needs, and even to the costly extravagancies of Qaddhafi's foreign policy. The development budget for 1976–1980 would require annual oil revenues of roughly $\$5 \times 10^9$, or substantially less than the $\$9.4 \times 10^9$ achieved in 1977[88]. This indicates that Libya might be inclined to reduce oil export yet further, to bring revenues closer in line with demand for foreign exchange. Since the traditional radicalism of Libya, in terms of pricing policies, will probably continue – after all, Libya has to defend its reputation as the country which demonstrated the reality of oil power to its fellow OPEC countries in 1970! – the desire to reduce output could reinforce the move towards setting production ceilings in order to push up the international price of oil.

## *Kuwait*

Kuwait's lack of resources is even more pronounced than that of Libya. With a population of about 1 million (over 60% of which are foreigners), and a total surface of 17 818 km$^2$ (of which only 3% is under cultivation)[89], Kuwait has little else to rely upon but its offshore fishing grounds (prawns and shrimps), service industries, and oil. The manufacturing sector is small – mainly food processing, construction, and petrochemical industries[90]; new industrial projects outside the oil and gas sector have been shelved. The government seems reluctant to push industrial development further, since this would inevitably further complicate the delicate problem of foreign workers. Besides, domestic markets are small, power supply from associated gas is limited by the oil conservation policy of the government, and pollution could rapidly become a serious problem in a territory as small as Kuwait[91].

Domestic consumption has already been developed fully — Kuwait is the Middle Eastern welfare state *par excellence*. The privileges do not accrue to the non-Kuwaitis, however, who often earn salaries two-thirds less than Kuwaitis. This duality is complicated by the fact that expatriates (often Palestinians) control influential positions in middle management and the media. Political unrest led to the dissolution of parliament in 1976, and the precarious population balance could become an element of potential instability.

Barring political change, the evolution of Kuwait's import requirements seems fairly predictable: they will not increase dramatically in real terms[92]. Given the modest but distinct successes in the past to diversify export earnings (a proof of the success of service industries, and also petrochemicals) it might be that the relative importance of oil exports could be reduced yet further from the present level. (Today, non-oil export revenues cover 23% of import requirements.) The imports showing the fastest growth over the last few years were armaments — the government has embarked upon a major effort to strengthen its armed forces[93]. Kuwait's foreign exchange requirements also include another security-related item — its aid. Since its independence, the country has recognized the need to dissolve the envy of other Arab states by largesse, and the generous and often very effective development assistance, as well as the politically motivated disbursements to front-line countries in the Israeli—Arab zone and the Palestinians also have to be seen in this context.

All told, Kuwait cannot hope to develop indigenous alternatives to the oil export industry. Two conclusions can be drawn from this: first, it seems in the country's interest to stretch the lifetime of its petroleum reserves, and to form a conservation policy, reducing output to levels which are close to foreign exchange requirements. Such a policy has been pursued since 1972, when the government first decided to impose production ceilings[94]. Second, Kuwait can invest money abroad to secure a rentier's income after oil reserves are exhausted. Such investments are constantly made, and already now account for a sizeable source of additional income[95]. Depending on world price levels, this appears to imply production levels around or below $2 \times 10^6$ b/d[Note 96]; but as was mentioned above, the bottom line of Kuwaiti oil production is not so much set by foreign exchange requirements than by the level of associated gas production necessary to keep the Kuwaiti economy running. This bottom line appears to be around $1.5 \times 10^6$ b/d.

## Saudi Arabia

Saudi Arabia combines typical features of the 'low absorbers' with some peculiar aspects. On the one hand, population is small (of the estimates ranging between four and eight million, the former figure is generally considered closer to the mark), and per capita income high (between $4400 and $8000). A very high share of foreign workers — between 30 and 40% of the population[97] — also provides justification to group the country together with the smaller Gulf sheikhdoms and Libya. On the other hand, the country covers a territory corresponding to that of the USA east of the Mississippi; and this provides opportunities for development non-existent in countries such as Kuwait. Mineral resources are an example: apart from the world's largest oil reserves (rivalled possibly only by Mexico) and large natural gas resources, there are iron ore, gold, copper, lead, zinc, silver and uranium deposits. Saudi Arabian planners even expressed the hope that mineral production excluding oil and gas could in the long run supply as much as 30% of GNP[98].

Agriculture is another area with considerable additional potential. At present, it employs about 50% of the population but contributes only 0.2% of GNP; large amounts of food have to be imported. The surface under cultivation and agricultural output could be expanded significantly, if water could be found, and the present Five Year Development Plan (1975–1980) foresees spending of $34 \times 10^9$ riyal on water development, and $4 \times 10^9$ riyal on other agricultural investments. Some major projects are under way, or already completed, such as the al-Hasa Irrigation Scheme, the Faisal Model Settlement Scheme, the Wadi Jizan Dam, and the Abdha Dam[99].

Industry is at present small, but expanding rapidly, and has been allocated $45 \times 10^9$ riyals during the Plan period — mostly in oil-related or energy-intensive industries[100]. However, as in the case of agriculture, manpower and infrastructure constraints are formidable obstacles: ports used to be so congested that Saudi Arabia deployed helicopters and airships to discharge freighters directly. About 13 000 km of new roads, a major expansion of port capacity, airports, electricity, and telecommunications are all major priorities in the present Plan[101].

The objectives of the Development Plan are much in line with those of other countries: dependence on oil exports is to be lessened by diversification of the economy into a broad industrial base, human resources are to be developed fully (at present illiteracy is still around 80%), and the standard of living is to be increased. All this, it has to be added, is supposed to be achieved in maintaining the old religious and moral values of Islam — a majestic contradiction whose resolution will be one of the most interesting and explosive aspects of Saudi Arabia's future evolution. *De facto*, these objectives imply a very high level of import dependence, for even the achievement of industrial and agricultural targets would still only result in limited import substitution, and create very small non-oil export revenues. A minor shift towards the export of processed hydrocarbons is conceivable, however, in the form of oil products, fertilizer, and natural gas liquids (NGL). Import figures that doubled between 1974 and 1975, then increased by over 50%, then again doubled between 1976 and 1977, indicate the rapidly expanding dependence on foreign goods and services for consumption and investment. This massive influx of wealth has, however, also resulted in severe over-heating of the Saudi economy, inflation rates reaching 40%, and widespread shortages[102]. This has led to a major reversal of economic policy: the government switched to a more conservative approach — expenditure was cut down, criteria for project assessment were tightened, and a general campaign was launched against wasteful spending. This policy was reinforced by the realization that the government was simply losing control over the economy.

The past few years have shown that Saudi Arabia could in the future spend much of its oil income at home, if prices did not increase further, but the money spent would not create any real productive wealth[103]. In terms of real assets to be created, the Development Plan with its $142 \times 10^9$ budget is clearly too ambitious, given the major, and by now well-known, bottlenecks existing in the Saudi Arabian economy and society. But even this budget assumes oil income based on production rates of $5 \times 10^6$ b/d at 1975 prices; thus, there would be ample income available for investment abroad, political use of oil money, and military purchases.

To assess the likely future level of oil exports from Saudi Arabia we need to look beyond the economics of the situation. Economic analysis indicates that production levels needed for domestic consumption and investment of petro-dollars are below probable future world import demand for Saudi oil. *How much* below, and how high an excess of production will be permitted, depends on political bargaining processes within Saudi Arabia, between Saudi Arabia and other OPEC countries, and between Saudi Arabia and the consumers. Even in the present regime considerable differences of views exist as to desirable levels of oil production, and the possible pace of Saudi evolution into the twentieth century. Some members of the royal family fear the social consequences of too rapid modernization; already now, foreign workers from other Arab countries (Yemen, Egypt) and beyond are considered with suspicion and treated with great caution. This, in turn, creates resentment. The contradictions between traditional Wahhabi patterns of social organization and modernization are increasingly evident: examples here are the role of women (confined, in spite of rampant labour shortages, to a few jobs such as teaching in girls' schools), and the armed forces (suspected as a potential threat to the existing order). Although the traditional political system has shown considerable flexibility, it is uncertain to what extent it can diffuse or absorb radical opposition.

Thus, there is the possibility that because of the risks of social change the monarchy will order a considerable slowdown in the effort to modernize Saudi Arabia. There could also be a reorientation towards traditional Wahhabi values. On the other hand, radical political changes and the evolution of a new type of regime cannot be excluded — indeed, it seems probable over the next 10 years, given the tensions and contradictions already in evidence. Failures in the attempt at development, growing social inequalities, unrest among foreign workers, and resentment at what is seen as a new form of exploitation and subjugation by the West, are already noticeable, and unlikely to diminish. The oil export policies produced by this type of social and political change are difficult to foretell, but it seems plausible to assume that they will be less 'rational' and 'internationalist' (in the sense of reconciling Saudi objectives with international responsibilities). Domestic change may also influence the direction of foreign policy within the Middle East, within OPEC, and towards the Western world; and these orientations also influence oil export policies. Here, too, the present circumstances appear as the optimum configuration from the viewpoint of industrialized consumers; a less favourable scenario is easily conceivable.

But even assuming no change in regime, the degree of expansion allowed by the Saudi government would probably be limited. At present, Aramco has not been authorized to expand installed capacity beyond present levels; the gas-gathering project under development can process gas associated with a production level of $12 \times 10^6$ b/d. Thus, ceilings will be set as Saudi production expands; and these technical and economic limits will be reinforced by political and strategic considerations in Saudi oil policy. A tightening of the supply—demand balance requiring Saudi production increases would expose the regime to pressure from other OPEC countries to let prices increase; thus, the dilemma of having to choose between its OPEC alliance and its main ally, the USA, might become more pronounced and difficult to manage. At the same time, such a situation would exacerbate a dilemma for the country itself: its growing dependence on the industrialized consumers and its stake in their economic well-being will push Saudi Arabia towards high production rates and a moderate price policy. On the other hand, such a policy could lead to depletion rates unacceptable even for a country with the reserves of Saudi Arabia, and to the risk of structural oil shortages on a more dramatic scale, if at a later point, than if Saudi Arabia followed a conservative depletion policy. Thus, Saudi Arabia's oil policy would have the responsibility of evolving a production profile over the period of energy transition which would avoid major disruptions. Given the intensity of pressures on the regime, and the fundamental weakness of the country, it seems highly unlikely that Saudi Arabia could be relied upon to pursue such a policy*.

Given this lack of incentive to produce crucially important quantities of oil, and the central position of Saudi Arabia among the exporters, the USA has greatly intensified its traditional links with the kingdom. The armed forces are re-equipped, trained and expanded largely with American assistance. American companies, above all Aramco, have taken over large parts of the development programme, and US personnel and citizens employed by Saudi Arabia number several thousands[104]. *De facto*, the US government has thus committed itself not only to the security of the kingdom, but also to a very direct and high stake in the survival of the present regime. The exact extent of this symbiotic relationship, which forms the pillar of Western international energy policy, at the present time, is difficult to assess, but some reports indicate the existence of a far-reaching understanding between the two states. This agreement, informal in character, and reached in discussions in the Joint Economic Commission, is said to cover provisions for the long-term investment of Saudi Arabian funds in the USA, a Saudi pledge to supply oil to the USA at a price which would not

---

* Some of these pressures operated in the 1979 crisis (see Postscript).

increase by more than 5% p.a. until 1984, and US commitments to the integrity and independence of Saudi Arabia under its present regime, as well as to efforts to reach a settlement of the Israeli–Arab conflict along the lines desired by the Arab governments[105]. While these reports have been denied by Saudi Arabia[106], and the accuracy cannot be assessed, it is certain that Saudi Arabia has cooperated in the establishment of a large US oil stockpile. At present, it is this close relationship between Riyad and Washington which keeps Saudi Arabian oil supplies at the levels required, and offers a reasonable prospect of supply stability to the Western world over the next years. This mechanism is political, not economic, and is dependent on the stability of the present regime in Saudi Arabia, while hardly in line with the narrowly defined economic interests of the country. Thus, its long-term reliability as an element of supply stability has to be doubted. Events during 1979 have underlined such doubts.

This country-by-country survey is completed in *Table 8.3* by an OPEC-wide table summarizing some figures and estimates about the medium-term development potential and the political direction of those countries. What results clearly from both the survey and the table are significant differences among them; and these differences should be kept in mind throughout the following points.

In essence, the problem of supply stability in the longer term consists of several layers with risks of different magnitudes: an important part of oil export can be considered relatively safe. It consists of (a) the minimum production of oil needed to keep producers' societies going (which involves, at least in some countries, substantial exports), and (b) import demand which cannot be paid from other exports or financed through borrowing.

In recent years, there has been a very rapid expansion of OPEC imports (see *Table 4.2*). These increases have quickly reduced the OPEC petrodollar surplus, and thus made the recycling problem quite manageable. The question is: will this increase continue? As our country-by-country survey indicates, the answers differ. A further increase in imports seems plausible in the food sector because of sluggish growth of the agricultural sector in most OPEC countries, which will be unable to keep pace with population growth and increased demands for a higher quality of diet. Another indication is the poor performance of exports other than oil in many OPEC countries; many of the new industries built up in the Persian/Arabian Gulf and elsewhere will have difficulties in competing in international markets[107]. Thus, export earnings from traditional and new industries will for quite some time be of limited importance. Finally, growing populations and rapidly rising expectations, together with the degree of underdevelopment still prevailing in many OPEC states, will, generally, create pressure for more imports. As noted before, most OPEC governments are vulnerable to pressure; even in the apparently strong, authoritarian regimes the rulers have to consider the demands of powerful military and civilian bureaucracies and the expectations and attitudes of the masses. These pressures for consumption translate into imports (e.g. arms, welfare, etc.). Given the size of the low

Table 8.3  Some Indicators of OPEC Resources and Political Stability

|  | Algeria | Iraq | Libya | Kuwait | Qatar | UAE | Saudi Arabia | Iran | Nigeria | Venezuela | Indonesia |
|---|---|---|---|---|---|---|---|---|---|---|---|
| (1) Population size | m | m | s | s | s | s | s–m | m–l | l | m | l |
| (2) Minority problems | s–m | l | 1[a] | 1[a] | 1[a] | 1[a] | 1[a] | m–l | m–l | m | m–l |
| (3) Dependence on foreign workers | s–m | s–m | 1 | 1 | 1 | 1 | 1 | s–m | s–m | m | m |
| (4) Level of education | m | m | m | 1 | m–l | m–l | s–m | s–m | s | m | s |
| (5) Development of health system | m | m | m–l | 1 | m–l | m–l | m | m | s | m | s |
| (6) Development of social security | m–l | m | m–l | m–l | m | m | m | s–m | s | m | k |
| (7) Power of army | m | l | m–l | s | s | s–m | m | m–l[b] | m–l | s | m–l |
| (8) Communist influence | s–m | s–m | s–m | s | s | s | s | s | s | s | s |
| (9) Role of Islam | m–l | m–l | l | m | l | l | l | l | m | — | m |
| (10) Panarabism | m–l | m–l | l | s–m | s–m | s–m | m | — | — | — | l |
| (11) Chances for medium-term stability | l | l | l | m | s–m | s–m | s–m | s | m | l | m |

Key: s = small  m = medium  l = large
a  Foreign workers
b  As of early 1979

Source: A. Gälli, 'Die politische und soziale Entwicklung der OPEC-Staaten,' in M. Tietzel (ed.), Die Energiekrise: Fünf Jahre danach, Neue Gesellschaft, Bonn 1978, pp. 73–100 (94); own assessments

absorbers, such pressures will be less important in comparison to total oil revenues, but they will play a role; in the high absorbers, demands for immediate consumption rather than acceptance of deferral in favour of austerity will be even stronger.

Successes in import substitution militate against a further increase of imports. Import substitution, as opposed to export industries, might have some impact on trade relations at least in the longer term, and import requirements of some OPEC countries (mostly relatively developed high absorbers such as Algeria or Venezuela) might increase less rapidly than in the past, once investments begin to bear fruit. Another, probably much more important factor militating against a continuation of the past boom are constraints in the absorptive capacity of the low absorbers: they have neither the needs nor the potential to step up imports of consumer and investment goods at the previous rate. Related to this, the flooding of OPEC markets with imports — of goods, but also of ideas, models and practices — might have exceeded the capacity of OPEC societies to absorb these. In Iran, the result of this (coupled with other specific factors) has been a powerful backlash against the whole strategy of modernization, and against the political institutions which had organized it. The same could happen in other OPEC countries — be they high or low absorbers. Although this might produce prolonged periods of instability and a temporary withdrawal from international trade, it seems unlikely that this would last for very long in high absorber countries where dependence on oil exports is great. The capacity to export and to absorb imports could, however, be affected negatively by such periods of instability (destruction of infrastructure, drastic change of course in development strategies, difficulties of re-establishing authority).

In low absorbers, such instabilities could have an even more profound effect on import demand: the urgency of resuming oil exports at high levels would be much less, the cushion provided by foreign exchange reserves and investment income from abroad much higher. Indeed, an adjustment of production rates of low absorbers nearer to domestic revenue needs as a permanent feature of a country's energy policy could well come as the result of such instability.

Finally, one factor should be mentioned which could influence import demand either way. Military expenditures are a very sizeable item in many of the import budgets of OPEC, yet one which obviously depends on circumstances unrelated to the oil policy of the country. Iran has shown how defence expenditures and imports can be affected negatively by political changes; on the other hand, the feeling of uncertainty and the vulnerability of their oil reserves might induce low absorbers to continue and even accentuate their military spending. This may be the effect of the Iranian crisis on Saudi Arabia.

On balance, then, the conclusion about future import demand and, more generally, oil revenue requirements, can be summed up as follows: Algeria, Iraq, Nigeria and possibly even Iran will be relatively stable sources of supply for most of their export potential. Together, they could account for exports of about 12 $\times 10^6$ b/d in 1985, and 8.5 $\times 10^6$ b/d in 1990 (rough estimates for 1985 and

1990: Algeria 1.1 and $1.0 \times 10^6$ b/d, Iran 3 and $3 \times 10^6$ b/d, Iraq 3 and $3 \times 10^6$ b/d, Nigeria 1.7 and $1.5 \times 10^6$ b/d). Stable supplies from the low absorbers will probably be around $10 \times 10^6$ b/d in 1985 and 1990 (Saudi Arabia $6 \times 10^6$ b/d, Kuwait $1.9 \times 10^6$ b/d, Libya $2 \times 10^6$ b/d). Even these quantities, which obviously fall far short of world requirements, cannot be considered 'safe': domestic instability and regional conflict could make these low estimates turn out to be optimistic.

The assumptions about export rates by low absorbers in the above figures already make some allowance for the use of oil revenues abroad, notably in the Arab world (partly for political reasons) and in the industrialized countries. In terms of supply stability, investments in the Arab world at current levels are probably almost indispensable for the lower absorbers as a means to influence developments and to secure vital foreign policy objectives. *Table 8.4* shows the foreign aid efforts of Saudi Arabia, Kuwait and Libya, and the significance of political objectives in those efforts. Supply stability could be much increased for industrialized countries, if the scope of aid by low absorbers was expanded significantly. If Arab OPEC countries took the lead in a major development effort for the whole Arab world, absorptive capacity would grow dramatically; such a scheme would create vested interests in the whole Arab world in high production rates, and the consequent political pressures would undoubtedly contribute to the stability of oil supplies to the consumers. Political obstacles are obviously large; but the possibility of such a development should not be ruled out totally. It depends to a large extent on the degree of cohesion and stability in the Arab world.

Investments in developed countries would appear to contribute much less to supply security. First, they are self-generating — income might be reinvested, and as such augment overall assets without further additions from oil revenues. Second, incentives to invest abroad are not all that attractive from a low absorber viewpoint, although this could be changed by appropriate measures of developed consumer governments (indexation of OPEC investments?). Finally, there must be concern about the effect on national independence of large-scale dependence on investment in the developed world.

Ultimately, the gap between revenue requirements and oil demand will therefore continue to exist — with all the implications for supply security. To this has to be added uncertainty about the future evolution of oil prices — the subject of the following chapters. The arguments in this chapter have largely been based on the assumption of stable real prices of oil — a majestic *'ceteris paribus'*. But leaving this aside for the moment, we can still conclude that only political steps can close the gap between supply and demand. This is also the present situation. Four political variables appear to be responsible for keeping supply from low absorbers, and principally Saudi Arabia, at the level required: the close relationship between Saudi Arabia and the USA (which reflects on most other low absorbers, as well); the interest of Saudi Arabia and other sheikhdoms in the Western economic system and in political cohesion; the ability to pursue

Table 8.4  Aid Disbursed in 1975 and 1976 by some Low Absorbers ($10^6$) (bilateral concessional disbursements only)

| | Kuwait | | | | Saudi Arabia | | | | Libya | |
|---|---|---|---|---|---|---|---|---|---|---|
| | 1975 ($10^6$) | (as % of total) | 1976 ($10^6$) | (as % of total) | 1975 ($10^6$) | (as % of total) | 1976 ($10^6$) | (as % of total) | 1976 ($10^6$) | (%) |
| Egypt | 451.6 | (50) | 107.2 | (29) | 948.9 | (53) | 496.8 | (24) | — | — |
| Syria | 176.0 | (19) | 2.7 | (0.3) | 242.2 | (14) | 189.9 | (9) | — | — |
| Jordan | 114.9 | (13) | 57.4 | (15) | 49.3 | (6) | 165.0 | (8) | — | — |
| Total Arab | 888.5 | (98) | 224.8 | (60) | 1 603.9 | (90) | 1 453.8 | (70) | 4.0 | (14) |
| Pakistan | 5.0 | (1) | — | — | 74.8 | (4) | 67.6 | (25) | 3.0 | (11) |
| Least developed countries | 60.7 | (7) | 42.5 | (11) | 269.2 | (15) | 418.4 | (20) | 8.7 | (32) |
| Total | 910.3 | (100) | 376.3 | (100) | 1 780.0 | (100) | 2 073.7 | (100) | 27.6 | (100) |

Source: Development Assistance Committee: 1977 Review Development Co-operation, OECD Paris 1977

strategies of 'international responsibility' against domestic pressures, which so far have been effective only in Kuwait and Libya; and the prevailing level of national and regional stability in the Arab Gulf countries. All this depends in the final analysis on the particular orientation of the present Saudi leadership; a reversal of policy by the Saudis could change the political factors signficantly, and fundamental domestic change in Saudi Arabia would create a different, and probably much less comforting situation. On balance, political and social transformation with periods of instability seem much more likely than a smooth evolution with unchanged energy policy priorities. It is hard to believe that Saudi Arabia's political institutions will survive unscathed the next 10 years of rapid social change.

CHAPTER NINE

# Price formation and evolution

The third fundamental energy policy objective to be considered concerns the price of imported energy. Conventional wisdom argues that importers of energy are interested in low prices, exporters in high prices; the European Community's interest, therefore, lies in low energy prices, given its large dependence on oil (and, increasingly, natural gas) imports. But this argument deserves some closer scrutiny. Attention is restricted to the price of internationally traded oil, which not only represents the bulk of present European energy imports, but also increasingly sets the price of other internationally traded forms of energy, particularly natural gas, in the longer run probably also coal.

To start our analysis two issues must be considered: the price of oil, and how it is determined. Arguably, no single figure is of greater importance for international economics than the price of crude oil. The quadrupling of prices in 1973/4 has led to a massive transfer of wealth and power on a scale comparable to few other instances in human history. But what *is* this price of crude oil?

There are two clearly distinct elements which make up the price of oil: cost and rent. Costs involve the outlay needed to produce, transport, refine and distribute a barrel of oil, and to return a profit on each of these operations. Since costs at each stage vary with a wide range of factors, average costs will be used to simplify calculations. The difference between the average costs and the market price of a barrel of crude can be considered as a rent — the surplus profit derived from particular advantages in the market. Such advantages can be based on technical factors, or on oligopolistic action. Examples of the first type of rent include quality differences (specific gravity, sulphur content), geographical location (which can reduce transport costs), production costs (influenced by field pressure, rock permeability, size, etc.) and technological differences (economies of scale, superior transport technologies etc.). Oligopolistic rents are surplus profits which can be reaped by the importing governments (which control the market for oil), the exporting governments (which control the resource), and by the oil companies (which manage the market). It is the latter type of rent which is crucial for Europe — for it accounts for most of the difference between average cost and the actual price of imported oil (*Figure 9.1* provides some examples).

---

\* This chapter represents a slightly modified version of an article 'The price of crude oil in the international energy market. A political analysis', published in the June 1977 issue of *Energy Policy*. Permission to reprint this article here is gratefully acknowledged.

Price formation and evolution 201

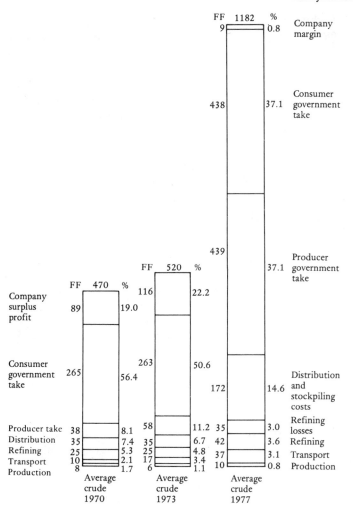

**Figure 9.1** *Breakdown of the cost of a ton of crude oil sold in France (figures in French francs and percentages) (Sources: J-M Chevalier 'Elements théoriques d'introduction à l'économie du pétrole, L'analyse du rapport de force', in Revenue d'Economie Politique, March–April 1975 and Bulletin de l'Industrie Pétrolière, Dec 28, 1977)*

The rent for the producer can be justified as a compensation for the loss of a non-renewable natural resource; this is the original, now all but forgotten, function of royalties. In principle, we could argue that this rent should then be equal to the cost of finding and developing an additional barrel to replace the barrel lost by production. This perspective, however, at best helps to justify demands for a rent, i.e. a surplus profit derived from oil operations in practical terms, though it will be next to impossible to establish a system which meets the theoretical requirement. In practice, whether a rent element can be secured or not depends on the structure of the market. The question of a 'desirable' price level is, therefore, hypothetical — as is Europe's interest in 'low' prices. Both ideas have to be replaced by concepts which take account of market structures. In this respect, Europe's interest lies not with any particular price level; nor is there a justified rent element which — added to costs — can determine the price of oil. The crucial factor is the *capability to influence price decisions*. This can be shown in another way. The real problem of the OPEC price increases in 1973/4 was not the new level, but the speed with which it came. Europe's interest in the price of international energy can thus be reformulated as an interest in avoiding rapid price increases. This can be secured only through some form of influence in the price decision-making. The same applies to the producers' desire to realize a surplus profit to compensate for the loss of a non-renewable asset.

To turn, then, to the second major issue under discussion: how are prices of oil decided? The huge rent element contained in present prices indicate that production costs play an insignificant role. International oil prices are, and always have been, set within a band of economic factors: the production costs at one end, and the costs of alternative sources of oil and energy at the other. At present, this band is very broad: production costs in the Persian/Arabian Gulf are in the range of 10 to 25 cents, in North Africa and Nigeria it is about 16 cents, including average rates of return on investment (figures refer to the early 1970s)[108]. The long-term supply price has been estimated at a maximum of 20 cents (in 1968 dollars) for the period of 1970–1985 in the Gulf[109]. The ceiling set by alternative sources is difficult to define, and indeed hardly exists in the short term. In the longer term, they appear to be between $3 and $10 for offshore oil, and above $20 for most alternative sources of energy such as synthetic and non-conventional oil, solar, and other renewables. For practical purposes, the price ceiling is, therefore, set rather by the capacity of the world economy to adjust to higher prices and the economic and political disruptions caused by oil price increases beyond a certain magnitude.

Most experts have so far focused on the *international oil industry* in their attempts to explain oil price movements over a longer period. Whether this industry is seen as essentially self-adjusting, and therefore amenable to competition (Adelman)[110], or an activity *sui generis* with particular rules and characteristics demanding some form of cartelization (Frankel)[111], or whether the focus is mainly on the large international firms which have dominated the industry since its beginnings[112], the concepts used and the perspectives chosen are essentially economic. Governments, and politics, enter this type of analysis at best at the fringes, and there is a tendency to explain government behaviour in terms appropriate for the oil industry (e.g. income maximization).

Here, a different line is pursued — not to invalidate the economic analyses already undertaken, but rather to supplement them; the main assumption is that political processes offer the chief explanation of price trends in the international oil market. Certainly, those processes unfold within a framework set by economic factors — the costs of production of crude oil (which set a floor to oil prices), and the cost of alternative sources of energy which could substitute for oil. But this band, within which oil price levels can move, is so large that it has lost much of its practical importance; while the price of alternatives to internationally traded oil did at times influence oil price developments in the past in the sense of a fairly effective ceiling, the enormous quantitative expansion of the international oil trade as a consequence of equally dramatic increases in world oil consumption has effectively removed the lid from oil prices. Short- and medium-term supply—price elasticities are so low that present oil prices still appear below their ceiling, and demand—price elasticities offer little more comfort.

Politics, then, determine the level of international oil prices for the time being: *the price of oil is a consequence of the relationship of forces in the international oil market.* This is the central assumption of our model[113].

If we take the international oil market as a system consisting of actors, and of the patterns of transactions and interactions between them, it might be useful first to attempt a classification and identification of the main types. Since the main transaction in the system is the exchange of crude oil (and products) against goods and services, or credit, this could be used as one variable. We would, then, distinguish between actors according to their international oil-trading position. The second variable used here is the stage of development reached by countries participating in the international oil market. On this basis, we can now identify major groups of actors: developed net importers and net exporters of oil, developing net importers and net exporters of oil (for reasons of convenience, we can call the developed importers 'consumers', and the developing exporters 'producers', bearing in mind that those terms are not altogether accurate). As a different, but also important type of actor, there are also the companies. We shall, for the purposes of this paper, ignore the distinction between socialist and capitalist countries, and between countries with a heavy degree of dependence on the international market, and those with low dependence.

Crude oil is exchanged between the different types of countries, at a certain price, mostly through the international oil companies. This price is closely linked to a globally formulated, centrally determined structure of crude oil prices (even discounts and credit terms are only deviations from those prices, and therefore presuppose them). Sometimes the price is simply fixed with reference to the global price, but in most cases it will be subject to bargaining processes. In so far as the terms of reference of those bargaining processes are the globally prevailing price levels, we are again directed towards the central price-setting process in the system. Such global price levels have existed since the early days of the international oil market, and their relation to production cost has at best been tenuous. This price is determined largely by the relationship of forces between three groups of actors: the oil companies, the developed net importers, and the developing net exporters (for reasons which will become clear later, the other two types of actors do not play a significant role in this process).

What factors determine the relationship of forces between these three groups? Chevalier's answer consists of two points: the long-term trend in costs of developing and producing the incremental barrel of oil needed to balance supply and demand; and the degree of evolution of social consciousness in the exporting countries[114]. Both explanations, however, appear unconvincing. His argument that long-term price trends changed fundamentally around 1970 towards rising prices is not necessarily correct. When he quotes the high development costs of North Sea and Alaska oil, he implies that these new reserves have to be developed for economic reasons. In fact, ample low-cost reserves are available in the Middle East; they would suffice to meet world requirements for some time to come[115]. The reason why these reserves cannot be exploited at the economically desirable rate is political[116]. Whether we have indeed entered an era of rising long-term costs is arguable – political reality affirms it, but economically speaking it might well be wrong. The second argument (the evolution of social consciousness) touches, as we shall see, an important point, but is too vague to serve as an explanatory category.

Three alternative sets of explanatory variables are suggested here: control over resources, patterns of interaction, and finally, influences from outside the international oil market. To assess the balance of force, the relative strength of an actor in the system, we have to evaluate his position using these three criteria. Once we have established relative strengths and weaknesses, we can proceed to analyse price interests of the prevailing actors, and draw conclusions about the development of oil prices.

Control over resources involves, primarily, crude oil exports and reserves, but also markets. Quantities of exports, market shares, size of reserves, are obvious indicators. One less obvious indicator is the ratio between oil imports/exports and domestic economic development, in other words: the importance of oil exports/imports for the functioning of the economy and society concerned. This, in itself, constitutes a balance if we take both exporters and importers: if the dependence of one side is greater than the other, this introduces an element of power in the system. Such an imbalance obviously exists now – as indicated by the accumulation of large oil revenue surpluses.

Many other resources are equally important in determining the strength of an actor in the international oil market. Among these are distribution and processing facilities, capital, and above all, technical and managerial know-how. In all but the last case, it is relatively easy to assess the distribution of control over these resources in quantitative terms; even for the last resource, a rough estimate of distribution is possible.

Distribution of resources goes a long way to explain the relative position of strength or weakness of an actor in the international oil market (and it now becomes clear why developed exporters and developing importers cannot play an important role in the bargaining processes – they lack resources). They do not, however, reflect fully the weight of an actor. A second crucial aspect is the degree of co-operation among actors, and in particular among actors of the same category. Co-operation can lead to a pooling of resources, and thereby contribute significantly to power. An analysis of patterns of interaction in the international oil market is, therefore, at least as important as an assessment of the distribution of resources. Two aspects stand out: the degree of co-operation or non-cooperation between actors of the same type, and the resulting overall structure of the system. The relevant structural patterns resulting from co-operation in the system could be termed *hegemonial and egalitarian*; those two patterns are ideal-type structures, with reality normally representing an intermediate pattern. In a hegemonial structure one actor, or one group of actors, dominates the system by occupying a centre position. This means the actor (or the group) interacts with the other actors in the system individually while the latter do not interact at all, or with much lower intensity. A less abstract way of describing such a structure is the old notion of 'divide and rule'. An egalitarian structure, on the other hand, is based on equally intensive interaction between all actors, or where all different groups of actors are equally well integrated. The latter case resembles the situation in industrial conflict, where both labour and management/capital are organized. In the case of the international oil market, with at least three groups of actors with substantial weight (developed importers, developing exporters, companies), an egalitarian structure will tend to produce alliance and alignment politics.

Hegemonial or egalitarian patterns can also dominate the structure within one group of actors — co-operation can be based on hegemony or equality. The latter is presumably more difficult to achieve, but potentially more stable. To a certain extent, control over resources also influences patterns of interaction: OPEC for instance, is basically an 'egalitarian' organization, but Saudi Arabia enjoys a special position because of her resource base.

External influences are the third set of variables influencing the distribution of power in the international oil market. Since this system can be seen as a sub-system of the wider international system, positions of rank, status, power and wealth of actors in the system affect their relative position in the international oil market as well. Similarly, structural patterns in the international system can affect relations in the oil market; e.g. between the USA and Saudi Arabia. Finally, values and ideologies have played an important role in shaping the international oil market, and the interaction of its actors — nationalism, neutralism, anti-imperialism, are obvious examples.

This briefly sketched model drawing together some explanatory variables affecting the relationship of forces in the international oil market tells us essentially which actors, or group of actors, dominates the market at any given time. Trends in the shift of resources, as well as in structural patterns, will also allow us to suggest the future development of the relationship of forces. While changes in prices might not always be perfectly synchronized with changes in this relationship, since the changes might take some time to be perceived by the actors in their full implications, in the longer run underlying shifts will have their impact on prices of crude oil.

If a rough balance exists in the international oil market, then price changes can be expected to be minor. Once we are dealing with a significant imbalance, then a crucial question for the resulting price changes is the interests and objectives of the dominant group of actors, or the dominant actor. The first part of our deliberations has dealt with the *capabilities* of actors to influence price levels in the international oil market; now we have to address *intentions*, or the price strategy pursued. Such strategies are complex; size and costs of reserves are only one aspect, though an important one: elasticities of alternative oil and energy supplies, and of oil demand, in the longer run, have to be assessed, and such estimates will influence price strategies for the exporters. Similarly, importers' strategies will be influenced by the cost of development of indigenous energy production, as well as by the economic advantages of low-cost energy imports.

But apart from these and other factors related directly to the international oil market, price strategies are influenced by a number of variables outside this market as such. These variables will be economic and political, and result from pressures within the actors' structure, as well as from outside. All these aspects tend to modify the fundamental interests of exporters and importers. In the case of an exporting country, the reserve position might mitigate the high-price strategy, while domestic pressures could tend to push up the price objective. Similarly, security considerations might tend to advocate higher price levels in importing countries, while consumer pressure would tend to favour lower prices. This picture could be further complicated by the differential impact of different price levels of actors of the same group, or from other groups. Exporters might moderate their price policies because of the relatively greater burden higher prices put on less developed countries or weaker industrialized countries. Alternatively, Saudi Arabia might oppose drastic price increases because this could tilt the regional balance of power in the Persian/Arab Gulf in favour of Iran and Iraq, which can utilize oil revenues more directly to increase their strength. This appeared to be one of the objectives of the Saudi Arabian production increases after the OPEC meeting in December 1976, which led to a price split in the international market[117].

The protection of high-price domestic energy production in spite of large disadvantages *vis-à-vis* oil imports could lead to disparities in the energy costs for different industrialized countries, and therefore to distortions in competitiveness[118]. The more important high-cost domestic production, the greater the interest in high international prices, if security reasons advocate the maintenance of domestic production. Importers should also be concerned about longer-term stability (or predictability) of international prices to allow domestic energy planning (an interest shared with the exporters).

Apart from such economic considerations, the oil price strategy of the developed net importers has political ingredients, as well. The importers might also be prepared to modify price objectives for the sake of alliance, because of international political considerations, and out of concern about the repercussions of low oil prices on the exporters.

One of the basic objectives of the oil companies is the long-term maximization of profits, and therefore a price level above production cost (including an average return on investment), to create additional investment capital to find new oil, and thereby ensure future profitability. This already implies a second objective: the self-preservation of the company in the longer run. The implications of this are not necessarily high, or rising, prices as the main objective: profits are a function of quantities supplied as well as of price per unit. In one sense, the company combines both the interests of the producer and the consumer. Given the fact that the oil companies possess few 'natural' assets in the international oil market (in the way oil fields are the natural assets of the exporters, and markets those of the importers), control of, or influence in, the market is of overriding importance. Price policies would then tend to become a function of the quest for control, while inversely the search for control would be a function of the price interests in the case of exporting and importing countries. This implies that the international oil companies would pursue a policy of stable or even lower prices if their position is to be strengthened at the market end, and of higher prices if it were to be expanded at the production end. The former is a strategy of replacing other sources of energy and the capturing of incremental energy demand; the latter implies either association with producer countries or development of alternative sources of oil and energy supply.

Let us now apply those theoretical concepts to the evolution of international oil prices since 1948. *Table 9.1* shows three major turning points in the price evolution of internationally traded oil – a fall in 1948 and 1949, a second fall in 1958–1959, and a rise around 1970. Each of these turning points marks the beginning of a different era in the international oil market (although changes are obviously in reality somewhat less tidy): an era of slowly rising prices during the 1950s, of falling prices throughout the 1960s, and of dramatic increases in the first half and again in the late 1970s.

How can these developments be understood in the light of the previous discussion? The first period until 1958–1959 could be called the era of the *majors*, for it was a group of eight large integrated oil companies which dominated the international oil market: Standard Oil of New Jersey (now Exxon), Royal Dutch/Shell, the Anglo-Iranian Oil Company (now BP), Mobil Oil, Gulf Oil, Texaco, Standard Oil of California, and the Compagnie Française du Pétrole. This dominating role was based on two factors: the share of those companies in the resources of the international oil market, and the structure of interaction in the market.

After the initial decline in oil prices in 1948–1949 (essentially a consequence of pressure by home governments of the companies, and of the integration of an enormously rich new oil-exporting region – the Middle East – into the international market)[119], the system reached a somewhat precarious equilibrium managed by the majors. They held a very large share of crude oil production and reserves destined for the international market (the Middle East, Venezuela), and also

Table 9.1 *The Evolution of International Oil Prices*

(A) Posted prices (in $/b, 1 Jan)

| | Saudi Arabian 34° API | Kuwaiti 31° API | Venezuelan 35° API |
|---|---|---|---|
| 1948 | 2.06 | (1.85) | 2.65 |
| 1949 | 1.81 | (1.60) | 2.65 |
| 1950 | 1.71 | (1.50) | 2.65 |
| 1951 | 1.71 | (1.50) | 2.65 |
| 1952 | 1.71 | (1.50) | 2.65 |
| 1953 | 1.81 | 1.67 | 2.90 |
| 1954 | 1.93 | 1.72 | 2.90 |
| 1955 | 1.93 | 1.72 | 2.90 |
| 1956 | 1.93 | 1.72 | 2.84 |
| 1957 | 1.99 | 1.80 | 3.05 |
| 1958 | 2.06 | 1.85 | 3.05 |
| 1959 | 1.90 | 1.69 | 2.84 |
| 1960 | 1.87 | 1.65 | 2.80 |
| 1961 | 1.80 | 1.59 | 2.80 |
| 1970 | 1.80 | 1.59 | 2.80 |
| 1971 | 1.80 | 1.68 | 2.80 |
| 1972 | 2.29 | 2.19 | 3.21 |
| 1973 | 2.59 | 2.48 | 3.21 |
| 1974 | 11.65 | 11.55 | 14.88 |
| 1975 | 11.25 | 11.15 | 14.66 |
| 1976 | 12.38 | 12.260 | 16.17 |
| 1977 | 12.09 | 12.37 | 13.99 |
| 1978 | 12.70 | 12.37 | 13.99 |
| 1979 | 13.34 | 12.83 | 14.69 |

Sources: C. Issawi and M. Yeganeh, *The Economics of Middle Eastern Oil*, Faber, London 1962; M. Field, 'Oil and the Middle East and North Africa', *The Middle East and North Africa 1976–1977*, Europa Publications, London 1976, pp. 73–108, Comité Professionnel du Pétrole, Pétrole 1975, Paris 1976; Petroleum Economist

(B) Delivered prices

Crude oil delivered to Japan, weighted average, landed price ($/b)

| Source of crude | 1958 | 1960 | 1962 | 1964 | 1966 | 1968 | 1970 | 1972 |
|---|---|---|---|---|---|---|---|---|
| Iran | 3.26 | 2.38 | 2.16 | 2.07 | 1.90 | 1.86 | 1.76 | 2.30 |
| Iraq | 3.23 | 2.40 | 2.26 | 2.19 | 2.02 | 2.06 | – | 2.50 |
| Kuwait | 2.97 | 2.35 | 2.07 | 1.98 | 1.85 | 1.87 | 1.80 | 2.37 |
| Saudi Arabia | 3.42 | 2.40 | 2.16 | 2.07 | 1.96 | 1.99 | 1.84 | 2.34 |

Crude oil delivered to six European markets, average landed price ($/b)

| Importer | 1953 | 1957 | 1961 | 1965 | 1972 |
|---|---|---|---|---|---|
| Belgium | 3.45 | 3.86 | 2.60 | 2.36 | 2.86 |
| France | 3.00 | 3.83 | 2.66 | 2.54 | 3.05 |
| Germany | 3.21 | 3.72 | 2.56 | 2.13 | 3.01[b] |
| Italy | 2.81 | 3.34 | 2.07 | 1.89 | 2.81 |
| Netherlands | 3.12 | 4.10 | 2.82 | 2.40 | 2.88 |
| UK | 3.19 | 3.78 | 2.70 | 2.46 | 3.02 |
| Weighted average | 3.08 | 3.75 | 2.56 | 2.27 | 2.95 |

a Government selling price (originally equalling 93% of posted prices)
b 1971
Source: N.H. Jacoby, *Multinational Oil*, Macmillan, New York 1975

controlled important segments of the international tanker fleet and the world refinery capacity. Similarly, the companies possessed a valuable though less easily quantifiable resource: knowledge. The petroleum industry in the developing exporting countries was still largely run by company men; the number of experts outside the USA and the majors was limited, and in the case of developing exporters, extremely small.

Of these resources, control over production and reserves was the most decisive. This control rested on the concession system which gave the companies full sovereignty over the oil resources in the main exporting countries in exchange for a bonus payment for the initial granting of the concessions, a dead rent during the period of exploration, and royalties and/or a share in company profits. This sovereignty was based on the observance of contracts, and therefore rested on somewhat shaky foundations; in fact, however, the contracts were kept in force by the underlying structure of the international oil system which we shall discuss later. (A summary of the control of the majors over resources in the international oil market can be found in *Table 9.2*).

The next strongest group of actors in terms of resources was in this period the importing countries (or more precisely, some industrialized importers, especially the USA, which had just joined this group)[120]. Their main asset was their control over markets, and over indigenous energy resources with which oil had to compete. It was pressure from these countries, and in particular from the USA through the Economic Co-operation Administration in charge of the Marshall Plan for Europe, which had led to the lowering of prices of Middle East crude oil in 1948—1949, and the establishment of a new, dual pricing system with two basing points — the Mexican Gulf and the Persian Gulf[121].

The structure of interaction in the market during this period was essentially hegemonial: the dominating elements were tacit co-operation of the majors, and the support of the US government for their regime in the international oil market. All majors were involved in joint concessions with interlocking directorates creating an informal web of communication and co-operation;

**Table 9.2** *Distribution of Resources in the International Oil Market: the Share of Eight Majors in Total Assets (excl. USA, socialist bloc)*

|  | Reserves of crude oil ($10^9$ b) | Production ($10^3$ b/d) | Refining capacity ($10^3$ b/d) | Tanker capacity[a] ($10^3$ dwt) | Product sales ($10^3$ b/d) |
|---|---|---|---|---|---|
| Absolute size of share of majors, 1953 | 75.9 | 4 576 | 3 844 | 10 014 | 1 299 |
| As % of total | (95.8) | (90.2) | (75.6) | (29.1) | (74.3) |
| Absolute size of share of majors, 1972 | 370.4 | 23 177 | 18 155 | 43 700 | 13 492 |
| As % of total | (71.1) | (74.0) | (50.8) | (20.0) | (58.3) |

a *Owned or managed. The figure including long-term chartered tonnage would be much higher; in 1972, their share in the world tanker fleet excluding socialist countries was 62.4%*
Source: *As for Table 9.1*

besides, their interest in avoiding price wars coincided and provided — in spite of competitive elements in their relationships — a basis for co-operative management[122]. The artificial, highly centralized price structure could, therefore, be maintained throughout the period, and posted prices even increased somewhat, helped by the upsurge in demand in Europe, the Korean war, and the closure of the Suez Canal in 1956. The USA (which is the home country of five of the eight companies, with a sixth, Shell, having a major US component), lent her support to the operations of the majors with her political, economic and military power; the other industrialized countries found themselves after the end of the World War in something of a clientele position. US power essentially also underpinned (together with the British presence in the Persian Gulf) the management system of the majors at the crucial point: the control over production and reserves. The Iranian failure to pull off the nationalization of her oil fields, and the fall of the government of Mossadegh, engineered with substantial support from the CIA, demonstrated the determination of the USA not to accept any radical reversal of the concession regime[123].

In this system, two dimensions of price conflict became apparent: conflict over the *size* of the surplus profit realized by the majors from their operations in developing countries, and conflict over the *distribution* of this surplus profit. The former was essentially a conflict between companies and importers, the latter between companies and exporters. In the first case, the role of the USA was crucial: she initiated the cuts of Middle East oil prices in 1948–1949, but also helped the majors to re-establish full control and price stability after this by protecting domestic oil production through voluntary import restriction schemes[124]. The high cost of domestic oil production in the USA, and the effective lobbying efforts of its operators, produced a US strategy favouring the fairly high level of international crude oil prices prevailing in the 1950s, including acceptance of two price rises in 1952/3 and 1957. Given the relatively small share of operating costs in total oil prices, and their stability, the evolution of posted prices gives a good indication of the evolution of the size of the surplus profit. It is worth noting, however, that in the latter half of this period the companies started to give importers rebates over posted prices, thereby reducing the size of the surplus profit.

The second dimension of conflict was between exporters and companies. The producer governments began to demand a higher share in company profits from their oil, and received it, at least to some extent. In the Middle East, this trend was started by the 50:50 profit-sharing agreement between Aramco and Saudi Arabia; it is noteworthy that for political reasons the US government helped this agreement to come about: it accepted a deduction of US oil company tax payments to host governments from their income tax liabilities in the USA, which meant effectively that the US taxpayer paid Aramco's taxes to the Saudi government[125].

At the same time, the improvement in the distribution of surplus profits between companies and producer governments was for the latter not as dramatic

**Table 9.3** *Distribution of Surplus Profit: Government and Company take, 1948 –1974 ($/b)*

|  | Middle East | | Venezuela | |
|---|---|---|---|---|
|  | Government take | Company take | Government take | Company take |
| 1948 | 0.34 | 1.52 | 0.85 | 0.70 |
| 1949 | 0.31 | 1.14 | 0.71 | 0.47 |
| 1950 | 0.37 | 1.20 | 0.61 | 0.57 |
| 1951 | 0.41 | 1.27 | 0.57 | 0.63 |
| 1952 | 0.53 | 1.11 | 0.76 | 0.62 |
| 1953 | 0.66 | 1.09 | 0.75 | 0.63 |
| 1954 | 0.72 | 1.08 | 0.74 | 0.66 |
| 1955 | 0.76 | 0.97 | 0.76 | 0.70 |
| 1956 | 0.78 | 0.97 | 0.82 | 0.76 |
| 1957 | 0.80 | 0.98 | 0.94 | 0.88 |
| 1958 | 0.79 | 1.05 | 1.01 | 0.55 |
| 1959 | 0.79 | 0.88 | 0.96 | 0.42 |
| 1960 | 0.74 | 0.87 | 0.99 | 0.45 |

|  | Eastern hemisphere | | | |
|---|---|---|---|---|
|  |  | Government take (average) | | |
|  | Company take (average) | Saudi | Kuwait | Iranian |
| 1960 | 0.565 | 0.750 | 0.765 | 0.801 |
| 1965 | 0.418 | 0.832 | 0.789 | 0.811 |
| 1970 | 0.336 | 0.883 | 0.828 | 0.862 |
| 1971 | 0.341 | 1.259 | 1.197 | 1.246 |
| 1972 | 0.283 | 1.437 | 1.409 | 1.358 |
| 1973 | 0.698 | 1.789 | 1.763 | 2.012 |
| 1974 | n.a. | 8.986 | 7.734 | 8.397 |

Sources: 1948–1960: As for Table 9.1. 1960–1974: D.A. Rustow and J.F. Mugno, OPEC, Success and Prospects, New York University Press, New York 1976.

as it might appear: the companies were able to secure various rebates and allowances from producer governments, thereby shifting the actual distribution to something like 32:68 in favour of the companies[126]. Nevertheless, the figures presented in *Table 9.3* show that the share of the producers in profits per barrel grew throughout the period, while the share of the companies fell — a first indication that shifts in the relationship of forces were beginning in the international oil market.

The equilibrium characterizing this period was indeed a precarious one: several factors of change began to gain momentum, and ultimately severely affected the international oil market. These forces, in order of importance, were the intrusion of newcomers, latent conflict among the majors themselves, and competitive oil policies among importing governments.

The newcomers were US oil companies faced with a competitive disadvantage *vis-à-vis* the majors as a consequence of higher production costs in the USA, and attracted by the huge profits which could be made with Middle Eastern (and Venezuelan) oil in the rapidly expanding markets of Western Europe, and indeed of the USA, where stock regulation to support the high domestic price structure produced a widening gap between indigenous supply and demand[127]. After the Second World War, a number of such 'independent' oil companies made their entrance in Venezuela and the Middle East. While in 1945, 28 US companies were engaged in oil exploration and/or production abroad, the number had increased to 190 in 1958[128]. The proliferation of internationally operating companies first led to heightened competition for concessions (and therefore improved terms for the producer governments), then, inevitably, this competition also moved downstream as soon as the independents were successful in their search for oil — they now had to force their way into markets.

Increased competition between independents and majors at all stages of the industry, however, destabilized the relationship among the majors, too. Their interaction at this point was a combination of collusive and competitive moves. While all had a stake in the maintenance of the prevailing system, each company also sought to profit from the rapidly expanding markets, and to increase its share. Moreover, the concession system left some companies with more crude oil than they could distribute through their own integrated channels, while others had to buy additional crude to satisfy the demand of their subsidiaries. Although this imbalance was eased by long-term supply arrangements among the majors, the difference of interest persisted, and it needed only an outside trigger to unleash some of the competitive tendencies within the oligopoly into price competition[129].

Finally, in the late 1950s and early 1960s, another set of actors appeared on the scene: the national oil companies of France, Italy and Japan. These companies were an expression of differences of interest among the importers: countries without their own majors saw an independent national attempt to secure sources of supply abroad as the best way to minimize costs and maximize benefits to their own economies. These national companies began to compete for concessions (introducing radically new concepts such as the joint venture or the contract agreement with producer governments, undermining the concession system altogether)[130] and to add to the volumes of oil traded internationally. At about the same time, another actor appeared on the market: the USSR started to export oil to Western European countries, contributing to oversupply.

Changes within the international oil market were partly affected also by changes in its international environment: the global rivalry between the two superpowers, the USA and the USSR — deadlocked at the central front in Europe through the nuclear stalemate — shifted more and more into the Third World, opening possibilities of independent action for developing countries. The rapid economic recovery of Western Europe and Japan led to greater political freedom of manoeuvre *vis-à-vis* the USA; while the Western alliance maintained a considerable degree of cohesion and leadership, some Western European countries and Japan began to develop their own energy policies, which found their manifestation in the already mentioned national oil companies.

The forces of change eventually led to a fundamental restructuring of the international oil market. This evolution began with a crucial US decision: the establishment of import controls in 1959, which virtually blocked a number of independents from the American market, and forced them to turn to Europe. The results were price reductions, first informally through discounts and rebates, then formally through a reduction in posted prices. This affected the income per barrel of the producer governments; their anger and frustration led directly to the foundation of OPEC in 1960.

The turning point of official price reductions in 1959–1960 was the beginning of a period of transition in the international oil market, which lasted until 1970. If we analyse this period in terms of distribution of resources, the most marked development is the reduction of the shares of the majors in several key sectors, mainly in favour of the independents and national oil companies, but also of the producer governments and their national oil companies (see *Table 9.2*). At the same time, the position of the producer governments was strengthened by a proliferation of know-how: voluntary and compulsory training schemes of the companies, and related efforts of the governments concerned, began to bear fruit, as did improved co-ordination and communication among producer countries. The weakening of the concession system also meant that producer governments expanded their control over production. In terms of patterns of interaction, the following observations can be made: the tacit collusion of the majors to maintain the established price structure continued, but its burden had significantly increased. The majors found themselves pressured from all sides – the producer governments demanded a higher share of the surplus profits, consumer governments wanted rebates and discounts, thereby demanding a reduction of the total size of the surplus profit, and the independents competed fiercely to secure market outlets. The existing surplus in the market forced the majors to step up competition among themselves. Nevertheless, they managed to achieve a rough balance between supply and demand at price levels not too far away from those of the 1950s, which still allowed substantial surplus profits since operation costs were way below the prices achieved.

A new element of interaction appeared with OPEC – the attempt to organize co-operation among producers. While OPEC managed to prevent further falls in posted prices, and to secure some further shifts in the distribution of the surplus profits in favour of producer governments, co-operation was still far from perfect, as the futile attempts to introduce production quotas in order to control the market and raise prices showed. The two programmes passed by OPEC in 1964 and 1965 ended in total failure[131]. This conclusion is corroborated by producer behaviour during the 1967 crisis when some Arab producers cut back supplies to the West: non-Arab producers, and in particular Iran and Venezuela increased their output, thereby blunting this attempt to use the 'oil weapon'[132].

A third structural change of importance has already been noted: the reduced cohesion of the industrialized consumer countries as a consequence of the declining hegemony of the USA in the Western alliance.

These changes combined to weaken substantially and continuously the

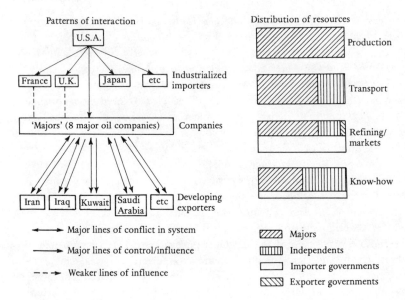

**Figure 9.2** *Patterns of interaction and distribution of resources in the international oil market, 1950–1960*

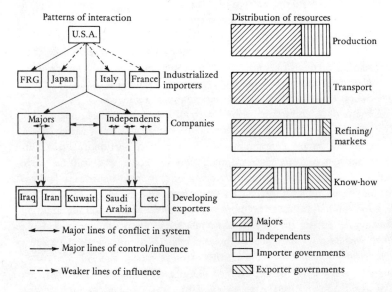

**Figure 9.3** *Patterns of interaction and distribution of resources in the international oil market, 1960–1970*

previously prevailing hegemonial structure of management in the international oil market. (*Figures 9.2* and *9.3* illustrate the changes in terms of control over resources and patterns of interaction.) To some extent, the centralized management of the majors remained, as witnessed by the stability of posted prices, and the relatively slow decline of delivered prices; US support for this system also continued, although the evolution of international politics had reduced the effectiveness of this support. Producer cohesion and co-operation was not yet strong enough to replace the existing management structure of the market, but sufficed already to insulate them from the decline in delivered prices, as *Table 9.3* shows: their income per barrel increased, although only slightly, and their share in the surplus profit grew substantially.

In fact, the conflict between producers and companies over the distribution of the surplus profit, and increasingly also over control of rates of production led to an interesting consequence: the establishment of a loose bipolar oligopoly between the majors and OPEC. The modest success of OPEC (phasing out of royalty expenses and of discounts) were negotiated between those two groups, without consultations with other producers or the independents[133]. A new alignment began to appear in the market.

The weakening of the old structure of management established after the Second World War created strong pressures towards the establishment of a new system of control. The transitional character of the 1960s was essentially a result of the entrance of a number of new actors in a situation providing strong incentives for such a proliferation. But this, in turn, led to developments which reduced competition: profit rates of the industry declined, the chances for striking bonanzas disappeared as the host governments pursued their interests more vigorously[134]. Once established, the newcomers had no longer an interest in price competition, but in the re-establishment of a stable, remunerative level of oil prices. Competition in this situation could only be introduced by the producers — and for them, as well, joint action promised much higher rewards.

In the negotiations in Teheran and Tripoli early in 1971, the predictable realignment of forces did indeed take place. By that time, a series of developments led to the establishment of OPEC control over prices and production, particularly the tight supply—demand situation. Libya's successful attack on the oil companies had led to substantial improvements in benefits derived from her oil exports, and apparently also a change in US policy. Previously, the USA had, because of her high-cost domestic oil production, suffered a competitive disadvantage *vis-à-vis* Western Europe and Japan who relied on cheaper oil imports. The Nixon government therefore seems to have abandoned the implicit support for prevailing price levels, and more or less openly encouraged OPEC to press for higher prices[135]. In Teheran and Tripoli, the producers effectively began to set prices, and to assume full control over their oil resources by threatening supply interruptions if the companies would not meet their demands. The unilateral setting of prices since 1973, and the actual embargoes and supply cutbacks in 1973—1974, only underlined this fundamental change. In terms of resources, OPEC had now acquired a very strong position: in 1974, it controlled 79.3% of the international trade in oil, and 64.5% of world oil reserves. It also began to move downstream into the tanker and refining business, and even to distribute oil to final consumers. The chief remaining assets of the companies lie in the intermediate stages of the oil business, and in their know-how. The consumers retain their ultimate control over markets and over alternative sources of energy which could replace oil in the future (in both cases, those assets are *de facto* shared with the companies). For a variety of reasons, the companies also were not unfavourably disposed to higher oil prices[136].

In terms of patterns of interaction, the crucial dimension was until 1979 clearly collusion among OPEC producers: even during a period of falling demand for oil, this collusion proved reasonably stable. The cohesion of OPEC has been enhanced by co-operation with the international oil industry which continued to fulfil the primary management function in the market — the balancing of supply and demand. This loose bipolar oligopoly is reinforced by the special status the former concession companies still enjoy: long-term supply contracts at reduced price levels, involvement in development projects in the host country, etc, which together offer attractive possibilities for profits, and the only way to maintain vertical integration of their operations.

Within the industry, cohesion was also strengthened after 1970 in several respects: there was no longer much incentive to break into the production stage, or into new markets, by lower prices. Competition was, therefore, shifted to the final distribution stage of the industry, where companies had in the past followed a strategy of maximizing outlets for crude oil available in large quantities and at low cost[137]. Emphasis was now placed on the profitability of all operations of the integrated companies; this was reflected in some cases even in a far-reaching restructuring of company presence in the industrialized markets[138]. The industry had now become fairly cohesive again in the upstream stage, as witnessed by the membership of the company negotiating group dealing with OPEC in Teheran and subsequent negotiations, which included the former independents. From the point of view of the companies, the shift in control towards the producer governments was certainly not desirable in so far as it implied a loss of their capacity to regulate the international oil market, but given the absence of any short-term alternative, they had to seek the most profitable and stable arrangement available to them. This meant co-operation with OPEC in the form of a bipolar oligopoly.

The industrialized net importers were now the main losers (a fact clearly borne out by the capital flows associated with the international oil trade, and even by changes in the distribution of the surplus profit: although the importing governments continued to secure a share through taxation of petroleum products, this share has been reduced considerably since 1970). Interestingly, this loss of control (which the consumers had exercised before 1970 through the oil companies by and large satisfying their interests) led to the establishment of an organization of the main industrialized importers (excluding France, which, however, has become something like a tacit partner): the International Energy Agency. It is noteworthy that now all major actors in the international oil market are organized: the egalitarian structure has evolved fully.

While the distribution of resources is at present favouring the developing net exporters organized in OPEC, the structure of the interaction reinforced this advantage further by fairly effective collusion of net exporters, and a loose alignment between exporters and the international oil industry growing out of a common interest in the maintenance of a system of oligopolistic management rather than competition, and the mutual profitability of such co-operation. The net importers originally confronted this bipolar oligopoly in an atomized and bilateral way, thereby failing to use effectively their resources. But even the present situation still seems to favour the exporter–company alignment because of a powerful imbalance between the need of the importers for sufficient oil supplies, and the need of the exporters for revenues: the latter is much lower, and can be secured through price increases, anyway, as long as the oligopoly works and price levels do not allow massive substitution of oil by other sources of energy. (the 1970–1978 market structure is described in *Figure 9.4*). The year 1979 ushered in an important new phase through the disintegration of OPEC co-operation and a further weakening of the companies (see Postscript).

The main axis of conflict in the system has clearly shifted since 1970: it lies now between producers and consumers, and mainly concerns the *size* of the surplus rent – or in other words, the price level. The developing oil exporters at present clearly command a very strong position both in terms of resources, and in terms of patterns of interaction. But the domination which OPEC has

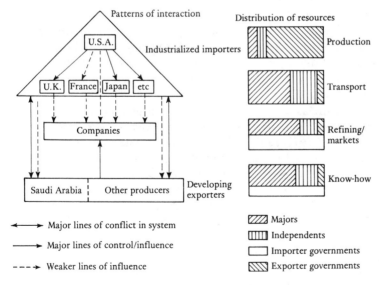

**Figure 9.4** *Patterns of interaction and distribution in the international oil market, 1970 to 1978*

established in the system poses problems – those of coalition politics. The cohesion of the developing oil exporters as a group, their co-operation in acquiring market control, depended mainly on identity of interest in the previously prevailing structure: higher prices, more market control. Now, OPEC has to sort out a new situation – one in which control has been established, and where differences of intention within OPEC will gain importance. These differences regarding preferable price levels are already becoming apparent; they reflect different resource positions, but also different political objectives outside the international oil market. This can complicate decision-making within OPEC, as the OPEC meeting in December 1976 demonstrated, and could even weaken OPEC cohesion altogether, as events in 1979 have demonstrated. This cohesion rested on Saudi hegemony within OPEC – a hegemony which was based on weak foundations (see Postscript).

The differences within OPEC first led to heated argument in Vienna, when the OPEC oil ministers met from 24 to 27 September 1975. The bargaining between OPEC 'hawks' (Iran, Iraq, Libya, Nigeria) and 'doves' (Saudi Arabia, supported by some other Gulf states) eventually ended with a compromise on the price of the marker crude, which was increased to $11.51/b. No agreement was reached on price levels for other types of crude oil, and market forces led to considerable fluctuations in production levels. On this question, an informal understanding was reached at the OPEC conference in Bali (28–29 May 1976). The price of the marker crude remained frozen, however, until December 1976, when the differences within OPEC over the desirable price level resulted in the first formal split in the organization: 11 of the 13 member states decided on a 10% price rise, to be followed by another 5% increase in June 1977; Saudi Arabia and the United Arab Emirates increased their prices by only 5% to $12.09/b (the other countries increased their prices – hypothetically, since only Saudi Arabia was actually selling the marker crude – in line with $12.70/b for Arabian Light). Saudi Arabia also announced its intention to raise production, which hitherto had been subject to a production ceiling of $8.5 \times 10^6$ b/d, in order to enforce her price policy. This split came to an end in June 1977, when a compromise was reached: Saudi Arabia and the UAE brought their prices in line with those of the other 11, while the other countries waived the price increase scheduled for the second half of 1977. Saudi Arabia had been unable to increase production significantly because of technical difficulties and bad weather in the Gulf; this no doubt was an important factor in the eventual compromise. More generally, it could be seen that even Saudi Arabia cannot fully dominate OPEC – in spite of her powerful position inside the organization. The split also has to be seen in its proper perspective – any interpretation of those events as the first sign of OPEC's decline would be misfounded. In fact, the opposite interpretation would appear to be closer to reality: OPEC's control over the international oil price has been so firmly established against any interference by other actors in the system that internal differences and divergencies of interest could come to the fore. In other words, OPEC had ceased to be on the defensive, or even in the sensitive period of establishing control — it was now clearly in charge of pricing decisions.

The decision in mid-December 1978 in Abu Dhabi to increase prices during 1979 in four stages at an annual average of 10%, with the level at the last stage of increase (1 October) amounting to a 14.5% increase over the one prevailing during 1978 (the new price of marker crude was to be $14.542/b by that time) only confirmed those conclusions. OPEC had again demonstrated its control over the world oil market. The year 1979 brought, however, a dramatic disintegration of OPEC cohesion, and supply interruptions and then lower exports from Iran ushered in a new phase in the world oil market — one of a fragmentary political order. Three OPEC conferences during 1979 proved increasingly clearly that OPEC had lost its grip, abdicating responsibility for oil prices to its individual member countries. As a result, prices more than doubled on average until the end of 1979.

This brief survey of the evolution of oil prices in the international market since the end of the Second World War appears to underline the usefulness of our model. The dramatic increase in the size of the rent, as well as the change in its partition between producer governments and oil companies, should be understood as the result of a fundamental shift in the balance of forces: OPEC asserted control over the bulk of internationally relevant oil resources and world oil trade, and at the same time profited from, and participated in, the dissolution of the hegemonial structure of interaction, and the establishment of a new, more balanced and egalitarian structure in which the resource advantage of the producers, and the power element inherent in the unequal distribution of benefits resulting from international trade in oil, gives the OPEC countries a decisive advantage. *Table 9.3 and Figures 9.2–9.4* illustrate this evolution.

CHAPTER TEN

# Prices and control in the international energy system: the prospects

As we have seen in the previous chapter, oil prices have historically been the result of complex bargaining processes in a field structured by the relationship of forces between the main groups of actors. An extrapolation of these forces into the future – hazardous as it may be – could provide some indication of the future evolution of control over world oil prices. This chapter will, therefore, assess the resources and strategies, the capabilities and intentions of actors in the international energy system. Again, the focus of the analysis will be on the three major groups of actors: the private oil companies, the producer governments, and the developed oil importers. The latter are presented through its pivotal power, the United States, and through their organization, the International Energy Agency.

## The private oil companies

Since 1970, any distinction between the majors and the independents as actors with distinct interests in the international oil market has been dissolved. The companies can, therefore, now be considered a fairly homogeneous force in the international oil market, notwithstanding continuing elements of competition. (See *Tables 10.1 and 10.2* for a description of the major companies.)

The private oil industry is transnational in character, and it *behaves* like an international force, i.e. independently of the national interests of the home

Table 10.1  *The Major Oil Companies in 1977*

| Company | Net income ($10⁶) | Capital expenditure ($10⁶) | Value of fixed assets ($10⁶) | Crude supply (10⁶ b/d) | Product sales (10⁶ b/d) | Natural gas sales (10⁶ ft³/d) |
|---|---|---|---|---|---|---|
| Exxon  | 2 423 | 3 569 | 20 538 | 5.343 | 5.266 | 10 488 |
| Gulf   |   752 | 2 068 |  8 332 | 1.687 | 1.669 |  2 470 |
| Mobil  | 1 005 | 1 285 |  9 043 | 2.370 | 2.299 |  3 240 |
| SoCal  | 1 016 |   899 |  6 199 | 3.403 | 2.455 |  1 505 |
| Texaco |   931 | 1 248 |  9 897 | 3.933 | 3.227 |  3 883 |
| BP     |   599 | 1 420 |  6 497 | 3.540 | 1.875 |    350 |
| Shell  | 2 640 | 4 377 | 16 038 | 4.847 | 4.676 |  6 674 |

Source: *Petroleum Economist*, May 1978

**Table 10.2** *Other Integrated Private Oil Companies ('Independents')*

| Company | Net income ($10⁶) | | |
| --- | --- | --- | --- |
| | 1975 | 1976 | 1977 |
| Atlantic Richfield | 350.4 | 572.5 | 702 |
| Cities Service | 137.7 | 217.0 | 210 |
| Continental Oil | 330.9 | 460.0 | 381 |
| Getty Oil | 256.7 | 258.5 | 309 |
| Marathon Oil | 128.1 | 195.8 | 197 |
| Occidental Petroleum | 172.0 | 183.0 | 215 |
| Pennzoil | 106.8 | 131.2 | 115 |
| Phillips Petroleum | 342.6 | 412.0 | 517 |
| Standard Oil (Indiana) | 787.0 | 893.0 | 1 012 |
| Standard Oil (Ohio) | 126.6 | 136.9 | 181 |
| Sun Oil | 220.0 | 356.0 | 362 |
| Union Oil of California | 232.8 | 268.8 | 334 |

Source: *Petroleum Economist*, March 1977, March 1978

country[139]. For example, during the supply crisis of 1973/4, even British Petroleum, half of whose assets are controlled by the British government, refused to give the UK preferential treatment as customer[140]. Thus, the industry and its strategies have to be studied separately.

In terms of control over resources, it is sufficient to quote some of the results of a Chase Manhattan Bank study of 29 international oil companies[141]. According to this study, these 29 companies accounted in 1976 for $29.8 \times 10^6$ b/d of crude oil production. Combined refinery throughputs came to $27 \times 10^6$ b/d, or 60% of total world refinery runs. In the same year, the group had total gross revenues of $\$271\,200 \times 10^6$, and a combined net income of $\$13\,100 \times 10^6$. *Table 10.3* provides a breakdown of these assets by area and by function; *Tables 10.4 and 10.5* give a more detailed look at the importance of the majors in 1976.

These figures clearly indicate that one of the major resources of the private oil industry is capital. Exxon, for example, has approved a four-year investment programme of $\$19 \times 10^9$ [Note 142], and together with Shell the company will spend $\$3.5 \times 10^9$ over six years in the North Sea alone[143]. Even more important than capital, however, is the technical know-how of the companies, and it is a measure of their strength that the leading US firms in terms of patents, Phillips Oil, received 329 patents in 1973 alone. The company held at that time 6477 patents in the USA, and 331 in 29 other countries[144]. Similarly, another US firm, Sohio (now linked to BP) earned 21% of its income before tax in 1974 from royalties of patented technology[145]. The leading role of the majors in terms of resources and know-how can also be gauged from the fact that the Soviet Union has been negotiating with BP on the transfer of offshore platform technology, and on permafrost development of oil[146]. To the extent that oil exploration and development is now moving into difficult areas, the biggest, best established companies stand to gain most, since only their capital and know-how can master difficult

**Table 10.3** *Investments in Fixed Assets by a Group of 29 Oil Companies, 31 December 1976[a] ($10^6$)*

| (A) Function | Gross assets | | Net assets | |
|---|---|---|---|---|
| | Total | of which USA | Total | of which USA |
| Production | 81 108 | 57 679 | 46 184 | 29 951 |
| Transportation | 22 992 | 13 447 | 17 256 | 10 379 |
| Refineries/chemical plants | 47 101 | 26 373 | 27 045 | 15 342 |
| Marketing | 24 683 | 11 131 | 14 724 | 7 095 |
| Other | 7 766 | 5 958 | 5 248 | 4 140 |
| Total | 183 605 | 114 588 | 110 493 | 66 907 |

| (B) Area | Gross assets | | Net assets | |
|---|---|---|---|---|
| | end 1976 | end 1975 | end 1976 | end 1975 |
| USA | 114 588 | 102 799 | 66 907 | 58 326 |
| Canada | 12 697 | 11 604 | 7 889 | 7 070 |
| Venezuela | 26 | 56 | 3 | 14 |
| Other Western hemisphere | 5 484 | 5 330 | 3 229 | 3 246 |
| Subtotal | 132 795 | 119 789 | 78 028 | 68 656 |
| Europe | 37 991 | 35 098 | 24 649 | 22 602 |
| Africa | 5 118 | 5 200 | 3 123 | 3 046 |
| Middle East | 1 649 | 1 540 | 828 | 825 |
| Far East | 6 052 | 6 116 | 3 865 | 3 898 |
| World total | 183 605 | 167 743 | 110 493 | 99 027 |

[a] Companies include those in Tables 10.1 and 10.2 (minus Pennzoil) plus Amerada Hess, Apco Oil, Ashland, Champlin Petroleum, Clark Oil & Refining, CFP, The Louisiana Land and Exploration Company, Murphy Oil, Petrofina, Skelly Oil, Superior Oil, and Tosco Corporation.
Source: Chase Manhattan Bank, Energy Economics Division, *Financial Analysis of a Group of Petroleum Companies, 1976*, New York 1977

**Table 10.4** *The Major Oil Companies in World Oil Production*

| Company | Crude supply ($10^6$ b/d) | | | % of total free world production | | |
|---|---|---|---|---|---|---|
| | 1975 | 1976 | 1977 | 1975 | 1976 | 1977 |
| Exxon | 5.411 | 5.576 | 5.343 | 12.4 | 11.8 | 11.3 |
| Texaco | 3.770 | 4.015 | 3.933 | 8.6 | 8.5 | 8.4 |
| Gulf | 2.040 | 1.801 | 1.687 | 4.7 | 3.8 | 3.6 |
| Mobil | 2.240 | 2.156 | 2.370 | 5.1 | 4.6 | 5.0 |
| SoCal | 3.025 | 3.525 | 3.403 | 6.9 | 7.5 | 7.2 |
| Shell | 4.786 | 4.732 | 4.847 | 10.9 | 10.0 | 10.3 |
| BP | 3.440 | 3.540 | 3.380 | 7.9 | 7.5 | 7.2 |
| | 24.712 | 25.362 | 24.963 | 58.4 | 55.2 | 53.0 |

Source: *Petroleum Economist*, May 1976, 1977, 1978; Comité professionnel du pétrole, *Pétrole*, 1977

Table 10.5  *The Majors in World Refining*

| Company | Refinery runs ($10^6$ b/d) | | | As % of world refinery throughput | | |
|---|---|---|---|---|---|---|
| | 1975 | 1976 | 1977 | 1975 | 1976 | 1977 |
| Exxon  | 4.331 | 4.359 | 4.348 | 8.2 | 7.7 | 7.4 |
| Texaco | 2.772 | 2.860 | 2.870 | 5.2 | 5.0 | 4.9 |
| Gulf   | 1.701 | 1.697 | 1.760 | 3.2 | 3.0 | 3.0 |
| Mobil  | 2.060 | 2.049 | 2.105 | 3.9 | 3.6 | 3.6 |
| SoCal  | 2.106 | 2.258 | 2.302 | 4.0 | 4.0 | 3.9 |
| Shell  | 4.258 | 4.191 | 4.238 | 8.0 | 7.4 | 7.2 |
| BP     | 1.560 | 1.780 | n.a.  | 2.9 | 3.1 | n.a. |
|        | 18.788 | 19.194 | n.a. | 35.4 | 33.8 | n.a. |

Sources: *Petroleum Economist, May 1977, 1978; BP Statistical Review of the World Oil Industry, 1976, 1977*

oilfield development. Already some smaller operators have been leaving the North Sea, making way for bigger companies[147].

Administrative and managerial expertise is possibly equally important for the functioning of the international oil market. The oil companies are still in charge of the day-to-day operations in producer countries even where they have been nationalized — their title has changed from owner to operator, but their role has changed little. And in the final analysis, it is still the majors which supervise the management of the international market, planning supply and demand balances in the long and short term, and operating the liftings, transport and refining, and probably no other organization could take over these functions from them[148].

A further asset of the companies is their control over markets and distribution networks. Without market outlets, the oil producers could not maintain present price levels, or at least would need to show much more effectiveness in co-operation among themselves than they have shown so far. Similarly, producers and consumers alike depend at present on the majors to provide the infrastructure of the international oil market, for although the amount of oil traded outside the companies' integrated network has been growing as a consequence of increased volumes of producer government oil entering the markets, this has not yet weakened the majors. Some of them have even established trading companies to join this particular oil activity[149].

There are several answers as to how the resources of the companies will be used, and to which strategy they will pursue. Overall, three different but not mutually exclusive strategies can be distinguished: *withdrawal, collusion with OPEC,* and *diversification away from oil.*

Control over producer countries by Western governments, a 'stable investment climate', was one of the factors which attracted the oil companies into the Third World[150]. Intent on maximizing profits in the *long run*, the majors are particularly interested in stability. The political volatility of the Middle East, therefore, encouraged the companies to develop alternative oil provinces, once the grip of the West on the Middle East countries began to slacken. In effect, alternative sources of oil could be considered an insurance policy: they allowed the companies to maintain their vertical integration by increasing the leverage against the maverick producer intent on challenging the status quo, e.g. by nationalization. Such supplies could be either totally neglected (as happened in Iran in 1951–1953, when the nationalized oil could not be exported for want of buyers), or pushed into a marginal position of supplying the amounts not provided for by rival producers (this happened to Iraq, which explains the slow development of its oil resources)[151].

Diversification of sources of supply began after the Second World War, mainly in parts of the Third World still under the control of the colonial powers. Royal Dutch/Shell were granted exploration rights in Nigeria in 1937, and the first well was drilled in 1951; the French national oil companies, RAP and BRP, began their search for oil in the French colonial empire at about the same time[152]. North Africa developed into a rich oil province after the disturbance of oil flows to Europe in 1956 (with the closure of the Suez Canal and the interruption of the IPC pipeline from Iraqi oilfields to the Mediterranean). In 1958, 17 oil companies held 77 concessions in Libya; six discovery wells were completed the following year[153]. After the foundation of OPEC, a new aspect of this strategy gained in importance: withdrawal from the Third World oil producer countries, and the development of safe new resources under the direct control of the developed countries. In the early 1960s, for example, Exxon decided on a $700 \times 10^6$ exploration programme outside OPEC areas. This search later led to important developments in Australia, Canada, Alaska, and the North Sea[154]. At about the same time, in the mid-1960s, the companies started to take an interest in synthetic crude oil production from oil shales, tar sands, and heavy oils[155].

This strategy of withdrawal has since been intensified, as investment patterns indicate. This does not imply a complete abandonment of the Middle East so much as a changing role of the companies in these countries; investment will be largely taken over by producer governments and the companies will function as suppliers of services and expertise, as well as operators of existing fields.

While the majors still appear to pursue this strategy — all of them, for instance, are involved in the North Sea development — the emphasis has changed. The original impetus towards diversification of supplies has given way to the view that these new production areas would be needed, anyway, to meet growing world demand for oil. In the meantime it has become clear that the companies' withdrawal from the major producer countries in the Third World will not shift the balance of power back to the companies: the natural advantage of the Middle East with its high concentration of the world's oil reserves at low production cost will prevent this; as long as the producers maintain a minimum of co-operation among themselves, there is virtually no way to undermine their position. Quite apart from the strength of OPEC, however, it is questionable whether the oil companies now have any *desire* to break OPEC — the reverse would appear closer to the truth[156].

For these reasons, the second strategy of close co-operation with the producers has largely come to overlap and supplant the first. The difference between the two strategies is that the former derives from, and aims at the maintenance of a position of strength, while the second emanates from weakness. This weakness stems from the main characteristic of the major international oil companies – their vertical integration[157], which means that oil is transferred from production to final distribution within affiliates of the same firm. If the companies wish to maintain vertical integration (and this has been one of their main assets in terms of profits and stability in the past), then they have to accommodate the producers[158]. The alternative would be attempts by the producers to set up their own distribution systems and market outlets: if this was paralleled at the consumer end by the appearance of strong national oil companies as buyers of the producers' crude, this could eventually ease the private companies out of their position in the market. In that case, there would be a scramble for supplies with great risks for the stability of the market – prices could oscillate wildly, and predictability might be totally undermined. Neither of the two would be in the interest of the private companies (nor of anyone else, for that matter).*

In order to continue informal vertical integration, the companies can rely on their main assets: access to markets, their technological and administrative skills, and their know-how in general, which make them necessary partners for the producer countries. The companies, in consequence, take on three functions for the producer countries:

(1) The management of production, exploration and development on a service agreement basis. The concessionaires maintain their role in day-to-day operations, but in a different legal framework. In exchange they receive a fee per barrel produced[159], and the right to lift certain amounts of crude oil on a long-term basis[160].
(2) They provide market outlets for the producers and continue to balance supply and demand through co-ordination of worldwide activities in the international oil market. The long-term purchase agreements of former concessionaires are one example of this, but equally the majors could conclude long-term agreements as buyers with producer countries in which they did not own concessions[161].
(3) They assist in the development programmes of the producer countries through joint projects in refining, petrochemical industries, etc.

As we shall see, this symbiosis between companies and producers rests on a mutual interest in maintaining price levels, and on the competitive advantage provided by vertical integration. If the companies were reduced to buyers of crude without special arrangements, they would have to compete for supplies. Whether this would result in a glut with falling prices and market instability, or a

---

* *This happened to some extent in 1979 when up to $8 \times 10^6$ b/d of OPEC exports were shifted to bilateral deals and spot transactions.*

continued tight market with high prices but a loss in their competitive advantage at the production stage, the companies would lose out — and so might the producers if they decided to forsake the 'safety net' provided by the companies.

On the other hand, conflicts between the producers and the companies have been reduced as a consequence of market developments, and now essentially concern the distribution of liftings on a worldwide scale and the siting of refineries. As long as the market for oil keeps expanding, or as long as OPEC countries are prepared to accept some cutbacks in liftings during periods of slack demand, and·work out among themselves some rough measure of burden sharing (as has happened repeatedly since 1973 with the Saudi willingness to see liftings fall substantially, and so to bear the brunt of the adjustment), conflicts over lifting rates should be manageable. Equally, the companies will be able to pass on the adjustment costs of increased volumes of refining at source to final consumers as long as the bipolar oligopoly of OPEC and the companies holds. Companies will be under pressure, however, from the consumers as well, and conflicting forces of industrial adjustment between old and new refinery industries will cut through the companies themselves, as well as affect the relationship between producer and consumer countries. Again, it can be expected that the relationship between OPEC and the companies will actually profit from its 'depoliticization': as in the case of price and supply security, the companies have been able to withdraw from the frontline of discussions, leaving those questions to governments to deal with.

The third long-term strategy of the companies is *diversification* away from oil into other sources of energy, but also away from energy. Since most of the alternative sources of energy are still plentiful outside the Third World (as well as in it), this allows what in effect constitutes a variant of the strategy of withdrawal. If successful, the balance in the international energy system would tilt away from oil, and away from the oil producers. The time frame of this strategy is much longer, however, than the original strategy of withdrawal, and its success is much more doubtful in terms of shifting market power in the international oil market. Thus, this strategy of diversification could equally be seen as another attempt to maintain integration, and thereby profitability and growth of the enterprise. The oil companies are turning into energy companies — this is how this strategy can be summarized[162]. The following examples illustrate this:

*Exxon*[163]:
>*Natural gas:* Worldwide sales in 1976 10 678 $\times$ $10^6$ $f^3$ /d.
>*Coal:* First coal mine opened 1970. Production total 1976 2.8 $\times$ $10^6$ t. Production to expand rapidly during the late 1970s, with capacity approximating 25 $\times$ $10^6$ t in 1980. Reserves of coal about 8.4 $\times$ $10^9$ t. R&D on coal gasification and liquefaction; pilot plant under development.
>*Shale oil:* Research since mid-1960s.
>*Tar sands:* Syncrude project in Alberta started in mid-1978 (31.25% share of Exxon affiliate in 125 000 b/d plant).

*Nuclear:* Uranium exploration since 1966 in many parts of the world; principally USA, Canada, Australia, South Africa and West Germany. Uranium production in 1976 1.5 × $10^6$ lb uranium oxide.

Fuel fabrication (market share in USA about 20%; orders of around $1 × $10^9$ at the end of 1974 — 75% US utilities, 15% Germany, Taiwan, Japan) in USA and Germany.

Enrichment: Joint feasibility study of commercial enrichment plant with General Electric, development of isotope separation based on laser technique in joint venture with Avco Corporation. Application for licence for reprocessing facility. Research in nuclear fusion.

*Solar energy:* R & D.

*Texaco:*

*Natural gas:* Worldwide sales in 1977 3882 × $10^6$ $f^3$/d.
*Coal:* 2 × $10^9$ t of reserves in USA; R&D on coal conversion.
*Shale oil:* R&D; acquisition of reserves.
*Tar sands:* Lease.
*Nuclear:* Exploration for uranium in USA.

*Socal:*

*Natural gas:* Worldwide sales in 1977 1505 × $10^6$ $f^3$/d.
*Coal:* Participation in coal industry through $333 × $10^6$ share in Amay, a major US coal and mineral producer; R&D on coal liquefaction.
*Tar sands:* R&D on synthetic crude oil production.
*Shale oil:* R&D on synthetic crude oil production.
*Nuclear:* Uranium production in USA to begin 1979; exploration in USA, Canada, and Australia.
*Geothermal energy:* R&D; lease of 300 000 acres in western USA.

*Gulf:*

*Natural gas:* Worldwide sales in 1977 2469.6 × $10^6$ $f^3$/d.
*Coal:* Production in 1977 8.5 × $10^6$ t. Coal ventures include acquisition of Pittsburgh and Midway Coal Mining Company in 1963, Peabody Coal (together with Kennecott in which Gulf has substantial share) in 1968 (meanwhile divested). Coal liquefaction pilot plant opened 1975. Very large coal reserves in USA (5 × $10^9$ t). R&D on coal gasification, including *in situ* gasification.
*Shale oil:* 50% of shale oil tract acquired in 1964; R&D in mining processes (*in situ* technology).
*Tar sands:* 16.75% share in Syncrude project in Alberta (see Exxon). R&D on *in situ* recovery of deep heavy oils.
*Nuclear:* Uranium production begins 1977; uranium reserves in USA over 100 × $10^6$ lb of uranium ore; uranium production in Canada in 1977 exceeds 5 × $10^6$ lb, of which about 50% share of Gulf.

1967: Gulf General Atomic (since 1973 joint venture with Royal Dutch/Shell) bought from General Dynamics: development of high temperature gas-cooled reactor, research in fast breeder reactor and nuclear fusion. Also shares in German HTR development (Hochtemperatur-Reaktorbau) and HTR fuel fabrication.

*Mobil:*
*Natural gas:* Worldwide sales in 1976 3092 $\times$ $10^6$ $f^3$/d.
*Coal:* Over 2 $\times$ $10^9$ t of reserves; R&D in coal gasification and liquefaction.
*Shale oil:* Substantial reserves; R&D in mining processes.
*Nuclear:* Uranium exploration.

*British Petroleum:*
*Natural gas:* Worldwide sales in 1976 368 $\times$ $10^6$ $f^3$/d.
*Coal:* Through Sohio (26% stake, rising to 54%, of BP) Old Ben Coal Corporation with annual production of about 10 $\times$ $10^6$ t. In 1976, acquisition of 50% stake in Clutha mines, Australia (annual coal production about 6 $\times$ $10^6$ t). Exploration in Australia, Southern Africa and Indonesia.
*Nuclear:* Uranium exploration in Canada and Australia.

*Royal Dutch/Shell:*
*Natural gas:* Worldwide sales in 1976 6657 $\times$ $10^9$ $f^3$/d.
*Coal:* Involved in coal production in USA and international coal trade; first deliveries to Europe made; share in Dutch coal terminal. In 1979 opening of coal mine in South Africa with annual capacity of 3 $\times$ $10^6$ t. Export project under consideration with Australian coal production; exploration in Indonesia. Total annual worldwide production in 1976 14 $\times$ $10^6$ t. Reserves in USA (1.5 $\times$ $10^9$ t) and Canada. R&D on coal conversion.
*Tar sands:* R&D on *in situ* production processes.
*Shale oil:* Participation in consortium to set up plant.
*Nuclear:* In 1969 10% stake in Urenco (European Enrichment Project). 30% stake in Interfuel (nuclear fuel production). In 1973 bought into General Atomics (see Gulf). 50% of shareholding in now abandoned US reprocessing plant.

*CFP:*
*Natural gas:* Worldwide sales in 1974 171.5 $\times$ $10^6$ $f^3$/d.
*Coal:* Production in South Africa.
*Nuclear:* Uranium exploration in Australia, Africa, Ireland, USA, Columbia. Uranium processing in France. Uranium production in SW Africa (10% in Rossing mine).
*Solar energy:* R&D into solar heating and photovoltaic technologies.

*Standard Oil of Indiana:*
  *Natural gas:* R&D on LNG transportation.
  *Coal:* R&D on coal liquefaction and gasification.
  *Shale oil:* R&D since 1964; ownership of shale tract jointly with Gulf.
  *Tar sands:* R&D on new mining techniques.
  *Other:* R&D on solar energy, fuel cells, oil gasification.

These examples of energy diversification only illustrate a general trend. For the time being, the scale of diversification is still relatively modest — for 15 companies analysed in an article by C.D. Russel, five had diversified only to an extent below 10% of their total activities, and six others to between 10 and 20%[164]. But the trend continues.

In addition to such horizontal diversification within the energy industry, many oil companies have also moved into other sectors. Obvious examples are petrochemicals and chemicals, which are to a large extent based on oil as a raw material. The oil companies are estimated to account for about 40% of total petroleum-based chemical industries in the world, and two subsidiaries of oil companies are among the top 20 chemical firms in the world[165]. Other activities are totally unrelated to energy: in the 1960s, for instance, Exxon expanded into leisure activities. Many oil companies have entered the mineral and mining industries; Socal and Gulf have been in urban development projects and real estate. Mobil has acquired large interests in the retail (Montgomery Ward) and packaging business (Container Corporation of America)[166].

How do these three strategies interlock? What are the oil companies' objectives? In effect, the strategy of close co-operation with the producers is a necessary precondition of the other two strategies, and therefore pivotal. For in order to develop alternative sources of energy, or to diversify away from energy altogether, the companies have to gain time, and increase their investment funds for the intermediate period of withdrawing from OPEC countries. It is for this reason that co-operation with producer governments becomes vital: this co-operation allows them to maintain present price levels, and thereby to secure 'adequate profit levels'. At the same time, since OPEC countries will put up most or all of the capital investment, some of the companies' capital which previously went to OPEC countries in order to assure a continuous flow of oil to consumers, will be freed. Finally, the high level of OPEC prices is a precondition for the profitability of alternative sources of energy in which the companies are investing: they have, therefore, every interest to co-operate with OPEC in order to prevent a fall in prices[167].

While it would be incorrect to assume that the majors have always been interested in maximizing prices (they are rather more concerned with the long-term *stability* of prices, and with price levels which guarantee *profits over a longer period*), from their viewpoint the present situation demands high and stable oil prices[168]. The high price is essential for forward planning and investments in alternative energy sources, as well as for political reasons: further drastic price increases would probably reawaken the distrust and suspicion of consumer countries and might increase pressure for anti-cartel actions. It is only high and stable prices which can guarantee the survival and growth of the large international energy firms; the risks of fluctuating prices with a dramatic fall and the break-up of OPEC as the opening event, would be incalculable.

A second basic set of objectives concerns profits themselves. Since higher prices alone do not suffice to increase the company profit margin, the companies in the past concentrated their profits at the production end, and accepted losses in refining and distribution in order to create outlets for their profitable crude[169]. As the OPEC countries increased their share of the rent at the producing end, the companies had to shift profits to downstream activities[170]. This necessitated rationalization of distribution networks, with whole national branches like Shell Italiana being sold[171]. In fact, the companies now have to rely on each and all of their operations making profits, since the former concentration of profits at the production stage now has to be spread over all the stages of the petroleum business.

For these reasons it seems clear that co-operation with the producers will continue to be of vital importance for the companies. In these circumstances, it is doubtful whether the producers really lost their leverage *vis-à-vis* the oil companies, as some sources close to the industry have argued[172]. The companies still have much to lose from a disruption of their relationship with the producers, and their interests with regard to oil prices appear to coincide largely with those of OPEC.

## OPEC

The strength of OPEC in terms of resources has already been mentioned: it rests primarily on the control of over 90% of international trade in oil, and about half of the total world oil production. OPEC has now also begun to move downstream, although its resources in such activities as refining and shipping are still comparatively minor. Also, the producers are still short of technical and administrative know-how. OPEC has made considerable efforts, however, in all these areas, and substantial progress has been made. The main instruments of diversification upstream and downstream (refining, distribution, shipping) have been the national oil companies. Here, partnerships with private oil companies have successfully transmitted oil technology, and the training programmes conceded as gestures of good-will towards the host countries have also been effective[173]. Overall, the strength of OPEC can be summarized as follows[174]:

(1) OPEC controls about 65% of the non-communist world's production, about 70% of its reserves, and about 90% of international trade in crude oil. In addition, OPEC oil has a large competitive advantage.

(2) The demand for oil is relatively inelastic to price changes, particularly in the short and medium term — a reflection of the vital importance of energy for the economies of developed and developing countries, and of the narrow limits on additional supplies of alternative energy sources. The implication is that OPEC is still in a position to raise prices for oil at any time and by considerable amounts, provided it has the political will and unity to do so. Price weaknesses observed repeatedly over the period 1975–1978 do not contradict this: they were the result of temporary excess supplies of oil over demand. This excess could have been eliminated quickly, and painlessly, by a reduction of production in Saudi Arabia and/or other low absorbers. That it did not happen was the result of essentially political decisions, not of market forces. Similarly, limited competition among suppliers during the same period did not reflect a weakening of OPEC's market power, but simply the difficulties of price differentials between various types and qualities of oil. While these differences between the marker crude and other crude oils were subject to market forces (and changes in these forces therefore demand adjustment of differentials)[175], the setting of the price level of the marker crude itself remained basically a political decision, and continued to offer a large variety of options for the producers.

(3) The OPEC members do not all need all the money they receive for oil exports: even the so-called high absorbers could reduce their spending somewhat without suffering much: international borrowing would be an easy alternative. In addition, countries like Saudi Arabia, Qatar and Abu Dhabi produce oil at a level which far exceeds their absorption capacity for revenues. There is, therefore, more than a kernel of truth in Sheikh Yamani's statement:

> 'In Saudi Arabia, we can easily cut down by about $5.5 \times 10^6$ b/d and live in comfort and meet our requirements inside and outside Saudi Arabia. In Iran, even with all the huge development programmes, they can easily cut down by $2 \times 10^6$ b/d, and will meet their requirements both internally and internationally. And this is also the case with many other countries. This is the difference between OPEC as a cartel and any other cartel... What we are doing right now is something foolish if I take it from the narrow angle of our selfish interest in Saudi Arabia. We are producing $5.5 \times 10^6$ b/d just to help the economy of the world, to help not to have another price increase.'[176]

As was pointed out under (2), this ability to reduce exports considerably is actually not needed, since increased revenues could always be provided by price increases, if the political will existed. The capacity to cut back offers, however, significant additional leverage which strengthenes the overall position of OPEC, and thereby paradoxically might even increase its ability and confidence to administer another price rise.

(4) The state with the largest exports, Saudi Arabia, is also the state which

needs money least. Saudi Arabia can therefore painlessly absorb a large part of fluctuations in market demand. This gives the country a key role within OPEC. At the same time, installed capacity and future potential give the Saudis a very strong position in the international oil market and therefore in international energy politics.

(5) The distribution of production rates in OPEC is still managed by the majors. They have no reason (except price) to single out any particular producer, and this helps in smoothing the impact of demand fluctuations for oil exports. This co-operation with the companies enables the producers to manage the international oil market effectively: an element of structural alliance of great importance.

(6) The OPEC countries have fairly close cultural and political similarities. All are Third World countries with a common commitment to a change in the international economic order in their favour; nine of them — or eleven if Nigeria and Gabon are counted — are Muslim, and seven are Arab countries.

On the other side, OPEC's position also shows several weaknesses. They are:

(1) The one-dimensionality of their power: on practically all other counts but oil they are weak and vulnerable, while the industrialized countries' power is based on a multidimensional strength — military, economic, technological, psychological, etc. This makes OPEC countries in many ways dependent on outside support from the developed world, as was shown in Part 2.

(2) Within OPEC, there exists substantial scope for disagreement. It is common wisdom, for instance, to distinguish between 'high' and 'low absorbers' among producers; the low absorbers are generally credited with an interest in less elevated price levels. This seems a somewhat simplified argument: first, it presupposes a prevalence of national over international interests even in the very long run, or, put differently, a possibility that too high oil prices might threaten producers with the loss of markets to other sources of oil energy. If one accepts, however, the present situation as a transition process away from hydrocarbons, then higher prices could also be in the interest of low absorbers, if they prevent serious adjustment crises. Second, price decisions are subject to a variety of influences other than those of oil economics: international and regional political considerations, general economic factors on the national and the global level. Nevertheless, differences about price levels could at any time break out among OPEC countries, even if they might not necessarily reflect structural differences of interest but political objectives. Quite apart from prices, however, there are a number of issues on which OPEC could fall out.

What will be OPEC's strategy in the future? From our analysis, it is clear that its overriding objective and concern has to be a perpetuation of its control over the international energy system. This market control depends, as pointed out in the previous chapter, on the strong resource position of OPEC, but above all on co-operation with the companies, and — crucially — among producer countries. According to a study by the US consultants Arthur D. Little[177], the main threat to OPEC would be a proliferation of consumer national companies buying oil directly from the producers. This would undermine the oligopoly of OPEC and private oil companies. Consequently, Saudi Arabia strongly warned the USA against the establishment of a US national oil company[178]. But co-operation among OPEC countries seems even more important for a continuation of OPEC strength — and it can be assumed that this will also be a priority for the strategies of the producers. Precisely because of its importance, a breakdown of OPEC co-operation seems thus unlikely.

This fundamental objective would also allow the producers to realize other goals which will figure in their strategy: stable or increasing earnings in real terms through price adjustments, the protection of surplus revenues against inflation and devaluation risks since these surpluses are seen as an expression of OPEC's willingness to produce oil in excess of their own revenue demand. These objectives could take specific shape in demands for pricing oil exports in SDRs or on the basis of a currency basket[179], for indexation of oil prices[180] and surplus revenue holdings in the West, and similar steps[181].

A third set of objectives concerns development of OPEC countries: demands under this heading will likely continue to be the transfer of technology on easy terms, the shift of industrial capacities into OPEC, particularly in sectors such as refining, steel, and fertilizers through a halt on further expansion in Europe in favour of a build-up of plants in producer countries, and the eventual opening of market outlets in the (industrialized) consumer countries to absorb the new industrial products from the exporters[182]. Further demands to help such exports appear likely as the new industrial production in OPEC seems to have little chance of being competitive; links between oil exports and steel exports, for example, could be one such demand.

Finally, OPEC will probably to some extent continue its support for LDC demands for a new international economic order. It seems doubtful, however, whether the producers would be willing and able to use much of their leverage to help achieve this goal[183].

In sum, the crucial aspect of OPEC's strategy will be its desire to maintain market control. Whether this means they will insist on their *exclusive* control over price and supply decisions, as they now do, seems, however, not certain: after all, even today the actual freedom of manoeuvre for the producers both with regard to price and to supply levels is severely limited by their vulnerability to economic crisis in the world economy, and their need for political support from the industrialized countries[184]. Thus, market control is even now *de facto* shared, although exercised exclusively by OPEC. It does not seem inconceivable that OPEC would be prepared to accept some formal arrangement of sharing control with other countries, if

(1) It becomes manifest that the difficulties involved in the transition problem are so great, and the risks involved so severe, that the producers prefer to

share their responsibility with other countries in a desire to prevent adjustment crises.

Or if

(2) Producers would prefer a safe share in market control which meets their basic concerns about sufficient oil revenues to diversify their economies and create new sources of income for the post-hydrocarbon era, over an uncertain future in which there might be an erosion of market power through the decline in OPEC's resource share in the international energy market.

Obviously, both conditions would also involve far-sightedness and the preparedness to sacrifice some short-term benefits.

The crucial variable affecting OPEC's market control over the coming years will be the degree of cohesion among exporters. Institutionally, OPEC could hardly be considered a strong supra-nation organization[185]. Of its four main working groups, the secretariat is the only permanent one. The Secretary-General is appointed for two years, but his secretariat is composed of members from national oil companies on loan, or from national administrations, which makes it difficult to develop much independence and freedom of action for the common objectives of the organization. OPEC is aware of the shortcomings of its secretariat, and has agreed on a reform programme. One suggestion is a 'Think Tank' attached to the secretariat, staffed by professionals and experts recruited both from rich and poor non-member and member states. The supranational element is even more absent in other bodies of OPEC. The Board of Governors, which oversees the work of the secretariat, consists of delegates from the member countries and meets only occasionally; the Economic Commission is likewise composed of national representatives. It acts as an advisory group on prices and price maintenance, and has so far had little actual impact on decisions. The real policy-making body of the organization is the Conference of Oil Ministers which has to approve decisions unanimously.

While solidarity of OPEC thus depends essentially on the degree of agreement among the main member states, the weak institutional framework can actually be an advantage: it gives governments a high degree of flexibility while still providing a degree of communication and co-ordination for the whole group.

Zuhayr Mikdashi has described OPEC not as a cartel (since there are no agreements on market sharing and production control), but as a 'structured oligopoly'. He notes[186]:

'... leadership in setting oil export terms belongs to the few. This leadership might be exercised by a country controlling large productive resources (e.g. Iran or Saudi Arabia) or by one that has the "right" perception of the balances of market forces at work, or the ability to act as a market "barometer", or as a catalyst (e.g. Algeria or Libya). In any case, the effectiveness of the co-operative bond is a function of the conditions and objectives of the member governments, of the policies and actions of other protagonists, and of the circumstances.'

Thus, it is the different policies and strategies inside OPEC which assume crucial importance. There, the position of Saudi Arabia gives this country a key role: since she could break or sustain OPEC price decisions over a wide range of the possible price spectrum by either flooding markets with additional production or curtailing exports, even countries like Algeria and Iran have at times taken a moderate view on prices to accommodate Saudi wishes[187]. Saudi political and economic objectives have, therefore, tended to be of great, but not exclusive importance in decisions made within OPEC. These objectives, in turn, reflect Saudi relations with the United States, and past differences over price levels between Riyad and Teheran were not unrelated to the different degree of co-operation of these two countries with Washington, but above all to Persian/Arabian Gulf politics[188]. Before the crisis in Iran, Saudi Arabia had substantially improved relations with this country. The revolution stopped, or at least interrupted, Iran's drive to become the predominant power in the Gulf, which had been a cause for some anxiety in Riyad. On the other hand, the destabilization of Iran worried Saudi Arabian leaders profoundly — as a 'writing on the wall' for its own future, and as a sign of growing Soviet encirclement. The Saudis were certainly concerned about supporting the Shah's regime within their capabilities, and this seems to have influenced their decision to accept and support the phased OPEC price rise decided in late 1978.

This leads us to a wider point: the future of OPEC will almost certainly be decided in the Gulf. Three major producers are concentrated here: Saudi Arabia, Iran, and Iraq. The three are linked through a web of conflictual and co-operational ties. There can be little doubt that since 1974 moves towards closer collaboration in the Gulf had strengthened OPEC, and the bargaining over prices has shown a willingness to compromise. Conversely, a serious disagreement over prices, and the breakup of OPEC, would have repercussions in the Gulf which could be very serious if prices fell dramatically. Thus, there will be a continuous interplay between Gulf politics and international oil affairs, the one being influenced by the state of the other, while influencing it at the same time.

In spite of subdued tensions within OPEC, its record of solidarity under difficult circumstances has been good. The organization survived, without major difficulties, a serious reduction in world demand outside the socialist countries from a peak of $50 \times 10^6$ b/d to about $43 \times 10^6$–$44 \times 10^6$ b/d at the end of 1975. Cutbacks occurred in all OPEC countries, with Saudi Arabia bearing the brunt of the adjustment. In dealing with the contractions of the oil market, the producers had two options: either to agree on a formal scheme of allocating supply cutbacks, or to accept lifting rates by the majors. The former way was tried, but failed to gain the support of Saudi Arabia and Qatar[189]. In effect, therefore, the OPEC members decided to leave the allocation of cutbacks to the companies; in order to increase their output, some members cut prices at the margin, either with or without formal approval of OPEC[190]. This was tolerated by Saudi Arabia, which provided the marker crude: had she decided to lower prices, OPEC price decisions would in effect have been broken. Some of the price adjustments, moreover, simply reflected necessary adjustments to changing price differentials between different crude oil types.

Although the debate within OPEC about pricing became quite vicious at times, it did not reach the point of being a fundamental danger to OPEC. This was made clear by the split in December 1976 and early 1977, which eventually was closed through a compromise between those countries which had accepted only a 5% price rise, and the '10 per centers'[191]. As it happened, differences were always settled — quite simply because the benefits each derived from OPEC solidarity were too great to be sacrificed. The costs of an OPEC break-up would be very serious indeed, if it resulted in a rapid price decline. Economically, the low absorbers, and Saudi Arabia in particular, might appear to be in the best position under such circumstances since they could produce enough oil at much lower prices to satisfy their needs for revenues. Saudi Arabia, however, and the other low absorbers are vulnerable politically and militarily, and an attempt to break OPEC, and bring down prices, could well lead to open or clandestine military force by Iraq and Iran. Thus, even the low absorbers would have much to lose if OPEC broke up — quite apart from the economic loss, and the loss of control in the market which might be difficult to recuperate, there would be very substantial political risks, which make it extremely unlikely that Saudi Arabia would ever risk a fundamental break with the rest of OPEC[192].

But as events in 1979 demonstrated, a disintegration of OPEC co-operation did not necessarily mean a fall in prices — provided it happened under circumstances of scarcity. Supply interruptions, and then lower exports from Iran, created a psychology of scarcity and insecurity — and prices increased dramatically, by more than 100% on average until late 1979. Prices were now set by individual governments; OPEC increasingly failed to impose even the semblance of a unified price structure. The 'price hawks' prevailed under these circumstances, for Saudi Arabia had lost the spare capacity and the political clout to prevent the price explosion. Announcements of individual production cutbacks for conservation purposes and the much lower exports from Iran sufficed to provide the necessary preconditions for a doubling of oil prices.

Until 1979, Saudi Arabia had committed herself *vis-à-vis* the USA to a policy of keeping oil prices stable, or at least making very modest increases. In other countries, however, such a further decline of the real value of oil exports, combined with a decline in volumes exported, exacerbated demands for price increases. This made the working out of compromises within OPEC almost impossible. Since the high absorbers' need for high revenues, and perhaps also for their development efforts, were frustrated by Saudi ability to keep prices frozen, a change in the underlying supply—demand equation towards scarcity was bound to cause another round of dramatic price increases, which even Saudi Arabia was unable to halt.

This fundamental weakening of OPEC is not without problems for the producers — in spite of dramatic revenue increases. One danger lies in the risk of a rapid rise of oil prices to a point where they exceed the point of equilibrium in the long-term energy outlook. Prices would then be likely to crumble dramatically. Another danger simply lies in the unpredictability and political instability of the present situation.

This could provide opportunities for a concerted effort of producers and consumers to establish order in the international oil market. Saudi Arabia might help to stop other OPEC members sliding into another round of dramatic price increases — the Saudis are likely to be more worried about the consequences of such a move on the world economy, and on their own position, than other OPEC members, provided, of course, that by then Saudi Arabia will still be governed by the present type of government.

## The United States

As in the past, the USA will play a central role in future international energy affairs. Partly, this is simply a consequence of the enormous resources still under the control of the USA: the largest market for oil, the largest volume of imports of any nation, still one of the greatest producers of oil and natural gas, and rich energy resources to replace conventional hydrocarbons in the longer run — shale oil, coal, uranium and the technologies for renewable energy sources. At the same time, the USA has the world's largest private oil industry with unparalleled know-how and resources. Equally important, however, for the future of the international energy system is the special role of the USA as the dominant power in the Western world and the international economy, and as one of two Superpowers. The immense importance of this world role, and the vital importance of the USA for other industrialized countries, economically, politically, and militarily, destine the USA for a leading (if not a leadership) role in world energy matters.

In some respects, the events of 1973/4 have arguably strengthened the position of the USA *vis-a-vis* other industrialized importers, notably Europe and Japan: they exposed the much greater vulnerability of the latter to energy supply interruptions, and gave America the chance to eliminate the relative disadvantage of a costly energy base for its economy, as well as to develop the potentially vast alternative energy resources available within its boundaries. On the other hand, its position in the international system turned the events of 1973/4 into a far more serious threat to the USA than to the other members of the Western alliance system: the successes of OPEC and OAPEC constituted a direct challenge to US predominance in international economics and politics. True, the years before had already seen a slow erosion of the United States' ability to impose its will on the international system. Politically, this was the result of the East—West conflict, economically that of the relative advance of other industrialized countries. To talk about multipolarity with these countries is one thing, to experience a successful challenge from a quite different direction, the Third World, another. After all, this challenge came from a group of countries whose economic position differed radically from that of the industrialized countries, and whose interests were consequently much less easy to accommodate. In essence, the question seemed to be whether the industrialized world, and its pivotal power, the United States, were prepared to share control over the international economy, or at least some of its vital parameters, with a number of newcomers who presented themselves as the spearhead of the whole Third World[193]. The stark choice therefore appeared to be either to accept a dramatic shift in wealth and political power effected by OPEC and OAPEC in 1973/4, or to try and 'roll back' this change. The basic issues at stake were security of supplies and the price of oil, and in addition the management of massive transfers of wealth; the initial reaction of the United States, as of other industrialized countries, did not sufficiently recognize the specific weaknesses and strengths of one another's positions. Ironically, as it turned out, the very asset of the West in this bargaining situation was its weakness: the ramifications of the action of the oil producers in the world economy through their impact on the industrialized economies were so widespread and long-lasting that they severely limited the freedom of manoeuvre of the producers who now had much to lose; on the other hand, the supposed strengths of the West, e.g. military power, was of very dubious value in this situation.

The initial response of the USA to the challenge was that of confrontation with the producers, and of assertive leadership with the other industrialized consumers. Both responses turned out to be of limited success. Interestingly, however, during the supply crisis itself, Washington followed a policy of appeasement of the Arab producers by working for a political settlement between Israel and the Arab world (the USA, however, had perfectly good reasons to do so even without the acute pressure from Arab oil exporters)[194], although threats of military intervention were used as part of the diplomatic process aimed at convincing the Arabs to withdraw their supply cutbacks[195]. After the crisis, however, the US strategy was directed to the fundamental issue: the shift in control over the international oil market. First steps date back to November 1973, when President Nixon announced his stillborn 'Project Independence' – a megalomaniac programme drawn up with little regard for reality, and with less for the problems and positions of America's allies[196]. The officially declared objective of this programme was to restore the energy independence of the USA through massive investments in alternative energy sources by 1980, and thus to regain US economic security against supply interferences. However, Project Independence also constituted a logical step in the attempt to break OPEC's power, and to reassert Western, and in particular US, dominance over the world energy system. A second tentative step in the same direction was made by Henry Kissinger when addressing the Pilgrim's Society in London (12 December 1973)[197]; this speech called for the establishment of an international energy action group to create incentives for increased energy production, to help conserve energy, and to co-ordinate energy R&D efforts. Although the principle of co-ordination and co-operation with the oil producers and other countries outside the OECD was professed, it was clear that the organization of a counter-cartel of consumers was the main thrust of the speech. As later events and US actions were to show — e.g. with regard to international price levels of oil — the objective was at the same time specifically American, and tailored to US interests rather than those of Washington's less well-endowed allies.

The speech in London led to the Washington Energy Conference in February 1974, which secured some important elements of the US objectives in setting up a co-ordinated Western response to the OPEC challenge[198]. The success, however, was only partial: France dissociated itself from major parts of the final communiqué, and in particular from the follow-up machinery to the conference. This machinery was developed eventually by the International Energy Agency (November 1974). With the IEA, Washington had to some extent achieved what it saw as a necessary precondition for reversing the changes of 1973/4 in the international energy system — yet, the Washington Conference also demonstrated the existence of significant policy differences among industrialized importers[199].

The Conference marked the end of a first, largely unilateral stage of Washington's response to the crisis — a stage dominated by lack of consideration for the interests of the more vulnerable European countries, and by attempts to push the other industrialized consumers into acceptance of the US approach to the oil producers. After that, co-operation began to play a more important role, consultations and concerted efforts increased, resulting in some modification of the US position. The scope of co-operation, however, was still limited to the industrialized countries — until the end of 1974, the USA refused to consider a dialogue with the producers within a multilateral framework, while at the same time pursuing its own brand of bilateralism with two crucial suppliers — Saudi Arabia and Iran, who were important from the viewpoint of Washington's strategic preoccupations[200]. Thus, the blocking of a broader dialogue contained not only elements of confrontation with OPEC, but also served to keep Europe and Japan outside an increasingly close relationship between the USA and key producer countries.

The Martinique summit between President Ford and President Giscard d'Estaing in late 1974 resulted in a further adjustment of the US policy towards a more open attitude towards the oil producers: Washington now accepted the idea of a tripartite conference between producers, industrialized consumers and developing countries in exchange for French *de facto* co-operation in the industrialized consumers' efforts to conserve energy, to develop alternative sources of supply, and to create a safety net to overcome the recycling problems associated with the vastly increased amounts of oil revenues[201]. But the USA continued to refuse a dialogue about anything but energy, and insisted that before the dialogue on energy could be opened the consumers had to agree on a comprehensive long-term energy programme, including measures to protect investments in alternative sources. Thus, the shift in the US position was at best partial — the strategy to break OPEC was still to be achieved, and an energy dialogue was seen in this connection as a useful way of putting pressure on OPEC with the support of other LDCs. Furthermore, the dialogue was to be held from a position of strength, particularly as far as the United States was concerned: they had to gain most from a protection of investments in alternative energy sources, i.e. a guaranteed minimum price for energy, which could have mobilized fairly rapidly, or so it seemed, the potentially enormous energy resources of the North American continent[202].

By the beginning of 1975, the central issues of the oil situation had become the producers' control over the international energy system and its price levels, and their support for a restructuring of the world economy. The acute crisis and its management problems in 1973 and 1974 had largely been overcome: the IEA had developed an emergency allocation scheme to cope with future supply interruptions, and the oil facility of the IMF had provided a mechanism to arrange for the recycling of surplus oil revenues to hard-hit consumers. Again, the USA had initially refused to search for a solution of the recycling problem including the producers, but had been convinced by other industrialized countries that this would be preferable. As a compromise, an additional $\$25 \times 10^9$ safety network, as a financial instrument of last resort, was drafted within the OECD[203] *. With these items out of the way, the US government now concentrated on the question of control in the international oil market. Initially, Washington had aimed at a substantial reduction of oil prices, and its strategy was designed to achieve this by breaking OPEC, if the producers could not be convinced to lower them willingly. The precondition for a change in the relationship of forces between the actors in the international oil market was a contraction of this market — only such a contraction provided reasonable hope of creating sufficient differences of interest, and enough pressure within OPEC to cause it to split. To achieve such a contraction, several developments had to be brought about[204]. The first was conservation of energy in the main importing countries, the second the development of alternative sources of supply within the industrialized countries. This latter development demanded the removal of restrictions and constraints on the exploitation of other sources of energy, but above all the provision of finance and the protection of investments in alternative sources of energy within the industrialized countries. Paradoxically, then, oil prices had to be kept up to bring them down — a paradox which is solved by the realization that it is *control* over prices, not the *level of prices as such*, which was the objective of US policy. In a major policy speech, Kissinger suggested a floor price for oil[205], an idea which was eventually accepted after much modification by the IEA.

In 1975, the attitude of the United States vis-à-vis the OPEC countries began to change. The reasons for this were many: it had become obvious that the development of alternative sources of energy in the USA was much more costly, time-consuming and difficult than had been expected originally: Project

* *This was never actually implemented since the US Congress refused to approve the scheme.*

Independence was quietly buried[206]. This meant continuing — and increasing — US dependence on OPEC; and this was corroborated by the growth of US oil imports since 1973. Besides, US energy policy had been caught in domestic conflicts between various branches of the Executive, and between Congress and President[207]. At the same time, analysts in Washington began to recognize the strength of OPEC, and to reassess the prospect of causing it to break up[208].

On the other hand, some of the pessimistic predictions of 1974 about the threat of international chaos as a consequence of the inability of the international financial system to cope with the vast surplus oil revenues turned out to be exaggerated: the private banking sector by and large proved capable of handling the much increased volumes of international liquidity. Some of the justification for the initial American strategy was thereby removed. At the same time, the quiet pressure from Europe and Japan for a more co-operative approach towards the energy problem, and the advantages which could be derived from the initiation of a negotiating process with OPEC, helped to change US policy — a change, which on the bureaucratic level was reflected in the eclipse of the influence of Assistant Secretary of State Enders, and the rise of the more co-operatively minded Charles Robinson, who replaced Enders in 1976[209]. By that time, US attention began to focus on the possibility of isolating the Soviet Union, which had next to nothing to offer in the North—South dialogue, and of thereby opening up a new area of international diplomacy, whereby the USA would clearly be able to push the USSR onto the defensive. On the other hand, a refusal by Washington to engage in the dialogue could have offered the Soviet Union the opportunity of making further advances in the Third World[210].

The reappraisal of the US policy came about only gradually, however. While Washington began to reconsider its attitude towards the energy issue (finding that differences between OPEC and the USA over the price question were not all that great), this issue had become inextricably intertwined with the wider issue of a 'new economic order' — a result of attempts by both the industrialized and the OPEC countries to enlist support from the Third World countries which did not export oil. Since the crisis in 1973/4 had, in one sense, been a challenge — a successful challenge! — to the established world economic and political hierarchy, it was hardly surprising that this should become the political substance of negotiations following the crisis; it was equally unsurprising that the support of the oil-importing LDCs should by and large go to the forces of change — although the alliance was not without strains.

Faced with demands for a fundamental restructuring of the international economic system, the USA again reacted, initially, by stalling. In the 6th UN General Assembly Special Session the USA hardly contributed anything, except some unacceptable last-minute proposals. The 29th General Assembly under the Presidency of the Algerian, Bouteflika, further exacerbated the confrontation between the USA and the Third World; the USA complained about the bloc voting at the UN, and threatened to withdraw from the organization. This deterioration in the climate, however, made both sides aware of the stakes at risk, and the undesirability of a confrontation which could only end with everyone worse off than before. Accordingly, a learning process began on both sides — in the case of the USA not least because of the opportunities offered by the dialogue, which were mentioned above. This change was reflected by the major speech prepared by Kissinger for the 7th UN General Assembly Special Session, setting out detailed proposals for a restructured world economy, which can be grouped around the following major points[211]:

(1) *Economic security:* items here were the co-ordination of economic policy among industrialized countries, the dialogue between producers and consumers of energy, and the stabilization of export earnings from raw materials.
(2) *Growth:* this was to be assured by better access of LDCs to world capital markets, and the transfer of technology.
(3) *Trade:* for the first time, the USA accepted generalized preferences.
(4) *Raw materials:* proposed were a world food reserve, and mixed 'fora' for raw materials on a case-by-case base.

Further steps forward were taken at the UNCTAD conference in Nairobi in May 1976, where the USA (in a speech prepared by Henry Kissinger and read by the US Representative at the United Nations, Daniel Moynihan) made a number of specific proposals to meet some of the demands of the LDCs, such as an International Resource Bank and an International Energy Institute. The spirit of this speech was constructive, although the host of proposals somewhat concealed the fact that on substantive issues, such as the integrated commodities fund, the debt problem of the LDCs, and the transfer of technology, the US position remained essentially unchanged.

This was hardly surprising — after all, the North—South dialogue involves a number of very direct conflicts of interest between North and South, between rich and poor, and involves a challenge to the world 'establishment'. It is the American multinational corporation, American political and economic influence in the Third World, which are the most obvious and most frequent targets of demands for a new international economic order by the LDCs — and to give up some advantages of the previously dominant position creates some dilemmas for the USA. At the same time, the shift in policy initiated in 1975 — although quite possibly tactical in character — will probably prove irreversible.

The new American administration under President Carter very much followed this logic: some further steps were taken to meet the demands of the LDCs, as will be seen later in the discussion of the Conference on International Economic Co-operation (see Part 4). The consistent element of US policy towards the South was still to make concessions where they were inevitable, while trying to defend US interests as much as possible. Inevitably, this also meant, by and large, an attempt to cling to the *status quo*. What was lacking in this policy was the preparedness to launch some bold and imaginative new ideas, although one might doubt whether any suggestion coming from the USA could count on much support from the Third World. The deadlock in North—South negotiations, which have become such an overriding aspect in the attempts to restructure the international order, is at least partly rooted in the psychology of mutual distrust and suspicion. In the Paris dialogue (the CIEC), the partial deadlock in the end came from an ironic reversal of positions: the USA insisted, together with the other industrialized countries, on a *quid-pro-quo* in the form of a permanent body of consultations on energy decisions. Thus, the situation had turned round full circle: while the USA initially refused to talk about North—South issues, and even about energy with the producers, it now had to accept that its energy objectives ultimately would have to be secured through a trade-off between concessions on North—South issues versus producer concessions on energy, where the industrialized countries had little to offer, let alone the means to impose. This irony, however, was not without its logic: from the US point-of-view the problem was still to secure some consumer influence over decisions concerning prices in the international oil market. A recognition of the distribution of power in the international oil market, which had been seriously miscalculated before, now led to the conclusion that such influence could not be imposed, only negotiated.

The mellowing of the US position was, of course, not unrelated to domestic events: oil imports, which had averaged about $6 \times 10^6$ b/d during 1973, had risen to about $8.5 \times 10^6$ b/d in 1977 — a sign of how little had been achieved by US domestic energy policy in that time. Oil and natural gas production had continued to fall, while energy demand had risen further (although at a rate much lower than before the 1973/4 crisis). The new administration had launched in April 1977 an ambitious energy programme[212], which recognized on the national level some crucial points made by OPEC internationally: the need to phase out dependence on oil and natural gas in a transition process to other energy sources, and the necessity to bring energy prices up to a level which would bring about such a transition. With this programme, the administration hoped to reduce oil imports in 1985 to about $6 \times 10^6$ b/d — an objective which was widely judged to be too ambitious even under the assumption that the programme would pass through Congress unscathed. What happened was of course a far cry from this: the modifications imposed by Congress further reduced substantially the value of the programme. Thus, in 1978 it had become evident that:

(1) The USA would continue to be heavily dependent on oil imports.
(2) Action to prevent an excessive increase of oil imports required measures in line with some key positions argued by OPEC.
(3) US import requirements would make an irreversible fall of demand for OPEC oil in world markets, and therefore the development of serious tensions within OPEC and the decline of producer control over the international oil market, quite unlikely.

Under these circumstances, the USA now directly depended on a close relationship with the producers — quite apart from the fact that this dependence would have existed even had the USA succeeded in reducing oil imports: Washington would still have to take into consideration the interests of its allies. The strategy pursued by the USA to achieve such a relationship was largely bilateral: the evolution of close alliances with Iran and, in particular, Saudi Arabia[213].

Undoubtedly, these special relationships also corresponded to wider strategic concerns of the USA — in the case of Iran, to the desire to contain Soviet expansion southward, and of ensuring regional stability in the Gulf area through a dominant Iranian 'guardian' of Western interests. Saudi Arabia provided the USA with an all-important ally in the Arab world, and in the attempts to settle the Arab—Israeli conflict. The latter, however, was not without problems since US and Saudi positions and interests were far from identical, and a smooth continuation of the alliance depended on the US capacity to bridge the gap between the Israeli position and that of the *majority* of the Arab world: Egypt alone was not enough.

In early 1979, some of the precariousness of the US international energy strategy had become evident. The revolution in Iran demonstrated the inability of the USA to control domestic developments in key producer states in the Gulf, which were vital to its entire energy strategy. The situation was certainly handled badly, but the underlying dilemma far exceeded tactical errors and misjudgements: how could the USA, how could the industrialized consumers in general, assure the domestic and regional conditions of stability in producer areas to maintain oil exports at sufficient levels, and thus avoid dramatic price increases?

For Saudi Arabia, the US failure to keep the Shah in power came as a major shock. US—Saudi ties had developed to the point where the USA — explicitly or implicitly — had committed itself to guarantee the survival of the regime in return for a Saudi Arabian policy of keeping prices down, production up, and petrodollars in US Treasury Bonds. At the same time, US efforts to settle the Israeli—Arab conflict had led to a serious split in the Arab world, which even Saudi Arabia could not easily heal. While trying to achieve reconciliation, the Saudis moved closer to the majority of the Arab world led by Iraq, which was opposed to the Egyptian—Israeli peace treaty negotiations. Regional policy considerations (concern about the threat to the regime in Iran and the need to align with the mainstream of Arab affairs), also affected oil price decisions in late 1978 and 1979.

Thus, in early 1979, the US relationship with both Iran and Saudi Arabia had come under strain; and the spectre of a domestic backlash and upheaval similar to the events in Iran in other Gulf producer states, in particular Saudi Arabia, was looming large. Yet there was no political alternative in sight for the USA but to rely on its special relationship with Saudi Arabia, and a continuation of close bilateral ties seemed crucial for the USA, and indeed for the whole Western alliance system. Yet the relationship with Saudi Arabia poses a number of problems. First, it is obviously vulnerable to a change in regime — and it must seem doubtful whether the USA could really prevent such a change in spite of its massive presence in the country. Second, the alignment with Saudi Arabia will draw the United States closely into the regional policy of Riyad. This has already helped to produce a change in the US policy towards the Israeli–Arab conflict; whether the different objectives of the partners in the Saudi–US alliance are reconcilable and will be reconciled, remains to be seen. Similarly, there could develop other regional problems where US and Saudi objectives are not necessarily identical. Again, it remains to be seen whether Washington would be able to steer clear of an involvement in regional conflicts against its own interests. Third, the USA now appears as a guarantor to assure the success of the development policy of Saudi Arabia. If progress towards economic diversification turns out to be slower than expected, and if some of the massive investments develop into white elephants, the alliance with the USA could become discredited as a new form of hegemonial imperialism and exploitation, with a violent backlash against a pro-Western orientation. This could be facilitated and exacerbated by a fourth problem, the social tensions and processes of change developing out of the rapid transformation of an extremely traditional, conservative and feudal society.

Apart from this special concern with some key producers, the US strategy in the international energy system can be expected to follow the lines developed since 1975: grudging acceptance of OPEC, possibly implying some willingness to come to some kind of institutionalized arrangement for co-operation along the lines sought during CIEC; co-operation with other industrialized countries, within the framework of the IEA and bilaterally, to maintain arrangements against supply interruptions and co-ordinate efforts to reduce dependence on oil imports in the longer run; and as much as possible a stabilization of international oil price levels. Domestic price increases needed to mobilize indigenous energy resources at least in the longer run could be achieved through increases and subsidies. The challenge for the USA, as for the other actors in the international energy system, will be the discontinuities which seem to be building up in this system over the next few years: US import demand between 1978 and 1985 can be expected to increase only slowly, to perhaps $9 \times 10^6$ b/d — Alaska oil has begun to make a contribution (about $1.2 \times 10^6$ b/d maximum production), and higher domestic prices and the imposed conversion of the US car fleet to much lower levels of petrol consumption are showing the first results in dampening energy and oil demand. On the other hand, this will only be a temporary break as long as the underlying trend towards high energy consumption is not halted: in this case, we might well see a period of coinciding additional supplies and low economic growth in the early 1980s, followed by a period of stable or even falling indigenous production of hydrocarbons in America and Europe, and steadily rising energy demand which is unlikely to be met sufficiently by other sources of energy than imported oil and natural gas. Thus dependence on OPEC exports will again increase.

The alternative for US international energy policy was a maximum exploitation of a period of Saudi predominance, thereby reaping the short-term benefit of lower import costs in OPEC and satisfying domestic political demands, or to forsake some of these benefits in concessions to producer countries (e.g. some price rises, market access etc.) in order to avoid the very substantial long-term risks involved in the previous strategy: the risk of sharp and dramatic changes of international energy prices, of consequent instability in the system with its implications for energy investments, and of a serious time-lag in the diversification away from hydrocarbons through energy conservation and alternative production. Apart from these risks directly related to energy issues, there was also the problem of domestic instability in some oil states, and such instability spilling over into regional tensions and conflicts. The USA opted for the easy alternative, and the result was the fragmentation and disintegration of Saudi hegemony in 1979. An international order for oil has to be reconstructed. The costs will now be higher than before — several years were missed. Yet one factor remains unchanged: it will still be the responsibility of the USA to take a leading role in international efforts to stabilize the oil situation.

## *The International Energy Agency*

Starting life as an instrument of the US strategy for control in the international oil market, the International Energy Agency has since developed a more conciliatory, co-operation-oriented role. This development has reduced the IEA's momentum: as the immediate crisis of 1973/4 receded, the differences between the industrialized consumer countries became more evident, and although co-operation among importer governments is still an important element in their overall strategy *vis-a-vis* the producers, difficult choices and alternatives faced by the importers, and in particular by the USA, have become evident at home, thereby also affecting the IEA.

Established in November 1974, the confrontational character of the IEA was initially difficult to deny. Its foundation coincided with a tough speech by Kissinger, urging the consumer countries to join forces, and ruling out any dialogue with OPEC at that stage[214]. The agreement setting up the Agency, which was signed the following day, mentioned the willingness of members to initiate a dialogue with OPEC, but in the prevailing atmosphere the hostile reaction of producers to this 'instrument of confrontation' hardly surprised anyone[215]. The fact that the IEA failed to organize all important industrialized consumers also demonstrated that it was very much an American initiative, to which other consumers had agreed as the unavoidable price to be paid for US commitments to an international approach and leadership in resolving the problems[216].

The core of the IEA is its Emergency Allocation Scheme which makes the organization something of an 'economic NATO'. This vital, security-oriented element of the IEA provided an area where the overriding interest in co-operation, and the need for US leadership, could find relatively easy agreement. But as the scope of the IEA was to expand into other areas, agreements — except on a technical level — turned out to be more difficult. The widely different energy interests of member countries (a function of their different resource-endowment), the pressures of domestic politics, and an ambiguous American strategy hovering between confrontation and dialogue with the producers, were all making consensus difficult to reach.

Apart from the fact that the success of a strategy of confrontation remained doubtful (it might have had the effect of uniting OPEC), the USA demonstrated that the consumers were not prepared to accept the necessary reductions in supplies: US imports of OPEC oil increased, rather than fell. A short-term energy programme worked out within the IEA, which foresaw a reduction of oil consumption in 1976 by 8% and in 1977 by 6% with respect to 1973, and a decline of imports by $2 \times 10^6$ b/d in 1975 (half of which was to come from the USA)[217], was only successful insofar as the recession brought down energy consumption, in any case. The IEA also began to monitor performance in member states with respect to conservation, and in its first report heavily criticized the USA — Washington came out bottom of the list[218].

The emphasis of the IEA then shifted more and more to long-term energy programmes and to the encouragement of R&D projects between member states. One key problem for long-term energy planning was the price issue. Although this was properly recognized by the IEA, its concept of an agreed minimum price for all IEA members was of doubtful value. In any case, the concept soon led to bitter disagreement among member states, as the energy-poor countries were reluctant to commit themselves to high prices, and thus to subsidizing indirectly energy production in the energy-wealthy members. The first group hoped to profit from an eventual fall in world oil prices; the second needed high prices to make the development of their resources profitable, to prevent competitive disadvantages *vis-à-vis* other industrialized countries, and to profit from high energy export prices. The initial idea of an energy floor price was abandoned and replaced by the concept of a *minimum safeguard price* (MSP), which was agreed upon in March 1975. The decision on the actual level of the MSP, however, turned out to be cumbersome, and only in January 1976 was it finally settled. Canada did not sign the agreement but promised that it would behave as if she was bound by it. Germany, Japan, Italy and Denmark then dropped their opposition against the proposed level of the MSP. France, of course, remained outside the agreement — in spite of the *de facto* co-operation with the IEA through the pursuit of parallel energy objectives, which was agreed between Presidents Ford and Giscard d'Estaing at their summit meeting in late 1975. The French wanted to keep their options open, for three principal reasons: they hoped to profit from falling world energy prices, they were still concerned about the confrontational aspects of the IEA, and they saw the MSP as prejudging a European Community energy policy (a worry shared by the EC Commission)[219].

The setting of the MSP at the level of $7/b for Persian/Arabian Gulf oil had originally been part of a US plan to turn the tables on the producers, and this was how the long-term energy programme was originally conceived. In the view of Thomas Enders, then Assistant Secretary of State, this programme was to achieve energy self-sufficiency in a reasonably short time, to restore the balance in the international oil market, and to give consumers some influence in oil price decisions[220]. Yet by the time it was agreed the MSP was hardly more than a paper figure: a realistic estimate of future energy prices in world markets would certainly be much higher. The other elements of the long-term programme (conservation efforts, removal of barriers obstructing the development of alternative energy sources, monitoring of member governments performances, and joint R&D efforts) were either of marginal importance only, or were simply caught in the crossfire of domestic politics which cared little for international obligations.

Thus, the IEA developed more into a useful but unexciting framework for consumer government co-operation on data collection, policy co-ordination, and for the facilitating of international R&D efforts (actually, most IEA-sponsored projects were one-country projects with financial participation by others). The progress made, and its limitations, were well demonstrated by the IEA Board meeting in October 1977, which brought agreement on a number of energy policy principles (notably the formulation of national energy programmes with import reductions as a central goal, and the commitment to raise domestic energy prices to levels which encourage conservation and development of alternative sources of energy), and on an oil import target for the IEA as a group in 1985 of $26 \times 10^6$ b/d (subsequently lowered in 1979 to $24 \times 10^6$ b/d). Finally, the Governing Board agreed on regular energy policy reviews by the IEA secretariat, which would monitor the policy achievements of member states; if these achievements were considered insufficient, the members promised additional steps[221].

While the rhetoric of the communiqué was laudable, it remained vague on several crucial points. Above all, the most specific commitment, that to an import target for 1985, was almost worthless since it was not composed of *national* targets (although national estimates must have played a role in the formulation of the group target). This offered member governments an easy way out of the responsibilities for their energy policy failures.

Apart from the useful contribution of the Emergency Allocation Scheme, the impact of the IEA appears limited. Two aspects, however, deserve attention. First, the evolution of the IEA demonstrated the decline of confrontational strategies in the international oil market. As the crisis-management aspects of the oil situation receded, they gave way to compromise rather than imposition in the attitudes of key industrialized countries *vis-à-vis* the producers, and the further the crisis waned away, the stronger became these elements in the IEA, and the more it lost its image of confrontation. Thus, the Agency could play a useful role in the Conference for International Economic Co-operation in Paris, where it was invited as observer to the energy group.

Second, the IEA has marked the beginning of direct and multilateral consumer involvement in the international oil market. Possibly the most interesting aspect of this is the information system built up by the IEA, which makes available data on international oil business transactions and on the oil industry to member governments[222]. Such information undoubtedly increases the capacity of consumer governments to influence events in the international oil market, and provide a check on the activities of the oil companies.

Two interesting conclusions can be drawn from these developments: first, the consumers show signs of organizing themselves. Although the vigour of the initial crisis management, and the assertive leadership of the USA have disintegrated somewhat, this new element persists. Second, there are several indications pointing towards a higher degree of direct government intervention in the international oil market from the consumer side. The two trends together may well provide the fundamentals of direct bargaining between producer and consumer governments over crucial issues, and the IEA could take on an important role in this connection.

Returning to the question asked at the outset of this chapter – the future evolution of world energy prices – it now appears likely that prices will continue to move upwards. The reasons for this lie not only in the energy transition facing the world; this is a long-term trend which strengthens the position of the oil exporters but cannot explain more than the proposition that world energy prices in 2000 will be higher than today. The actual setting of oil prices will probably continue to be effected by OPEC; this, at least, seems the conclusion from our attempt to assess the future distribution of bargaining power in the international energy system. All three groups analysed show some international weaknesses; their co-operation is fragile and rests on permanent processes of adjustment and compromise. This is least pronounced for the companies, and most obvious in the case of the importers. But the problems of OPEC are by no means small, either. The year 1979 showed that domestic instability of a key OPEC country could seriously destabilize the precarious balance in the international oil market. Pressure towards discontinuities in the system might well build up, and could result in new crises and dramatically wrong decisions. This leaves the importers in an awkward position: apart from the security guarantee provided by the IEA, the achievement of its main external policy objectives will remain outside its influence. At the same time, the present system of control and its serious fragmentation in 1979 leave major problems unsolved.

## Notes to Part three

1. A discussion of the vulnerabilities of industrialized importers can be found in E.N. Krapels, *Oil and Security, Problems and Prospects of Oil Importing Countries*, IISS, London 1977 (Adelphi Paper No. 136), Pt. 1
2. See E.K. Faridany, *LNG: 1974–1990, Marine Operations and Market Prospects for Liquefied Natural Gas*, Economist Intelligence Unit, London 1974 (Quarterly Economic Review Special No. 17), p.41
3. H.P. Drewry Shipping Consultants, *The Seaborne Transportation of Liquefied Natural Gas, 1977–1989: Trades, Costs, and Revenues*, Drewry, London 1977, p.25
4. The producers have declared their intention to ship 50% of their LNG exports in their own carriers (E.K. Faridany, *LNG: 1974–1990, Marine Operations and Market Prospects for Liquefied Natural Gas*, Economist Intelligence Unit, London 1974 (Quarterly Economic Review Special No.17), p.46). At present, they are still far away from this target; the financial implications, and the orders which would have to be placed to meet the targets, are spelled out in P.H. Drewry Shipping Consultants, *The Involvement of Oil Exporting Countries in International Shipping*, Drewry, London 1976, pp.56–57. According to this, OPEC would have to order about 2 407 000 m$^3$ of LNG carrier capacity at a total cost of about $2880 x 10$^6$ until 1980
5. It is interesting to note that during the 1973/4 embargo. Algeria did not interfere with LNG deliveries to the USA (E.K. Faridany, *LNG: 1974–1990, Marine Operations and Market Prospects for Liquefied Natural Gas*, Economist Intelligence Unit, London 1974 (Quarterly Economic Review Special No.17), p.14).
6. E.N. Krapels, *Oil and Security, Problems and Prospects of Oil Importing Countries*, IISS, London 1977 (Adelphi Paper No. 136) Pt. 1
7. P.R. Odell, *Oil and World Power, Background to the Oil Crisis*, 3rd edn., Penguin, Harmondsworth 1974, pp. 142–143
8. Agreement on an International Energy Programme, Article 19
9. Agreement on an International Energy Programme, Articles 19 and 62
10. E.N. Krapels, *Emergency Oil Reserves Programmes in Selected Importing Nations*, Royal Institute of International Affairs, London 1977 (mimeographed paper)
11. See, for example, G. Lenczowski, *Oil and the State in the Middle East*, Cornell University Press, Ithaca 1960, pp.321–337; W.Z. Laqueur, *The Struggle for the Middle East: The Soviet Union and the Middle East, 1958–1970*, 2nd edn., Penguin, Harmondsworth 1972, pp.152–153; B. Shwadran, *The Middle East, Oil, and the Great Powers*, Wiley, New York 1959, p.539; G.W. Stocking, *Middle East Oil, A Study in Economic and Political Controversy*, Allen Lane, London 1971, pp.458–459; H. Lubell, *Middle East Oil Crises and Western Europe's Energy Supplies*, Johns Hopkins University Press, Baltimore 1963, pp.8–16
12. R. Stobaugh, 'The oil companies in the crisis', *Daedalus*, Autumn 1975, pp. 179–202
13. *The Financial Times*, 21 April 1974, 9 Dec. 1975; H.P. Drewry Shipping Consultants, *The Tanker Crisis, Causes, Effects, and Prospects to 1985*, Drewry, London 1976, pp. 11–36
14. *The Financial Times*, 21 April 1974, 9 Dec. 1975
15. *Middle East Economic Digest*, 18 April 1975, p.15; see also *The Economist*, 8 Feb. 1975, 5 April 1975
16. H.P. Drewry Shipping Consultants, *Shipping Statistics and Economics*, No.88, Feb. 1978, p.27
17. *The Financial Times*, 28 March 1978
18. *The Economist*, 17 May 1975
19. *The Financial Times*, 5 April 1975
20. H.P. Drewry Shipping Consultants, *The Involvement of Oil Exporting Countries in International Shipping*, Drewry, London 1976, pp.127–136; *Petroleum Intelligence Weekly*, 5 April 1976; *The Economist*, 21 Feb. 1976
21. H.P. Drewry Shipping Consultants, *Shipping Statistics and Economics*, No.88, Feb. 1978, p.2
22. *The Financial Times*, 28 Feb. 1977, 28 March 1978; *The Economist*, 27 Aug. 1977
23. H.P. Drewry Shipping Consultants, *The Trading Outlook for Very Large*

Tankers, Drewry, London 1975, quoted in *Petroleum Economist*, Sept. 1975, pp.341–343; *The World Tanker Outlook, BP Briefing Paper No. 77/02*, BP, London 1977; A.G. Bailey, *World Tanker Prospects, 1977–1981*, Tilney & Co., Liverpool 1977

24 *Middle East Economic Survey*, 14 Nov. 1977; Comité professionnel du pétrole, *Pétrole 1977*, CPdP, Paris 1978

25 *Middle East Economic Digest*, 22 March 1976

26 *Le Monde*, 22 Jan. 1975

27 *The Financial Times*, 16 Feb. 1976

28 H.P. Drewry Shipping Consultants, *The Involvement of Oil Exporting Countries in International Shipping*, Drewry, London 1976, Sec. 2, Pt. 4

29 *The Financial Times*, 16 Feb. 1976. Typical costs are given as $100/dwt for the resale of a new tanker, $50/dwt for a second-hand purchase, and $200/dwt for new tankers (H.P. Drewry Shipping Consultants, *The Involvement of Oil Exporting Countries in International Shipping*, Drewry, London 1976, p.54)

30 *The Financial Times*, 24 Feb. 1976

31 *Le Monde*, 22 Jan. 1975. See also *Middle East Economic Digest*, 5 Sept. 1975, p.5

32 H.P. Drewry, Shipping Consultants, *The Involvement of Oil Exporting Countries in International Shipping*, Drewry, London 1976, pp. 53–54

33 E. Garnett, *Tanker: Performance and Cost, Analysis and Management*, Cornell Maritime Press, Cambridge, Maryland 1969, p. 14

34 Data from *Facts and Figures*, American Petroleum Institute, 1971, p.263

35 M. Willrich and M.A. Conant, 'The international energy agency: an interpretation and assessment', *American Journal of International Law*, April 1977, pp.199–223 (205)

36 See *The Financial Times*, 21 June 1976; *The Middle East and North Africa, 1977–1978*, Europa Publications, London 1977, pp. 334, 537

37 *The Middle East and North Africa, 1977–1978*, Europa Publications, London 1977, p.537

38 *The Middle East and North Africa, 1977–1978*, Europa Publications, London 1977; *The Financial Times*, 21 June 1976 (Survey: Iran); *The Financial Times*, 1 July 1975; *The Economist*, 12 July 1975

39 *The Financial Times*, 25 Feb. 1976 (Survey: Kuwait), 25 Feb. 1977 (Survey: Kuwait)

40 This argument was widely used in economic analysis after 1973/4 – see, for example, E.R. Fried, 'World market trends and bargaining leverage', in J.A. Yager, and E.B. Steinberg, *Energy and US Foreign Policy, A Report to the Energy Project of the Ford Foundation*, Ballinger, Cambridge Mass. 1974, pp.231–265

41 *Europe Oil Telegram*, 22 Dec. 1977; *Petroleum Economist*, Aug. 1977, p.316

42 Y. Sayigh, 'Arab oil politics, self-interest and responsibility', *Journal of Palestine Studies*, Spring 1975, pp.59–73 (64–65)

43 *International Herald Tribune*, 6 Sept. 1976

44 *Middle East Economic Survey*, 20 June 1974. This can be seen largely as the explanation of the Saudi insistence on no, or very low, price increases

45 *International Herald Tribune*, 3 March 1976

46 *The Financial Times*, 25 Feb. 1976 (Survey: Kuwait)

47 *The Middle East and North Africa, 1977–1978*, Europa Publications, London 1977, p.215

48 'Atlas économique et politique mondial', *Nouvel Observateur*, Paris 1977, p.17; *The Middle East and North Africa, 1977–1978*, Europa Publications, London 1977, p.220

49 *The Middle East and North Africa 1977–1978*, Europa Publications, London 1977

50 'Atlas économique et politique mondial', *Nouvel Observateur*, Paris 1977, p.18

51 *The Middle East and North Africa, 1977–1978*, Europa Publications, London 1977, pp.223–224; on the development strategy of Algeria, see G. Destanne de Bernis, 'Les problèmes pétroliers Algériens', *Etudes Internationales*, Dec. 1971; J.-M. Chevalier, *Le nouveau enjeu pétrolier*, Calmann-Levy, Paris 1973, pp.174 *seq.*; B. Etienne, *L'Algérie, Cultures et Révolution*, Seuil, Paris 1977, Ch. 8; *The Financial Times*, 12 May 1978

52 *The Middle East and North Africa, 1977–1978*, Europa Publications, London 1977, p.225; see also *The Middle East and North Africa, 1978–*

1979, Europa Publications, London 1978, p.220; *The Financial Times*, 12 May 1978
53 *The Middle East and North Africa, 1977—1978*, Europa Publications, London 1977, p.219
53a *The Economist*, 17 June 1978, p.98
54 This is explicit in the Algerian development plan formulated for hydrocarbon development for the period up to 2005: oil income for the thirty years is estimated at \$95 x 10$^9$, and natural gas sales are thought to render \$156 x 10$^9$ Four-fifths of the total sales would be abroad (*The Financial Times*, 12 May 1978)
55 Algeria has up to now mainly taken a 'hawkish' position on prices
56 *The Middle East and North Africa, 1978—1979*, Europa Publications, London 1978, p.342
57 *The Middle East and North Africa, 1977—1978*, Europa Publications, London 1977, p.342
58 'Atlas économique et politique mondial', *Nouvel Observateur*, Paris 1977, p.129; see also *The Financial Times*, 12 Sept. 1978
59 *The Financial Times*, 12 Sept. 1978
60 K.A. Hammeed and M.N. Bennett, 'Iran's future economy', *Middle East Journal*, Autumn 1975, pp.418—432 (425), quoting from *Employment and Income Policies for Iran*, ILO, Geneva 1973; see also M.H. Pesaran, 'Income distribution and its major determinants in Iran', in J. Facqz, (ed.), *Iran: Past, Present, and Future*, Aspen Institute for Humanistic Studies, New York 1976
61 *The Middle East and North Africa, 1977—1978*, Europa Publications, London 1977, p.342. Since 1976 growth has slowed down considerably, and has been estimated at 17% for 1976/7, and even lower for 1977/8
62 *The Financial Times*, 12 Sept. 1978
63 *Neue Zuercher Zeitung*, 3 Aug. 1978; *The Financial Times*, 12 Sept. 1978; *The Middle East and North Africa, 1978—1979*, Europa Publications, London 1978, pp.342—343
64 *The Middle East and North Africa, 1977—1978*, Europa Publications, London 1977, p.348
65 *The Financial Times*, 28 July 1975
66 *The Financial Times*, 28 July 1975; *The Middle East and North Africa, 1977—1978*, Europa Publications, London 1978, p.353. There is little evidence that manufacturing exports have made much progress, rather on the contrary. Non-oil exports are still predominantly agricultural products and traditional goods such as cotton and carpets
67 *The Middle East and North Africa, 1977—1978*, Europa Publications, London 1977, p. 383
68 G. Mitchell, 'Iraq's battle of endurance supported by Soviet backers', *New Middle East*, April 1973, pp.39—42
69 *The Middle East and North Africa, 1977—1978*, Europa Publications, London 1977
70 'Atlas économique et politique mondial', *Nouvel Observateur*, Paris 1977, p.128
71 *The Middle East and North Africa, 1977—1978*, Europa Publications, London 1977, p.385
72 *The Middle East and North Africa, 1977—1978*, Europa Publications, London 1977
73 *Arab Oil and Gas*, 1 June 1978
74 *Le Monde*, 17—18 June 1976 (Survey: Iraq)
75 *Le Monde*, 17—18 June 1976 (Survey: Iraq)
76 *Africa South of the Sahara*, Europa Publications, London 1975, p.613
77 *The Financial Times*, 14 Feb. 1977 (Survey: Nigeria)
78 *The Financial Times*, 14 Feb. 1977 (Survey: Nigeria)
78a *The Financial Times*, 29 Aug. 1978
79 *The Financial Times*, 29 Aug. 1978
80 *The Financial Times*, 29 Aug. 1978
80a *The Financial Times*, 30 Aug. 1978
80b *The Financial Times*, 30 Aug. 1978
80c *The Financial Times*, 30 Aug. 1978
81 *The Financial Times*, 30 Aug. 1978
82 'Atlas economique et politique mondial', *Nouvel Observateur*, Paris 1977, p.154
83 *The Middle East and North Africa, 1977—1978*, Europa Publications, London 1977, p.537
84 J.A. Allan and K.S. McLachlan, *Agricultural Development in an Oil Economy, The Case of Libya*, School of Oriental and African Studies, London 1976 (mimeographed paper delivered at the Conference on the Impact of Oil on the Social and Economic Life of the Oil Producing Countries, University of Sussex, Institute for Development Studies, 28 May 1976)
85 *The Middle East and North Africa, 1977—1978*, Europa Publications, London 1977, p.537
85a *Neue Zuercher Zeitung*, 16 March 1978

86 J.A. Allan and K.S. McLachlan, *Agricultural Development in an Oil Economy, The Case of Libya*, School of Oriental and African Studies, London 1976 (mimeographed paper delivered at the Conference on the Impact of Oil on the Social and Economic Life of the Oil Producing Countries, University of Sussex, Institute for Development Studies, 28 May 1976); J.A. Allan, K.S. McLachlan and E.T. Penrose, *Libya: Agricultural Economic Development*, Frank Cass, London 1975, Ch.4; *The Middle East and North Africa, 1977–1978*, Europa Publications, London 1977, p.539

87 J.A. Allan and K.S. McLachlan, *Agricultural Development in an Oil Economy, The Case of Libya*, School of Oriental and African Studies, London 1976 (mimeographed paper delivered at the Conference on the Impact of Oil on the Social and Economic Life of the Oil Producing Countries, University of Sussex, Institute for Development Studies, 28 May 1976); J.A. Allan, K.S. McLachlan and E.T. Penrose, *Libya: Agricultural Economic Development*, Frank Cass, London 1975

87a *Neue Zuercher Zeitung*, 9 March 1978

88 *The Middle East and North Africa, 1977–1978*, Europa Publications, London 1977, p.471

89 *Atlas, Buechergilde Gutenberg*, Frankfurt/M. 1976, p.135

90 *The Middle East and North Africa, 1977–1978*, Europa Publications, London 1977, p.471

91 *The Financial Times*, 25 Feb. 1976 (Survey: Kuwait)

92 *The Financial Times*, 25 Feb. 1976 (Survey: Kuwait)

93 H.W. Maull, 'The Arms Trade with the Middle East and North Africa', in *The Middle East and North Africa, 1977–1978*, Europa Publications, London 1977, pp.114–120 (119)

94 *The Middle East and North Africa, 1974–1975*, Europa Publications, London 1974, p.468

95 *The Financial Times*, 25 Feb. 1976 (Survey: Kuwait). Income from official reserves alone has been estimated at $\$1 \times 10^9$, or about 17% of total government revenues in 1975

96 *The Financial Times*, 25 Feb. 1976 (Survey: Kuwait). But note that the realization of a large LPG export scheme planned to come on stream around 1980 would require an oil production level of about $3 \times 10^6$ b/d. Kuwait is the most vulnerable producer with regard to domestic effects of low oil production – apparently, problems of insufficient natural gas supply (which is associated with oil production, and therefore depends on production levels of oil) already caused problems during exceptionally low oil production periods in 1976: natural gas supplies electricity and energy for all vital functions of Kuwait society, including desalination

97 Committee on Energy and Natural Resources, US Senate, *Access to Oil – the United States Relationship with Saudi Arabia and Iran*, GPO, Washington 1977, p.37; A. Gaelli, 'Die politische und soziale Entwicklung der OPEC Staaten', in M. Tietzel, (ed.), *Die Energiekrise: Fuenf Jahre danach*, Verlag Neue Gesellschaft, Bonn 1978, pp.73–100 (83)

98 *The Middle East and North Africa, 1975–1976*, Europa Publications, London 1975, p.600

99 *The Middle East and North Africa, 1977–1978*, Europa Publications, London 1977, pp.603–604; *The Financial Times*, 21, 28 March 1977 (Survey: Saudi Arabia)

100 *The Middle East and North Africa, 1977–1978*, Europa Publications, London 1977, pp.605–606, 608; *The Financial Times*, 28 March 1977

101 *The Financial Times*, 12 Jan. 1976 (Survey: Saudi Arabia), 21, 28 March 1977 (Survey: Saudi Arabia)

102 *Arab Oil and Gas*, 16 March 1976; *International Herald Tribune*, 3 March 1976; *The Financial Times*, 21, 28 March 1977 (Survey: Saudi Arabia)

103 For the first time in a number of years, the Saudi Arabian budget was in deficit in 1975/6—an indication of the loss of control over spending. For reports on colossal waste of money in defence spending and in the behaviour of the royal family, see *Afro-Asian Affairs*, 15 June 1978, 29 March 1978, 27 April 1977

104 See *The Middle East*, May 1978, p.30; *Afro-Asian Affairs*, 24 Aug. 1977; *Le Monde*, 27 Jan. 1977; Committee on Energy and Natural Resources, US

Senate, *Access to Oil — The United States Relationship with Saudi Arabia and Iran*, GPO, Washington 1977
105 *Afro-Asian Affairs*, 18 May 1977; *International Currency Review*, Vol. 9, No. 2, 1977, pp.5–18. On the Joint Economic Commission, see S.D. Hayes, 'Joint economic commissions as instruments of US foreign policy in the Middle East, *Middle East Journal*, Winter 1977, pp.16–30
106 *The Financial Times*, 20 Sept. 1977
107 *Petroleum Economist*, April 1977, pp.125–127; J. Bedore and L. Turner, 'Die Industrialisierung der oelproduzierenden Laender im Mittleren Osten, Plaene, Probleme, Auswirkungen', *Europa Archiv*, No.16, 1977, pp. 538–546; J. Bedore and L. Turner, 'Saudi and Iranian petrochemicals and oil refining: trade welfare in the 1980s?, *International Affairs*, Oct. 1977, pp. 572–586; *The Financial Times*, 17 April 1978; see also L. Turner and J. Bedore, 'The trade politics of Middle Eastern industrialization', *Foreign Affairs*, Winter 1978/9, pp.306–322 (308–311)
108 See M.A. Adelman, *The World Petroleum Market*, Johns Hopkins University Press, Baltimore 1972, Ch. 2, particularly p. 76. For production cost calculations for Persian/Arab Gulf oil in income tax assessments, see M. Field, 'Oil in the Middle East and North Africa', in *The Middle East and North Africa, 1975–1976*, Europa Publications, London 1975, pp.69–102 (90). According to these figures, production costs in the Gulf in 1975 were between 7 and 60 cents per barrel, the latter figure being for offshore oil; the average cost was about 20 cents
109 M.A. Adelman, *The World Petroleum Market*, Johns Hopkins University Press, Baltimore 1972, p.76
110 M.A. Adelman, *The World Petroleum Market*, Johns Hopkins University Press, Baltimore 1972, p.76
111 P.H. Frankel, *Essentials of Petroleum*, 2nd edn., Frank Cass, London 1969
112 See, for example, E.T. Penrose, *The Large International Firm in Developing Countries: The International Petroleum Industry*, Allen and Unwin, London 1968
113 This approach builds on the work of J.E. Hartshorn,*Oil Companies and Governments*, Faber, London 1969,

and J.-M. Chevalier, *Le nouvel enjeu pétrolier*, Calmann-Levy, Paris 1973 (under the English title *The New Petroleum Stakes*, Allen Lane, London 1975). Chevalier's theoretical approach is developed at length in 'Elements théoriques d'introduction à l'économie du pétrole, l'analyse du rapport de force', in *Revue d'Economie Politique*, March–April 1975, pp. 231–256
114 J.–M. Chevalier, 'Elements théoriques d'introduction à l'économie du pétrole, l'analyse du rapport de force', in *Revue d'Economie Politique*, March–April 1975, pp.243–249
115 This point is made in M.A. Adelman, 'Is the oil shortage real? oil companies as OPEC tax collectors', *Foreign Policy*, Spring 1972, pp.67–109
116 Ultimately, the contradiction is one between the principles of the nation-state, national sovereignty, and a global economic perspective which would favour the exhaustion of cheap energy resources before the world turns to the more expensive alternatives
117 *International Herald Tribune*, 18 Jan. 1977
118 Such a situation existed in the late 1960s, and has led some analysts to argue that the US government secretly pushed for oil price increases to offset the economic disadvantage *vis-à-vis* the low-cost energy economies of Japan and Western Europe. See P.R. Odell, *Oil and World Power, Background to the Oil Crisis*, 3rd edn., Penguin, Harmondsworth 1974, p.195, and also V.H. Oppenheim, 'Why prices go up: the past: we pushed them', *Foreign Policy*, Winter 1976/77, pp.24–47
119 M.A. Adelman, *The World Petroleum Market*, Johns Hopkins University Press, Baltimore 1972, pp.138–139; J.-M. Chevalier, *Le nouvel enjeu pétrolier*, Calmann-Levy, Paris 1973, p.36
120 The USA became a net importer of petroleum in 1948
121 M. Field, 'Oil in the Middle East and North Africa,' in *The Middle East and North Africa, 1976–1977*, Europa Publications, London 1976, pp. 73–108 (84); see also M.A. Adelman, *The World Petroleum Market*, Johns Hopkins University Press, Baltimore 1972, pp.138–139, and G.W. Stocking, *Middle East Oil, A Study in Economic*

and *Political Controversy*, Allen Lane, London 1970, pp.339–402. The Marshall Plan Administration financed very large crude oil purchases for Western Europe; the authorizing act had requested that such purchases be made at the lowest competitive prices

122 E.T. Penrose, 'Vertical integration with joint control of raw-material production: crude oil in the Middle East,' in E.T. Penrose, (ed.), *The Growth of the Firm, Middle East Oil, and other Essays*, Allen and Unwin, London 1971

123 C. Tugendhat and A. Hamilton, *Oil, The Biggest Business*, 2nd edn., Eyre Methuen, London 1975, Ch. 14; G.W. Stocking, *Middle East Oil, A Study in Economic and Political Controversy*, Allen Lane, London 1970, pp. 144–181

124 M.A. Adelman, *The World Petroleum Market*, Johns Hopkins University Press, Baltimore 1972, pp.150–155

125 P.R. Odell, *Oil and World Power, Background to the Oil Crisis*, Penguin, Harmondsworth 1974, pp.29–30

126 M. Field, 'Oil in the Middle East and North Africa', in *The Middle East and North Africa, 1975–1976*, Europa Publications, London 1975, p.84

127 J.-M. Chevalier, *Le nouvel enjeu pétrolier*, Calmann-Levy, Paris 1973, pp.44–45; H. Elsenhans, 'Entwicklungstendenzen der Weltoelindustrie', in H. Elsenhans, (ed.), *Erdoel fuer Europa*, Hamburg 1974, pp.7–42 (14–15)

128 G.W. Stocking, *Middle East Oil, A Study in Economic and Political Controversy*, Allen Lane, London 1970, p.404

129 G.W. Stocking, *Middle East Oil, A Study in Economic and Political Controversy*, Allen Lane, London 1970, pp.413–415. The implications are discussed in E.T. Penrose, 'Vertical integration with joint control of raw-material production: crude oil in the Middle East', in E.T. Penrose, (ed.), *The Growth of the Firm, Middle East Oil, and other Essays*, Allen and Unwin, London 1971

130 The joint venture model was initiated by the Italian national oil company ENI in agreements with the Egyptian national oil company and the Iranian NIOC in 1957; the Iranian national oil company also introduced the service contract with another national oil company, the French Elf/ERAP (1966)

131 Z. Mikdashi, *The Community of Oil Exporting Countries: A Study in Governmental Co-operation*, Cornell University Press, Ithaca 1972, pp.112–133

132 H.W. Maull and O. Ursachen, *Perspektiven, Grenzen*, EVA, Cologne 1975, Ch. 1

133 Z. Mikdashi, *The Community of Oil Exporting Countries: A Study in Governmental Co-operation*, Cornell University Press, Ithaca 1972, pp.143–145

134 N.H. Jacoby, *Multinational Oil*, Macmillan, New York 1975, p.248, gives the following figures for rates of earning of US direct investments abroad (as % of net income to net assets):

|  | Petroleum | Mining and melting | Manufacturing | Other |
|---|---|---|---|---|
| 1955 | 30.2 | 13.1 | 13.0 | 10.2 |
| 1960 | 13.6 | 13.1 | 10.5 | 9.1 |
| 1965 | 12.5 | 15.6 | 11.9 | 10.9 |
| 1966 | 11.7 | 16.8 | 10.9 | 9.8 |
| 1967 | 11.8 | 17.1 | 9.3 | 9.2 |
| 1968 | 12.3 | 16.3 | 10.4 | 9.7 |
| 1969 | 11.1 | 14.4 | 12.4 | 11.3 |
| 1970 | 11.0 | 11.9 | 11.6 | 11.1 |
| 1971 | 12.5 | 8.1 | 11.9 | 11.7 |
| 1972 | 9.9 | 6.2 | 14.1 | 12.2 |

135 See M.A. Adelman, 'Is the oil shortage real? Oil companies as OPEC tax collectors?, *Foreign Policy*, Spring 1972, and V.H. Oppenheim, 'Why prices go up: the past: we pushed them', *Foreign Policy*, Winter 1976/77
136 For details on the Teheran and Tripoli Negotiations, see F. Rouhani, *A History of OPEC*, Praeger, New York 1971, pp.15–25, and J.-M. Chevalier, *Le nouvel enjeu pétrolier*, Calmann-Levy, Paris 1973, pp.95–102
137 See J.-M. Chevalier, *Le nouvel enjeu pétrolier*, Calmann-Levy, Paris 1973, pp.229–232; *Petroleum Economist*, May 1976, pp.182–184; *The Financial Times*, 14 March, 13 June, and 2 July 1975
138 See, for example, the case of Gulf Oil: *The Financial Times*, 2 July 1975
139 M.Wilkins, 'The oil companies in perspective', in R. Vernon, (ed.), *The Oil Crisis*, W.W. Norton, New York 1976, pp.159–177
140 A. Sampson, *The Seven Sisters*, Hodder and Stoughton, London 1975, p.263; R.B. Stobaugh, 'The oil companies in the crisis', in R. Vernon, (ed.), *The Oil Crisis*, W.W. Norton, New York 1976, pp.179–202
141 The Chase Manhattan Bank Energy Division, *A Financial Analysis of a Group of Petroleum Companies, 1976*, New York 1977
142 *The Sunday Times*, 16 May 1976
143 *The Financial Times*, 7 May 1976
144 Phillips Petroleum Company, Annual Report, 1973
145 BP, Annual Report, 1974
146 *The Financial Times*, 21 July 1978; *International Herald Tribune*, 20 April 1976
147 *The Financial Times*, 5 April 1975
148 P.R. Odell and L. Vallenilla, *The Pressures of Oil, a Strategy for Economic Revival*, Harper and Row, London 1978, pp.40–41
149 *The Economist*, 3 April 1976
150 N.H. Jacoby, *Multinational Oil*, Macmillan, New York 1975, pp.123–124
151 T.R. Moran, 'New deal or raw deal in raw materials', *Foreign Policy*, Winter 1971/2, pp.119–136
152 H. Elsenhans, 'Entwicklungstendenzen der Welterdölindustrie', in H. Elsenhans, (ed.), *Erdöl für Europa*, Hoffman and Campe, Hamburg 1974, pp.7–47; see also P.R. Odell and L. Vallenilla, *The Pressures of Oil, a Strategy for Economic Revival*, Harper and Row, London 1978, Ch. 2
153 G.W. Stocking, *Middle East Oil, A Study in Political and Economic Controversy*, Allan Lane, London 1970, p.411
154 A. Sampson, *The Seven Sisters*, Hodder and Stoughton, London 1975, p.177
155 A. Sampson, *The Seven Sisters*, Hodder and Stoughton, London 1975, pp.177–179; H. Elsenhans, 'Entwicklungstendenzen der Welterdölindustrie' in H. Elsenhans, (ed.), *Erdöl für Europa*, Hoffman and Campe, Hamburg 1974, p.40
156 P.R. Odell and L. Vallenilla, *The Pressures of Oil, a Strategy for Economic Revival*, Harper and Row, London 1978, pp.40–41
157 E.T. Penrose, *The Large International Firm in Developing Countries: The International Petroleum Industry*, Allen and Unwin, London 1968; B. Dasgupta, 'The changing role of the major international oil firms', *Journal of the Middle East*, Jan. 1975, pp.1–31 (4–6)
158 J.-M. Chevalier, *Le nouvel enjeu pétrolier*, Calmann-Levy, Paris 1973, pp.233–235; A. Sampson, *The Seven Sisters*, Hodder and Stoughton, London 1975, pp.229–318; *The Sunday Times*, 16 May 1976
159 P.R. Odell and L. Vallenilla, *The Pressures of Oil, a Strategy for Economic Revival*, Harper and Row, London 1978, pp.40–41. Fees for managerial and technical assistance range between 20 and 35 cents per barrel (see also *The Financial Times*, 25 Feb. 1976, and 30 March 1976). Companies also often receive discounts on the official selling prices, which seem to be worth up to 50 cents per barrel
160 Under the ARAMCO–Saudi agreement, amounts to be lifted are about $7.5 \times 10^6$ b/d, with the companies having to pay certain penalties if these export volumes are not met. In Kuwait liftings of BP will be 450 000 b/d until 1980, while Gulf will take 500 000 b/d, with a possibility of renewal after 1980 at a level not below 800 000 b/d for the two companies together (*The Financial Times*, 25 Feb. 1976, 30 March 1976)

161 This has already happened in the case of Kuwait: see, for example, *Le Monde*, 21 Jan. 1975. Here Shell has a long-term purchasing agreement for 310 000 b/d, Exxon for 100 000 b/d, and similar contracts exist for Arabian Oil (a Japanese firm) and Aminoil *The Financial Times*, 25 Feb. 1976

162 See J.-M. Chevalier, *Le nouvel enjeu pétrolier*, Calmann-Levy, Paris 1973, Ch. 5; *Hearings before the Permanent Subcommittee on Investigations, Committee on Government Operations, US Senate*, 22 Jan. 1974, GPO, Washington 1974 (Current Energy Shortages Oversight Series, Pt. IV), pp.540–546; Ford Foundation, *A Time to Choose, America's Energy Future*, Ballinger, Cambridge, Mass. 1974, Ch.9; C.D. Russel, 'Diversification des activités des sociétés pétrolières', *Revue de l'Energie*, April 1977, pp.269 seq

163 Details on companies are derived from sources quoted in note 162, and on company annual reports for the period 1974–1977

164 C.D. Russel, 'Diversification des activités des sociétés pétrolières', *Revue de l'Energie*, April 1977

165 C. Tugendhat and A. Hamilton, *Oil: The Biggest Business*, 2nd edn., Eyre Methuen, London 1974, Ch. 30

166 C. Tugendhat and A. Hamilton, *Oil: The Biggest Business*, 2nd edn., Eyre Methuen, London 1974, p.367; *The Financial Times*, 2 July 1975; annual reports of companies

167 H. Elsenhans, 'Entwicklungstendenzen der Welterdölindustrie', in H. Elsenhans (ed.), *Erdöl für Europa*, Hoffman and Campe, Hamburg 1974, p.19; J.-M. Chevalier, *Le nouvel enjeu pétrolier*, Calmann-Levy, Paris 1973, pp.233–235

168 *International Herald Tribune*, 9 Dec. 1974. One problem of falling prices for the companies would be large losses on stocks: *The Sunday Times*, 16 May 1976. See also P.R. Odell and L. Vallenilla, *The Pressures of Oil, a Strategy for Economic Revival*, Harper and Row, London 1978, pp.32–33

169 This is, of course, a book-keeping exercise since in a vertically integrated company the only way to assess profits is to deduct total costs from total revenues. The profits which could be realized on the basis of OPEC oil were, however, so large that inefficient refining and distribution operations could be accepted. See E.T. Penrose, *The Large International Firm in Developing Countries: The International Petroleum Industry*, Allen and Unwin, London 1967

170 This was already achieved to some extent in 1975, when the majors of the USA managed to shift their profits from operations abroad to operations in the USA. This appeared to include more than just the higher profitability of domestic production. See *Petroleum Economist*, May 1976, pp.182–184, and J.-M. Chevalier, *Le nouvel enjeu pétrolier*, Calmann-Levy, Paris 1973, pp.229–232

171 See C. Tugendhat and A. Hamilton, *Oil: The Biggest Business*, 2nd edn., Eyre Methuen, London 1974, p.171; *The Financial Times*, 14 March 1975, 2 July 1975, 13 June 1975

172 *Petroleum Economist*, July 1975, p.2

173 B. Dasgupta, 'The changing role of the major international oil firms', *Journal of the Middle East*, Jan. 1975, p.15; for the training programmes of the companies, see G. Lenczowski, *Oil and States in the Middle East*, Cornell University Press, Ithaca 1960, Ch. 12; G.W. Stocking, *Middle East Oil, A Study in Political and Economic Controversy*, Allen Lane, London 1970, pp.134–136

174 M. Field, 'OPEC's strength in depth', *Middle East International*, Nov. 1975, pp.11–13

175 R.J. Deam, J. Leather and J.G. Hale, *Understanding Energy, A Rational Basis for Planning Alternative Strategies*, Energy Research Unit, Queen Mary College, London 1977 (mimeographed paper)

176 *Middle East Economic Survey*, Supplement, 25 Oct. 1974, p.9

177 *The Sunday Times*, 25 April 1976

178 *The Sunday Times*, 25 April 1976

179 See *The Financial Times*, 21 Feb. 1975, 4 April 1975; *The Financial Times*, 18 July 1978, 19 July 1978

180 *The Financial Times*, 19 July 1978

181 See also Part IV, Chapters 11 and 12 on CIEC and the Euro-Arab dialogue, where these demands were voiced

182 See Part IV, Chapters 11 and 12

183 L. Turner, 'Die Rolle des Oels im Nord-Sued Dialog', *Europa Archiv*, No. 4, 1977, pp. 102–112

184 P.R. Odell and L. Vallenilla, *The Pressures of Oil, a Strategy for Economic Revival*, Harper and Row, London 1978, Ch.8
185 Z. Mikdashi, 'The OPEC process', in R. Vernon, (ed.), *The Oil Crisis*, W.W. Norton, New York 1976, pp.203–216 (200–207); Z. Mikdashi, *The International Politics of Natural Resources*, Cornell University Press, Ithaca 1976
186 Z. Mikdashi, 'The OPEC process', in R. Vernon, (ed.), *The Oil Crisis*, W.W. Norton, New York 1976, p.208
187 A. Hamilton, 'OPEC's next step', *Middle East International*, June 1976, pp.8–10
188 L. Turner and J. Bedore, 'Saudi Arabia–Eine Geldmacht', *Europa Archiv*, No.13, 1978, pp.397–410
189 *The Financial Times*, 24 Feb. 1974
190 M. Field, 'Oil in the Middle East and North Africa', in *The Middle East and North Africa, 1977–1978*, Europa Publications, London 1977, pp.77–113
191 M. Field, 'Oil in the Middle East and North Africa', in *The Middle East and North Africa, 1977–1978*, Europa Publications, London 1977, pp.77–113
192 E.T. Penrose, *The Role of OPEC in the World Oil Industry*, SSRC, London 1978. (Paper delivered at SSRC Conference on the International Context for Energy Decision-Making in Britain, 18 July 1978.)
193 P.R. Odell and L.Vallenilla, *The Pressures of Oil, a Strategy for Economic Revival*, Harper and Row, London 1978, Ch. 4; see also R. Vernon, (ed.), *The Oil Crisis*, W.W. Norton, New York 1976
194 K. Knorr, 'The limits of economic and military power', in R. Vernon, (ed.), *The Oil Crisis*, W.W. Norton, New York 1976, pp.229–243 (230). See also H. Maull, *Oil and Influence: the Oil Weapon Examined*, IISS, London 1975 (Adelphi Paper No. 117)
195 *Arab Report and Record*, 1–15 Jan. 1974, 16–31 Jan. 1974
196 R. Corrigan, 'Project independence by 1985', *Energy Policy*, March 1975, pp.73–74
197 *Keesing's Contemporary Archives*, Vol. 1974, pp.26 293 and 26 294
198 Bo Widerberg, *Oil and Security*, SIPRI, Stockholm 1974, App. 7
199 M. Willrich and M.A. Conant, 'The International Energy Agency: an interpretation and assessment', *American Journal of International Law*, April 1977, pp.99–224
200 Committee on Energy and Natural Resources, US Senate, *Access to Oil–the United States Relationship with Saudi Arabia and Iran*, GPO, Washington 1977
201 *Strategic Survey 1974*, IISS, London 1975, p.25
202 P.R. Odell and L. Vallenilla, *The Pressures of Oil, a Strategy for Economic Revival*, Harper and Row, London 1978, Ch.4; the US strategy is outlined in T.O. Enders, 'OPEC and the industrialized countries: the next ten years', *Foreign Affairs*, July 1975, pp.625–637. See also the influential article by W.J. Levy 'World oil cooperation or international chaos', *Foreign Affairs*, July 1974, pp.690–713
203 *Strategic Survey, 1974*, IISS, London 1975, p.24
204 T.O. Enders, 'OPEC and the industrialized countries: the next ten years', *Foreign Affairs*, July 1975
205 *Keesing's Contemporary Archives*, Vol. 1975, pp.27 479 and 27 480
206 See, for example, *Petroleum Economist*, Feb. 1976, pp.43–45; S. Hoffmann, 'Les Etats Unis: du refus au compromis', *Revue française de Science Politique*, Aug. 1976, pp.684–695 (690)
207 *Le Monde*, 17 July 1975; *The Economist*, 9 Aug. 1975; *International Herald Tribune*, 3 Sept. 1975
208 *International Herald Tribune*, 29 April 1975
209 S. Hoffmann, 'Les Etats Unis: du refus au compromis', *Revue française de Science Politique*, Aug. 1976, p.690
210 S. Hoffmann, 'Les Etats Unis: du refus au compromis', *Revue française de Science Politique*, Aug. 1976, p.691
211 S. Hoffmann, 'Les Etat Unis: de refus au compromis', *Revue française de Science Politique*, Aug. 1976, p.689
212 *Time*, 2 May 1977
213 Committee on Energy and Natural Resources, US Senate, *Access to Oil–the United States Relationship with Saudi Arabia and Iran*, GPO, Washington 1977
214 W. Hager, 'Die Internationale Energieagentur: Problematische Sicherheitsallianz fuer Europa', in

W. Hager, (ed.), *Erdoel und internationale Politik*, Piper, München 1975, pp.87–114 (96–97)
215 *Arab Report and Record*, 16–30 Nov. 1974
216 W.L. Kohl, 'The International Energy Agency: the political context', in J.C. Hurewitz, (ed.), *Oil, the Arab–Israeli Dispute and the Industrialized World: Horizons of Crisis*, Westview, Boulder, Colorado 1976, pp.246–257
217 *Neue Zuercher Zeitung*, 21 Nov. 1975; *Keesing's Contemporary Archives*, Vol. 1975, pp. 27 480 and 27 481
218 *International Herald Tribune*, 22 Nov. 1975
219 M. Willrich and M.A. Conant, 'The International Energy Agency: an interpretation and assessment', *American Journal of International Law*, April 1977, pp.99–224; *Neue Zuercher Zeitung*, 3 Feb. 1976; *The Financial Times*, 31 Jan. 1976; *Le Monde*, 21–22 Dec. 1975
220 *International Herald Tribune*, 2 Feb. 1976; see also T.O. Enders, 'OPEC and the industrialized countries: the next ten years', *Foreign Affairs*, July 1975
221 *Neue Zuercher Zeitung*, 8 Oct. 1977; *Le Monde*, 7 Oct. 1977
222 M. Willrich and M.A. Conant, 'The International Energy Agency: an interpretation and assessment', *American Journal of International Law*, April 1977 p.221

PART FOUR

# Towards a global management of energy problems: Elements and proposals

Up to this point, our analysis can be summarized in two straightforward statements:

(1) Europe will continue to be heavily dependent on imported energy from a small group of developing countries, and this dependence will continue to pose difficult problems of supply security, supply availability, and cost.
(2) To secure its vital interests in the energy field, Europe will have to search for international co-operation. Neither supply security, nor long-term availability of energy imports at reasonably stable or at least only slowly changing prices will be possible unilaterally: world energy interdependence has reached a level where different vulnerabilities, asymmetries in power and leverage, and above all difficulties in harmonizing the time-frame of national energy policies with each other (the necessity of oil importers to reduce dependence on oil imports as opposed to the need for exporters to turn a depleting asset into self-sustained, permanent economic growth) simply demand common efforts of management.

Co-operation will, therefore, be the *leitmotiv* of the following chapters; it might thus be useful to take a closer look at this concept itself. The first distinction to be made is the different *directions* of co-operation. The first direction of co-operation for Europe will continue to be common policies with other industrialized countries. With regard to security of supplies, remaining problems appear minor: existing mechanisms seem broadly adequate. Any improvement would also primarily concern the industrialized world; the defensive character of the IEA agreement on emergency sharing (or any similar scheme) practically excludes possibilities of co-operation with OPEC supply security, although co-operation in other areas will undoubtedly reflect upon OPEC's willingness to consider supply interruptions, and therefore have consequences for supply security, as well. Commitments to avoid such interruptions for political reasons would also be valuable, for any embargo and/or cutback would distract attention from more important underlying energy issues, and also distort temporarily the forces working towards energy transition: prices might oscillate sharply, demand growth could be constrained temporarily, and both would inevitably also change perceptions about the problems. The danger would lie in over-reactions, deceptive crash programmes leading to waste of resources, and a deterioration of the awareness of common problems for both exporters and importers.

The real issues, according to this argument, are the long-term stability of supplies and their price. Here, co-operation basically concerns the relationship between exporters and importers of energy, although concerted efforts by the industrialized countries (and of each industrialized country nationally!) are obviously also necessary: in the absence of such concerted action, the failure of one nation to reduce oil imports could have grave consequences for the international oil market, and thus affect everybody; this is the essence of the challenge. However, as the analysis in Part 3 has shown, even the industrialized countries taken together cannot expect to regain control over supply and price of their energy imports: this control will remain with OPEC. Co-operation between the industrialized countries and the exporters is therefore crucial to secure European international energy policy objectives.

The second question to be posed is: what are the alternatives to co-operation? As far as the dimension of industrialized importers among themselves is concerned, the alternative would be competition — competition for scarce and expensive energy imports. Whatever form such competition takes (bidding up prices, scramble for 'preferential' access to oil, efforts to foster exclusive 'special relationships'), the risks will be very serious for the fabric of Western co-operation on other issues: the deterioration of intra-Western relations will undoubtedly affect trade policies (as the weaker countries try to push their way back into access to oil imports), and quite possibly also co-operation in other areas, such as efforts to stem proliferation of nuclear weapons.

With regard to co-operation with OPEC, the alternatives appear at least as risky, and even less promising[1]. Efforts to break the cartel, to regain control over oil production and prices, are likely to end in failure, and would thus only set back the chances of success for a co-operative approach. The other alternative, a disentangling of industrialized countries from energy dependence, is simply not realistic for most of them even in the very long run. The United States could probably do it — but the result of its success would only be accentuated imbalances within the Western alliance, without any policy alternative to co-operation[2]. At best, it could turn the market situation around, and put OPEC temporarily under pressure; but even then, a strategy to break up OPEC, and to roll back prices, would be unwise, given the traumatic implications of such a move on OPEC societies and economies, on regional stability in the Middle East, and on exporters' attitudes when their supplies will be needed again.

The third aspect to be considered is the *character* of co-operation: what kind of co-operation do we want, what kind do we offer? A first possibility would be to select the most important oil exporters and 'co-opt' them — thus dividing the OPEC 'haves' from the 'have-nots'. This might be a useful way to encourage OPEC moderation and avoid abuses of OPEC power — but the risk is that this strategy is bound to run into trouble as soon as it works to the disadvantage of the co-opted countries. This is inevitable if the divisions within OPEC reduce its political leverage. Besides, co-operation will depend on few key countries, and thus increase the vulnerability to domestic political changes in such states.

In the long run a safer strategy seems to be one designed to maximize mutual longer-term advantages in broad co-operation, thus spreading the risks, and enhancing moderation (multilateral co-operation tends to eliminate elements of irrationality, and to induce caution). This already determines the choice on a fourth aspect: the *bilateral* versus the *multilateral* approach. As pointed out already, bilateralism implies competition between industrialized countries; besides, energy prices are not amenable to bilateral arrangements.

Long-term stability of supplies lends itself more readily to bilateral arrangements. A consumer offering attractive economic assistance, development aid, technical co-operation, and trade preferences might achieve a higher degree of supply stability than one unable to offer the same benefits. This will not solve the overall problem, however: if a supply squeeze develops, and prices increase so as to force rationing of imports, the weakest countries will suffer first and most — but the damage will spread to stronger countries through international trade, finance, and politics. No country can isolate itself from the reverberations of an energy crisis — as even Eastern Europe and the Soviet Union found out after 1973/4.

Bilateralism was greatly supported in the aftermath of the 1973/4 crisis: European countries and Japan rushed to conclude oil-against-technology deals with some producers, often including arms transfers on a large scale[3]. Gradually, however, the emphasis shifted to agreements of co-operation in more general terms. The much increased interaction between importers and exporters in the development of exporter economies implied a substantial increase in influence of the consumer countries. Western, in particular US, citizens are needed to train the armies of exporter countries, to fly their planes and man their new sophisticated military equipment; they build and supervise the new industries, the infrastructure, and the services developing along the shores of the Persian/Arabian Gulf. The close associations of a producer country with one particular importer, e.g. the relationship between the USA and Iran, and above all between the USA and Saudi Arabia, led to an uneasy symbiosis[4]. This new form of bilateralism is probably necessary to stabilize today's precarious energy situation (even at times of substantial surplus capacity and market gluts a cutback of Saudi oil production to, say, $5 \times 10^6$ b/d would cause tension and put pressure on prices), but, as indicated, it is not without dangers[5]: the exclusive relationship between Washington and Riyad could be detrimental to America's allies when there is a supply squeeze; the relationship is vulnerable to tensions and conflicts in the Persian/Arabian Gulf, and above all to changes in the regime in Saudi Arabia; it reduces the freedom of manoeuvre of the US government and risks drawing the USA into regional conflicts against its interest and desire; and it could result in a backlash from other Arabs against this new type of 'imperialism' and foreign domination at a time when a co-operative Saudi Arabian regime might be even more needed than today.

After the 'second oil crisis' in 1979, bilateral attempts to secure oil supplies through government-to-government contacts reappeared in force. As a result, the structure of the international oil market seemed about to change once more substantially — this time by reducing the importance of the established oil companies. This new wave of bilateralism seems not without problems and illusions — above all the illusion that countries can isolate themselves against an energy crisis while others suffer its impact.

Overall, then, the conclusion is that bilateral co-operation is risky and ultimately insufficient. Bilateralism is not harmful *per se*, if it is limited. But the essential factor has to be a multilateral co-operation on the major issues in order to avoid transition crises. We will therefore have to attend to existing multilateral efforts before we conclude the analysis with a number of specific suggestions and ideas. The following is a survey of some such efforts: the North—South dialogue (Conference of International Economic Co-operation); the Euro—Arab dialogue; and the Mediterranean policy; the Lomé Convention, and the negotiations between Iran and the European Communites.

CHAPTER ELEVEN

# The North-South dialogue

When OPEC turned the tables on the industrialized countries and acquired control over world oil prices and supplies, this represented a dramatic shift in power towards LDCs in an important subsystem of the international order. The success of the oil exporters could not exist without reverberations in other LDCs, and it is not surprising to find in retrospect that the debate about a new international economic order was a result of the 'new international energy order' forced by OPEC. Historically, energy issues are thus linked to the wider conflict about global redistribution and development; ever since, the two issues have moved in tandem in many international fora – as indeed they ought to, since energy *is* a crucial factor in overall economic and social development and welfare.

The first strand of the debate, the dialogue on energy, had originally been demanded by OPEC countries – an irony which is explained by the fact that such initiatives were taken long before 1973. After the turning point in 1973/4, it was the industrialized countries who took the initiative in calling for international energy co-operation. However, not without disagreement: initially, the USA hoped to succeed with a strategy of counter-cartelization designed to break up the unity of OPEC. Three major developments finally brought about a modification of the US policy: the realization that OPEC's cohesion was much greater than originally presumed; the growing importance of OPEC exports for the present and future supplies of energy to the USA (not to mention other industrialized regions); and the awareness that to become independent of OPEC for energy would be much more time-consuming and difficult than originally expected – in fact, it soon became apparent that for a very long time it would be virtually impossible[6].

The demand for a new international economic order – the second strand of our theme – was first made systematically by the 6th Special Session of the UN General Assembly in 1974. Also in the same year, the French Foreign Minister Michel Jobert launched the idea of an energy dialogue. The two demands – one from the LDCs, one from the industrialized world – were quickly moulded together by the former, but above all by the oil exporters: the Saudi Arabian oil minister Sheikh Yamani proposed an enlarged conference of oil exporters, developed countries and the rest of the LDCs, and a Ministerial Conference of 68 developing countries in Dakar (February 1975) put this position forward formally as a position of the Third World. Significantly, this position was supported at the OPEC summit meeting in Algiers in March 1975: the final declaration established the principle that an energy dialogue between OPEC and the industrialized world had to include on an equal footing a dialogue on North—South questions in general. The declaration also put forward certain fundamental positions which were to accompany the whole evolution of the dialogue: the demand that oil prices ought to be indexed, and that energy concessions by OPEC would have to be balanced by concessions made by industrialized countries to assist the development of all developing countries. More particularly, the OPEC heads of state expected the industrialized world to:

(1) Take measures to stabilize the prices of LDC commodity exports.
(2) Launch an effective food programme for LDCs.
(3) Transfer technology.
(4) Reform the international monetary system and international financial institutions in such a way as to safeguard the interests of the Third World.

The declaration also called for developed countries to agree to the construction of a major part of the refining, petrochemical and fertilizer plants in OPEC countries, and to protect the value of OPEC surplus investments in these countries[7].

In spite of general acceptance of the idea of a broad dialogue by both the key countries, Algeria and the USA, their positions were still far apart: the USA approved, but only on the condition that industrialized countries strengthened co-operation within the IEA by agreeing on a programme to develop alternative energy sources and to set a floor price on oil; Algeria, on the other hand, linked the conference to general issues of development, refusing to treat these as a mere appendix to discussions on energy. This difference was not sufficiently reduced to secure the success of a first preparatory meeting for the dialogue in Paris (7—15 April 1975). Broadly speaking, the cohesion of the seven developing countries (Algeria, Saudi Arabia, Iran, Venezuela, Brazil, India and Zaire) blunted all attempts by the developed countries (the USA, the European Community and Japan) to focus the agenda on energy issues[8]. Further modifications of the US position since May 1975 enabled the French President, however, to call a second preparatory meeting (13—16 October 1975). This time, the representatives agreed on the composition, the timing, and above all the agenda of the dialogue: the Conference of International Economic Co-operation (CIEC) was to have four commissions dealing with energy, raw materials, development, and financial questions. Each commission would comprise 15 countries — 10 developing and five developed countries; each (as well as the whole Conference) would have two chairmen, one from the developed, one from the developing countries. Altogether, 27 countries were to participate — eight from the industrialized world, and 19 from the LDCs (soon to become known in the Conference jargon as G-8 and G-19): the eight developed countries were the European Community, as represented by the Commission, the USA, Japan, Canada, Spain, Sweden, Switzerland and Australia. The LDC members of the CIEC were Algeria, Saudi Arabia, Indonesia, Iran, Iraq, Nigeria and Venezuela for OPEC, and Argentina,

Brazil, Cameroon, Egypt, India, Jamaica, Mexico, Pakistan, Peru, Yugoslavia, Zaire and Zambia. Several international organizations (UN, UNCTAD, IEA, OECD, OPEC, the FAO, GATT, UNIDO, UNDP, the IMF and the World Bank) were also admitted to discussions, but not to vote in the commissions. Members of the Conference who wanted to take part in the work of a commission in which they were not represented could send an observer but without the right to speak and vote. A ministerial meeting at the beginning of the CIEC would be charged with setting out the general direction of the working commissions within the framework of the understanding reached by the preparatory meeting, and with concluding the work at the end of the CIEC[9].

The consensus on the content of the dialogue was fairly vague — any subject relating to the overall themes of the Conference could be raised in the commissions. This concealed substantial disagreements between the parties — while the inclusion of issues other than energy was clearly a success for the LDCs, the industrialized countries had managed to keep the agenda for the commissions very general. Thus, the Conference could open as planned on 16 December 1975 in Paris — but the initial declarations of the ministers already highlighted differences between the two groups, and also amongst one another.

The two major items in the four-day meeting were the composition of the working groups, and their chairmen. The eight co-chairmen of the commissions formed, together with the two Conference chairmen, a steering group for CIEC, and consequently could exert considerable influence. The latter had been already selected beforehand (the Canadian Foreign Minister Allan MacEachen, and the Venezuelan Minister for International Economic Affairs, Manuel Pérez Guerrero) — but the commission chairmen and the members of the commissions had to be chosen by the meeting. After considerable debate, the result was as follows:

(1) *Energy Commission:* Saudi Arabia and USA as chairmen; other members are Algeria, Brazil, Canada, the European Community, Egypt, India, Iran, Iraq, Jamaica, Japan, Switzerland, Venezuela, Zaire.
(2) *Raw Material Commission:* Japan and Peru as chairmen; other members are Argentina, Australia, Cameroon, the European Community, Indonesia, Mexico, Nigeria, Spain, USA, Venezuela, Yugoslavia, Zaire, Zambia.
(3) *Development Commission:* Algeria and the European Community as chairmen; other members are Argentina, Cameroon, Canada, India, Jamaica, Japan, Nigeria, Pakistan, Peru, Sweden, USA, Yugoslavia, Zaire.
(4) *Financial Commission:* the European Community and Iran as chairmen; other members are Brazil, Egypt, India, Indonesia, Iraq, Japan, Mexico, Pakistan, Sweden, Saudi Arabia, Switzerland, USA, Zambia.

The Energy Commission first met in February 1976, and held further sessions in March, April, June, July, September, October and November of 1976, and had final sessions in preparation for the Ministerial Meeting concluding CIEC in April and May 1977. Work was divided in two stages — the analytical first phase, and an action-oriented second phase[10].

## The analytical phase of CIEC

Work focused on four major subject areas, namely

(1) A general analysis of the energy situation.
(2) The price of energy.
(3) The availability and supply of energy.
(4) International co-operation in the energy field.

During this phase, seven papers were presented by the group of nineteen LDC countries, and sixteen papers were submitted by the group of eight industrialized countries.

In the general analysis and the supply/demand debate, Saudi Arabia led the way for the LDC participants. Presenting a substantial statistical assessment of past developments and future scenarios, broadly based on Western sources, the Saudis added their own interpretation, which constituted a vigorous attack on the industrialized world. In particular, the Saudi representative made the following points:

(1) The industrialized countries accounted for a disproportionately large share of world energy consumption; they covered their demand by drawing excessively on the oil wealth of OPEC countries, thus causing a huge waste of resources needed for the future.
(2) The industrialized countries had prevented the refining of oil in OPEC countries, and thus profited from the processing of this raw material in their own countries.
(3) The developed countries were deliberately fudging the prospects of alternative sources of energy to justify a further heavy claim to OPEC exports.
(4) The price rises in 1973/4 had been salutary in opening up the possibilities of eliminating waste, and alleviating the burden on oil in covering world energy demand.

From this analysis, the G-19 paper drew the following conclusions:

(1) The burden of meeting the core of world energy demand had to be shifted to other energy sources such as nuclear energy, hydro-electricity, geothermal, and solar energy to arrive at a conservation of oil for 'noble' uses in the long term.
(2) In the medium term, the possibilities of coal had to be developed much more forcefully.
(3) Natural gas, which had to be flared in the past to meet oil import demand of the industrialized countries, had to be put to use rapidly; the price increases in 1973/4 offered the possibilities to do so.
(4) Much more vigorous conservation efforts were needed by developed countries so as to make available oil and gas for the future demand of the Third World[11].

Against this, the Eight argued that the G-19 analysis was understating the potential of world oil reserves; proven reserves were no indication of ultimately available supplies: as exploration and production technology advanced, further resources would become available, which would quadruple the quantity of proven reserves. The Eight refused, however, to look beyond 1985, arguing that beyond that point the probability of technological break-throughs, and the uncertainties about growth rates, made any forecast meaningless[12].

Thus, the assessment of the general energy outlook revealed already substantial differences; it also demonstrated that the Eight were playing for time, refusing to be drawn into an analysis which would require urgent action and concessions towards the LDCs.

On prices, the Nineteen wanted to group work around two separate aspects: price levels, their impact on alternative sources of energy production, and replacement costs on the one hand, and the purchasing power of oil and accumulated oil revenue surpluses on the other. In their analysis the Eight preferred to look at the impact of energy prices on the world economy, presenting papers and arguments about the relationship between oil prices and growth, inflation, unemployment etc. The fundamental difference in positions with respect to oil prices thus surfaced quickly; it was to remain unbridged to the very end of the dialogue[13].

Finally, the analytical phase also discussed some aspects of the situation of the oil-importing developing countries; here, the USA contributed the ideas of an International Energy Institute and an International Resource Bank — ideas which had been launched by the State Department at the UNCTAD meeting in Nairobi in May 1976. These discussions did not progress very far, however.

## *The action-oriented phase of CIEC*

At this point, a meeting of senior officials (8–10 July 1976) concluded the first phase of CIEC, and decided to open the action-oriented stage. A work programme was finally agreed by the co-chairmen of the dialogue; this defined the following areas of work for the Energy Commission:

(1) Energy resource availability and supply and development.
(2) Energy prices and the purchasing power of energy export earnings.
(3) Technical, scientific and financial co-operation in the energy field.

A total of nine proposals were put forward by the Eight, while the Nineteen submitted one joint proposal on the first item, and sub-groups of the Nineteen one proposal each on the other two topics. In November 1976, three contact groups were established within the Energy Commission to facilitate progress. This was abandoned, however, in favour of one single contact group during the last stage of commission work in April and May 1977.

On the question of energy availability, and supply and demand development, much common ground was developed. The Nineteen were able to present a common view, and this view was broadly reconcilable with the positions adopted by the industrialized countries as a result of two major outside events: President Carter's energy policy statement (20 April), and the London summit meeting (7–8 May 1977). Carter's energy programme accepted the need for drastic measures to conserve oil and gas, since otherwise the world would face a severe supply/demand imbalance (reconcilable only through dramatic price increases) in the 1980s. He also accepted the notion of domestic energy price increases in order to stimulate oil and gas production, as well as an expansion of the contribution of other sources of energy. Thus, the US position was now that energy prices had to reflect true replacement costs in order to bring supply and demand into balance. Third, Carter's policy contained some indexation of oil prices in the provision that prices for newly discovered US oil would be allowed to rise to 1977 OPEC price levels *in real terms.* Thus, the USA now, domestically, favoured some of the steps demanded by OPEC on the international level. The London Summit recognized the fact that the world was going through a transition period, in which oil and gas would eventually be replaced by other sources of energy. Equally, the Summit acknowledged the particular problem of the oil-importing developing countries, and endorsed the idea that the World Bank should play an expanded role in helping these countries develop their indigenous energy resources.

In the area of *energy co-operation* differences persisted, however. They even gained a new dimension in that OPEC countries now refused some of the proposals put forward by oil-importing developing participants of CIEC. Thus, there was no common position of the Nineteen on this topic; the paper submitted by the LDCs was actually carried by Brazil, Cameroon, India, Jamaica and Zaire. It suggested an international energy co-operation programme with a mixture of short-, medium- and long-term measures to (a) alleviate and ultimately resolve the structural problems of the oil-importing LDCs, and (b) provide for effective diversification of energy sources in OPEC countries. More specifically, the proposal contained:

(1) A special IMF oil facility for oil-importing LDCs from 1977 to 1982 with an annual size of $4 \times 10^9$ SDR, to be contributed by developed and OPEC countries; interest rates for this facility should be subsidized. This was meant as a short-term measure.
(2) In the medium and long term, provisions for the transfer of technology and information by developed countries to oil-importing LDCs, and for financial contributions by OPEC and the industrialized countries at a level of $300 \times 10^6$ SDR p.a. from 1978 to 1980; thereafter, financing of the development of LDC energy resources outside OPEC should be expanded through the World Bank, which was asked to provide $5 \times 10^9$ SDR p.a. in the period 1980 to 1984 in addition to its other lending operations. The transfer of technology would also help OPEC countries to develop a diversified energy structure; one particular source of energy aimed at in this connection was

nuclear, where Iran in particular insisted on access to technology and equipment.

Even more controversial than this proposal was the one submitted by the European Community: it concerned continuing consultations on energy issues. The paper suggested:

(1) Exchange of information on the energy market, in particular with regard to all aspects of the future evolution of supply and demand, exploration and exploitation policies, prospects of introducing alternative energy sources, security of supply and stability of markets, and energy investment plans.
(2) Consultations on energy prices within the framework of the impact of energy prices on the world economy, balance of payment problems, surplus financial assets of exporters and inflation, and the problems of oil exporters purchasing power and the prices and costs of various energy sources.
(3) The creation of a forum capable not only of serving as an international clearing house for energy information and data, but also of providing technical and operational assistance in the development of traditional and new energy sources (an idea taking up the US suggestion of an International Energy Institute); lastly, it was submitted that an additional mechanism was needed to ensure adequate, 'action-oriented consultations'[14].

The USA submitted a more detailed proposal on its idea of an International Energy Institute, and also revived the suggestion of an International Resource Bank. The IEI was conceived as a kind of non-governmental Third World complement to the IEA: a body where in-depth analysis of LDC energy problems could prepare and supplement assistance in technical, operational and financial aspects of developing the indigenous energy resources of the LDCs. The IRB could be a mechanism which primarily provided the necessary financial backing for such projects.

On prices, differences within the Energy Commission also persisted, as already indicated. The two proposals submitted came from Japan for the Eight, and from Egypt, Iran, Iraq and Venezuela for the Nineteen. The Japanese suggested essentially that energy prices take into account:

(1) Energy supply and demand.
(2) The cost of existing alternative sources of energy.
(3) The range and cost of developing new energy sources.
(4) The impact on the world economy, bearing in mind the problems of oil-importing developing countries and developed countries heavily dependent on imports.
(5) The concern of oil-exporting countries with the purchasing power of their export earnings.

The proposal of the LDCs, on the other hand, took the following factors into account:

(1) The competitive standing of oil *vis-à-vis* other sources of energy.

(2) The purchasing power *of the unit value* of energy exports (including accumulated export earnings) should be protected through indexation against inflation and currency depreciation.
(3) Two possible sources of adjustment of oil prices as *independent* from each other: the protection of purchasing power (implying a stabilization in real terms of the price per barrel), and the competitive position of oil (implying the possibility of price *increases* in real terms for each barrel)[15].

## Results of the Energy Commission

On 2 June 1977, the CIEC ended in confusion: the result of 18 months of discussions was a short communiqué, listing areas of agreement and disagreement in the four working commissions. For energy the final Ministerial Conference agreed on:

(1) Conclusion and recommendations on availability and supply of energy in a commercial sense, except for the purchasing power constraint.
(2) Recognition of the depletable nature of oil and gas. Transition from an oil-based energy mix to more permanent and renewable sources of energy.
(3) Conservation and increased efficiency of energy utilization.
(4) The need to develop all forms of energy.
(5) General conclusions and recommendations for national action and international co-operation in the energy field[16].

The notion of 'energy transition' contained in item (2) could be orderly or chaotic. But to achieve an orderly transition process, urgent efforts were needed to conserve energy, to develop alternative sources as quickly as possible, and to enhance recovery rates from existing oilfields. Above all, an orderly transition required OPEC price decisions to be in line with bringing about the necessary efforts in conservation and alternative production, such as to prevent major disruptions by rapid price adjustments. This seems to have been agreed in the dialogue – it does address the crucial issue in the energy field, but provides hardly more than a very general outline for a co-ordinated effort on a global scale. Besides, this agreement concealed in fact a kind of package deal with mutual, but only vaguely defined responsibilities, whose discharge was left entirely to national decisions.

Agreement on point (1) only concerns commercial energy transactions: the OPEC countries made it clear that embargoes or changes in production rates were entirely questions of national sovereignty. A number of agreed constraints were seen to influence availability of exports – e.g. the domestic requirements of OPEC, and the replacement costs of oil produced. No agreement was reached on the demand of the Nineteen that availability should also be subject to the protection of the purchasing power of a barrel of oil exported by OPEC.

The general conclusions and recommendations required special attention, which should be given to the structural problems of oil-importing developing countries, and the need for an International Energy Co-operation Programme within the overall framework of an International Economic Co-operation Programme (here again, the basic trade-off demanded manifests itself: oil against development concessions!). The specific recommendations addressed themselves to:

(1) Effective measures of co-operation between developed and developing countries with regard to transfer of energy technology and technical assistance.
(2) An expanded role of the World Bank in the development of energy resources in oil-importing LDCs.
(3) A review by the World Bank in co-operation with the International Development Agency and the IMF about the possibilities of increasing the capital flow into LDCs for the expansion of indigenous energy supplies; this should not affect the other activities of the Bank.
(4) The possibility of other international and regional financial institutions contributing to such efforts.
(5) Steps to facilitate the rapid integration of OPEC's refining, petrochemical, fertilizer and other downstream industries and their products into the expanding global industrial community — at an economically practical rate.
(6) The fact that CIEC had studied the possibility of an International Energy Institute, or an expansion of energy activities under UN auspices. Expeditious international consideration and appropriate steps of achieving the objectives mentioned above were recommended.

The list of *disagreements* in the Energy Commission reads as follows:

(1) The price of energy and the purchasing power of energy export earnings.
(2) Accumulated revenues from oil.
(3) Financial assistance to bridge external payment problems of oil-importing developing countries.
(4) Recommendations on resources within the Law of the Sea Conference.
(5) Continuing consultations on energy.

The disagreement on *price* had always been present within the dialogue: the industrialized countries were unwilling to accept the notion of indexation of oil prices and accumulated surplus earnings, and disputed OPEC's assessment of the factors to be taken into consideration in pricing decisions. As early as February 1976, the Commission could not even agree on the methodology to use in analysing the price issue. In the end, the reciprocal demands — that oil prices be adjusted to protect the purchasing power of OPEC versus the demand that pricing decisions should be considered in respect of their impact on the state of the international economy — remained unreconciled.

On *accumulated surplus earnings*, Saudi Arabia had submitted a proposal for the Nineteen in the Financial Commission. This proposal stipulated the protection of the real value of such surpluses invested in developed countries through indexation, through guarantees of compensation against currency devaluations, and safeguards against confiscation, freezing, and similar measures by developed countries against oil exporters' assets in their countries. The paper also demanded assurances for the prompt transferability of investments and income out of developed countries. This was not accepted by the Eight.

On *financial assistance* for oil-importing countries to alleviate short-term balance of payments problems, agreement in principle did not produce consensus on specific measures such as the ones suggested by the LDCs in the Energy Commission. In fact, the issue produced substantial disagreement among the Nineteen. Iran, in particular, had rejected this idea outright — as had other OPEC members, when the oil-importing countries suggested that 'adequate differential and remedial' action be taken to alleviate the structural problem of their economies (this latter proposal apparently aimed at a two-tier oil price). Intra-OPEC differences were not unrelated to this: after the OPEC meeting in Doha, which had led to a split in the oil price structure between the 'five per centers' and the 'ten per centers' (December 1976), Iran boycotted all progress in the CIEC Energy Commission — suspicious about the way Saudi Arabia had tried to direct discussions towards compromises acceptable to the other chairman of the commission, the USA. The Eight were also unwilling to accept these proposals of financial assistance.

The disagreement on the question of *sea-bed resources* and their exploitation was hardly surprising — this subject was at the same time negotiated at the Law of the Sea Conference, and had to be left to it. Since the problem is extremely complex, CIEC could hardly be expected to make a breakthrough with some broad agreement on principles.

Disagreement on *continuing consultation* was the major setback for the industrialized countries: on this they were most eager to receive concessions from the oil exporters. After all, continuing consultations were the only realistic way for the industrialized countries to regain some influence over the decisions crucially important for their future development, namely the price and availability of oil. That the CIEC ended without any concrete achievements on this topic (in spite of considerable willingness by some exporters and other developing countries to consider a continuation of the Energy Commission), is explained by reasons both within and outside the Energy Commission. First, they relate to the split within OPEC, and within the developing world: the moderates headed by Saudi Arabia and India (and, in the Ministerial Meeting, to a lesser extent Iran, who until then had also extended its boycott to co-operation issues) were apparently willing to make concessions, but had to draw back when they came under pressure from the more radical participants among the LDCs, who preferred energy co-operation to be under the auspices of the UN. A compromise proposal by the Saudis was unacceptable to the Eight. In addition, the European Community proposal apparently damaged the atmosphere in the Energy Commission considerably: the unfortunate differentiation between 'energy co-operation' and 'energy consultation' awoke old suspicions among the LDCs about the ultimate objectives of the industrialized countries — was this not just a pretext to regain some control over decisions rooted in the painfully acquired sovereignty of OPEC (and other developing) countries? There was no doubt that OPEC was totally unprepared to even consider giving up some of its freedom to set oil prices formally (what went on behind the scenes between Saudi Arabia and the USA, was another matter). And this position was fully supported by the LDCs, partly because the European Community proposal lacked any inherent *quid-pro-quo* of some generosity: it was clearly geared only to its interests, and the interests of the industrialized countries.

The same also applies to the general outcome of CIEC — for at that stage the argument about consultation had become immersed in the bundle of unresolved issues produced by the four working commissions[17]. On this level, too, the industrialized countries failed to produce significant and generous concessions, which could have created the pressure and support for a trade-off between energy and development concessions. One of the victims of this last stage was the US proposal for an International Energy Institute: assistance to the oil-importing developing countries was in principle amenable to agreement between OPEC and the industrialized countries. The atmosphere had been damaged, however, by the European Community proposal, and besides some OPEC countries were apparently unenthusiastic about this idea. In this way, the idea of the IEI, in spite of the considerable support it had from various sides, was rejected at the dialogue.

That the fundamental trade-off between development concessions and energy concessions did not materialize, seems a missed opportunity. Even on the most crucial point in the energy discussions — the continuation of co-operation in an institutionalized form — a compromise appeared possible. After all, even consultations about prices were at that point part of international affairs — if only in bilateral talks between the USA and key oil exporters. The theoretical sovereignty of OPEC with regard to pricing decisions was, in practical terms, also somewhat constrained: the impact of oil price levels on the state of the world economy *was* a factor taken into consideration at least by Saudi Arabia, if not by other exporters — and they all had a stake in it. It was, however, premature at this point to hope for a definite institutional framework for regulating and managing energy questions; a more promising strategy would have been to aim at such institutionalization step by step, starting with the least controversial aspect: the need to assist the oil-importing developing countries in developing their own energy supplies.

The results of the other three commissions at the end of CIEC clearly underlines the conclusion that a fundamental trade-off, a breakthrough in North–South relations, did not take place. Roughly summarized, the achievements and failures can be characterized as follows[18]:

(1) *On raw materials:* acceptance was reached on the principle of the establishment of a Common Fund for commodities, although the exact details of the Fund were to be negotiated later within UNCTAD; agreement also was

established on international co-operation in marketing and distribution of primary commodities; on research and development to help producers of natural products competing with synthetics; on measures to assist LDCs to develop and diversify their indigenous natural resources; on improving the Generalized System of Preferences; and on the need to identify areas for special and more favourable treatment of LDCs in international trade negotiations (Tokyo Round). These agreements, however, were more than balanced by the serious discrepancies which continued to exist between the two sides. Even on the Common Fund, positions were still far apart, as the further slow evolution of negotiations was to demonstrate; the agreement in principle was almost meaningless. The indexation of raw material prices, measures relating to the interests of LDCs in world shipping, as well as some aspects of compensatory financing, local processing of raw materials, and diversification, all were deadlocked in the discussions. The same was true for the control over the production of synthetics. Finally, the LDCs did not secure 'differential', i.e. more favourable treatment for their exports to industrialized countries, and therefore refused in turn 'unimpaired' access to their raw materials.

(2) *On financial questions:* agreement was reached on a number of non-controversial questions regarding LDC access to capital markets, of private foreign investment, and co-operation among developing countries. No consensus was reached on measures against inflation, the special treatment of oil exporters surplus revenues (a point already touched upon in the energy context), and some crucial aspects of private foreign investment (e.g. criteria for compensation in case of nationalization, and transferability of profits).

(3) *On development:* agreement was reached on measures to help the infrastructural development of LDCs, particularly in Africa. An increased share of LDCs in the world's total industrial production was accepted, as was the implementation of several UNCTAD recommendations on the transfer of technology, and the 22 recommendations of the World Food Conference 1974. Equally, agreement was reached on the need for additional financial flows to the LDCs. Here, the major sticking point was the demand of the poorest countries for an immediate and general debt relief and procedures for the initiation of a dialogue on debt rescheduling. Other areas of disagreement included access by LDCs' industrial products to Western markets, assistance measures for the industrialization of the Third World, and a code of conduct for transnational corporations.

Some of the areas of disagreement (in particular debt relief and the Common Fund) have been the subject of further discussions in other fora, and some progress was made. This illustrates two major aspects of the CIEC: on the one hand, its constant interplay with other international organizations such as UNCTAD, the IMF, GATT, and a number of other bodies. This oscillation of subjects between CIEC and other institutions could be, and certainly was at times, used as a means to gain time and postpone unpleasant decisions; on the other hand, it gave the dialogue in Paris a kind of 'Court of Appeal' function on questions concerning the new international economic order, which it could not really discharge. At the same time, the point mentioned illustrates that it was unrealistic to see CIEC as a qualitatively new and different framework for great advances in North—South negotiations; progress was bound to be piecemeal and sluggish, given the stakes involved and the very intractability of many issues: the expectation that the blueprint for the old economic order could be exchanged against a new one, was simply naive since neither existed — even the old order could at best be seen as a complex web of structures full of internal contradictions and forces of change in diverse directions. The term 'order' is here clearly misleading. This is true even more for the new international economic order which could hardly be more than a dynamic process of constant adjustment and reforms in specific areas. Thus, a fundamental problem of CIEC was that it was plagued from the beginning with *exaggerated expectations*[19] — exaggerations about the possible achievements within the dialogue in general, but also about the degree of autonomy of CIEC, which could not really be considered as a negotiating forum, given its very limited autonomy. Thus, CIEC not only passed topics and items onto other organizations (often to find that they were returned rapidly!); it also registered faithfully the repercussions of decisions taken elsewhere: examples here are the OPEC price decision in Doha late in 1976, the London Summit Meeting in 1977, or the European Community Councils.

Another fundamental difficulty encountered by CIEC was the large gulf of *conceptual differences*[20] about the idea of a new international economic order. As Amuzegar has pointed out, a different analysis lies at the root of these divergencies; but they are also reflected in such divergencies as *dirigisme* versus market forces, incrementalism verus fundamental reforms, etc.

Furthermore, each side *overestimated its bargaining power*[21]. Leverage was less than might have been possible, if each side had represented a united bloc: internal differences within the Eight and the Nineteen clearly existed, yet both pretended to be united. The willingness and ability of OPEC to use its muscle within the framework of the dialogue was clearly limited; yet the LDCs as a group relied on its leverage for major progress across the board. On the other side, the Eight were long handicapped by the unwillingness of the three key countries, USA, Japan and Germany, to move on substantive issues; yet in the end they had to accept the need to accommodate the LDCs, and thus made concessions which — if produced at earlier stages of the Conference — could have had much more impact.

Fourth, several *organizational aspects* of CIEC created difficulties[22]. For one, each side represented a larger grouping, and therefore felt constrained to hold to positions formulated jointly, and in the case of the LDCs within a much broader context. Yet this meant that participants often felt unable to depart from maximalist positions — thus negotiating freedom was seriously limited, sometimes against the better knowledge of the participants themselves. This was further exacerbated by the principle of unanimity, which was adopted for the discussions. Thus, any decision or recommendation could be prevented by the objection of one member country of the Conference. Finally, the inadequacy of the CIEC secretariat contributed much to holding up the work in the commissions.

Although some of the blame for the disappointing outcome of the dialogue can be attributed to the LDCs themselves (e.g. absence of effective leadership, holding to maximalist positions when progress might have been possible), most of the fault lies with the developed countries. The West intended to cause a rift between the members of OPEC and the other LDCs, drawing the latter to their side with proposals for oil price stabilization and development of their indigenous resources; somewhat unrealistically, they hoped at the same time to gain substantial concessions on energy, while being able themselves to avoid major concessions on other issues. If that could not be achieved, the intention was simply to get away with as few concessions as possible, each concession matched by something given from the LDCs.

In a manner reminiscent of the worst 'Cold War' syndrome, the industrialized countries saw the demands for a new economic order[23] as a major threat. This, in turn, only helped to deepen the mistrust on the other side, to whom stalling by the developed countries seemed to confirm the arguments of the radical members. Thus, a mutual psychological deadlock overshadowed the Conference; by the time the developed countries dropped some of their prejudices and aversions against the suggestions of the LDCs, time was running short.

It was hardly surprising that the tactics of the developed countries did not work: there was more to gain by the oil-importing developing countries staying with OPEC in the hope that its pressure would secure some major concessions from the industrialized countries, than from supporting the latter, who plainly were unwilling to make concessions of a substantial nature on issues other than energy. This policy of oil-importing LDCs ('NOPEC') probably paid off, although the gains in the end were limited. Yet given the right policy incentives by the industrialized countries, it seems doubtful how solid this alliance between oil exporters and oil-importing LDCs will be: support by OPEC for the demands of the LDCs in general appears only qualified, and there were some indications that several OPEC members were only concerned with avoiding pressure by oil-importing developing countries. This seems particularly true for Saudi Arabia. One can presume that the proceedings of the Energy Commission were informally co-ordinated between Washington and Riyad. The Saudi—US alliance thus became a vital element in the dialogue, which was, however, not particularly conducive to better results for the LDCs in general. Opposition from other OPEC countries against a trade-off of energy concessions against the offers of the developed countries finally led to failure in several crucial aspects of the dialogue: Saudi Arabia had to accept that she had moved too far outside the mainstream of LDC thinking, and under these circumstances she preferred to fall back in line. As on other occasions, Saudi Arabia's position within OPEC was proven to be strong, but ultimately not unassailable.

Had the industrialized countries developed more imaginative concrete proposals, it might have been possible to constructively channel the possibility of 'NOPEC' pressure on OPEC into substantial concessions on energy matters. To this end, however, major initiatives on other North—South issues were clearly required.

Ultimately, this failure to propose an imaginative package deal capable of attracting the support of the NOPEC participants indicates a failure of leadership. Ideological blinkers and preconceived notions, domestic constraints, and the problems of accepting a major transfer of resources at a time of economic crisis, all help to explain this failure, but they do not excuse it. More to the point seems the difficulty of transition faced by the US government; the new President took over only in the last stage of CIEC, and although the Carter administration clearly showed more flexibility on North—South issues than the previous one, it was not yet sufficiently installed, or sufficiently prepared, to contribute some major initiatives.

Japan also was hampered by elections in late 1976 — but in any case, she had not yet shaken off her traditional foreign policy of keeping a low profile, and seemed unwilling and unable to take the lead. In this situation, Europe had an opportunity to develop initiatives — and the preconditions seemed to be established with the participation of the European Commission on behalf of the Nine: the Communities thus talked with one voice. Moreover, the European Community appeared in a favourable position to exert such leadership: it had developed something like a cohesive development philosophy in the Lomé Convention, which had found much support among LDCs; it had a better 'feel' about how to handle North—South issues than the USA, which tended to oversell the virtues of private business and market forces; and it enjoyed a psychological advantage vis-à-vis the USA as a power without military and hegemonial ambitions, as a 'civilian superpower' of great cultural attraction for the LDCs.

Yet such expectations were probably too optimistic — they failed to recognize the deep divisions within the Communities on development issues, and the limits on Europe to act independently. Besides, the organization of Community participation in the dialogue was, as Wolfgang Hager has pointed out, hardly conducive to creative contributions by the Nine[24]. Thus, the European Community long played the familiar game of 'wait and see' — in that case, waiting for the arrival of Carter. Only in the last stage of the CIEC discussions did the Nine act, and produce some proposals which were partly adopted in the final meeting of CIEC. These initiatives, however, hardly contradict the above analysis: they represent little more than palliatives designed to prevent the dialogue ending in utter frustration[25].

In spite of these shortcomings, the Nine did at times manage to play a mediating role, and the differences within the Communities undoubtedly also helped to moderate the objections of Germany, at least to some extent. The mediating position of the Nine reflected the logic of the energy situation: the European Community was much more dependent on oil imports from OPEC than the USA and would continue to be so. On the other hand, it could not seek direct negotiations with producers: quite apart from the political, economic and military dependence on the USA, prices, long-term stability of supplies, and even oil supply security have to be dealt with on a global scale. These issues have, therefore, to be addressed in fora which bring together all those with sufficient power to upset an agreement reached by any group excluding them. Ideally, it should also include those without such leverage, but who are substantially affected by the decisions made (i.e. the oil-importing LDCs).

Finally, the European position responds to the logic of the international economic and political system. The USA, as the dominant global power, preoccupied with the global balance of East—West forces, was more inclined to opt for confrontation with the Third World, in an attempt to contain and divide LDCs and to stabilize US domination of the international hierarchy. Europe, on the other hand, was much more vulnerable to pressures from the Third World, and can be more flexible in adjusting to a restructured international system since her obligations and interests in the existing order were not as clearly defined.

But Europe depends on the USA in many ways; she cannot risk confrontation, and therefore must avoid, at any cost, choosing between the USA and the Third World. This can best be achieved by trying to persuade the USA towards a greater accommodation of Third World demands. A gradual realization in Washington that the changes since 1973 had been deeper and more lasting than initially recognized helped to shift US policy; besides, the US administration realized that the North—South dialogue in effect excluded the socialist group from an area of international affairs — its upgrading therefore placed the USA in a very favourable position *vis-à-vis* the USSR. The new US government under President Carter also showed more understanding and greater flexibility towards demands from the Third World, although the change was gradual, and not as substantial as some LDCs had hoped. But it was now generally recognized that a continuing effort to create and sustain momentum in the North—South dialogue was in the best interests of the West. However, there has been little effort to defuse some of the potential for future conflict and instability, and to provide sustained demand for industrial production on a global scale and realize growth targets needed to smoothen the adjustment of industrialized societies.

Yet it would be wrong to consider the result of CIEC as totally negative. Undoubtedly, the concrete achievements were rather limited: a general acceptance of a Common Fund, an almost equally general promise to increase development aid (although some industrialized countries, notably Japan, made specific pledges), and finally $\$1 \times 10^9$ for a Special Action Programme for the poorest countries. Yet, in the energy field, CIEC was the first occasion on which producers and consumers of internationally traded energy came together to discuss a whole range of problems concerning energy. In fact, all three central aspects of the energy problem (price, supply stability, and supply security) were discussed. Moreover, the associated status of the IEA meant that even the oil companies had (through the consultation mechanism of the IEA) some indirect, if tenuous, link with the negotiations. The failure to institutionalize the Energy Committee has, therefore, to be considered as a missed opportunity. The reasons for this failure have been mentioned above; they lie in the link between the energy issue and the larger North—South conflict. While this connection has to be recognized and dealt with politically, it is unfortunate in that it might well prevent solutions which are in the interest of all parties in the international energy system. OPEC's concern that an institutionlized Energy Committee would reduce its freedom to decide price levels without outside interference may be considered somewhat short-sighted since in the long run wrong price decisions are bound to have serious repercussions on the oil exporters themselves. Part of the problem here arose from the concern of the producers that non-oil LDCs might pressure them into a price policy they would consider too low. Such a concern, however, could be resolved only by a split price system. For the impact of high oil prices on development and the desirability of high price levels for global energy reasons are unreconcilable in any sensible way except by some form of differentiation between developing and developed oil importers.

Failure to institutionalize a consultative mechanism between producers and consumers might not be a major setback for on the positive side, the Energy Committee can point to a number of agreements between the two sides. The need to phase out hydrocarbons over the next decades, the necessity for stringent conservation and the development of alternative energy sources, the preferential treatment for non-oil developing countries, and the desirability of a growing role of the producers in downstream operations are cases in point. Such agreements mark a substantial evolution of positions and indicate the value of co-operation between producers and consumers within a tripartite framework. The question is no longer whether a different approach will be chosen eventually, but rather: how rapidly will this co-operation take on more stringent forms in order to avert the risks and dangers of serious confrontation in the international energy system.

CHAPTER TWELVE

# The Euro-Arab dialogue

During the crisis in 1973/4, the European Community foreign ministers agreed to a common position on the Middle East conflict. While developments towards such a position had been under way since 1972 (no doubt partly triggered off by the rapidly increasing power of the Arab oil producers), it was the first time that the Nine had officially pledged the Community's support of certain principles for a Middle East settlement[26]. This declaration of 6 November 1973 could be considered as the starting point of the Euro–Arab dialogue, and its basic problem: the potential for conflicting policy directions of the Community with regard to the Arab world, and to the USA, its most crucial ally.

The 6 November declaration was widely perceived as pro-Arab, and caused considerable anger in Washington, where the government apparently was surprised by the European initiative. The already tense relations between Europe and the USA (the European refusal to co-operate in the US airlift to Israel, and the pronounced differences in reaction to the Middle East crisis, had soured relations in the 'Year of Europe') deteriorated further; mutual reproachments about lack of consultation and understanding of the other side's position were exchanged. Kissinger saw the European initiative as hampering his own mediation efforts in the Middle East.

After the Middle East crisis had begun to subside, the USA turned to the problem of economic security in a more fundamental way – first sketched out in Kissinger's speech on 12 December 1973 in London, in which he called for an energy action group of the industrialized consumers. Only a few days later, on 14 and 15 December, the European Community held a summit meeting in Copenhagen, at which – rather unexpectedly – four Arab foreign ministers appeared, suggesting a comprehensive dialogue between Europe and the Arab world. The European response, in terms of its energy policy, to the crisis was affected initially by the difficult choice between co-operation with the producers or joint action with the USA which at that time strove to build up a strong counter-cartel, and to break the oil producers' new power. The dilemma split the Community, with France dissociating herself from the results of the Washington meeting, and the others in the Community following the US lead. However, shortly after the Washington Conference, and its apparent decision to construct a consumer organization before any overtures were made to the producers, the Community agreed to the Euro–Arab dialogue[27]. After further US–European tensions, the Community accepted on 12–22 April 1974 a formula which allowed it to pursue both the Atlantic and the Euro—Arab dialogue; the Nine promised informal consultations with the USA before the European Community arrived at decisions.

The Euro—Arab dialogue began officially in Cairo on 18—20 June 1974, after preliminary soundings between the Community and the Arab League. The participants in the dialogue on the European side were not the Commission, but the member governments; discussions were carried out within the framework of the Davignon political committee, the foreign policy co-ordination machinery of the European Community, sometimes with a representative of the Commission. The Arab side included the 19 Arab League member states and the Palestine Liberation Organization. The first preliminary meeting in Cairo accepted the procedures suggested by the European Community — initial diplomatic contacts, joint working groups to examine potential fields of co-operation, and finally a joint ministerial-level conference. A second preparatory meeting between the French Foreign Minister and the Head of the European Community Commission, and the Secretary General of the Arab League and the Kuwaiti Foreign Minister confirmed the Cairo agreement, and set November 1974 as the date for the first meeting of the joint commission[28]. The meeting had to be postponed, however: the Arab side demanded that the PLO participate in the discussions, while some European Community countries (Germany, Britain, Netherlands, Denmark) insisted that only states should be admitted. US objections also may have helped to create the delay. In February 1975, a compromise was found: each delegation would be free to nominate its members, obscuring their national origin in one large delegation. At the same time, the procedure of the dialogue was to be reversed, with the working groups being advanced, and the General Commission postponed[29].

The meeting of the working groups in Cairo on 10—14 June 1975, however, again ran into problems of a political nature: the European Community had in May concluded an association agreement with Israel within the framework of its Mediterranean policy. This led to strong criticisms by Arab countries, and to intimations that the Euro—Arab dialogue might have to be postponed again[30]. In the event, however, the Arab side accepted the position of the European Community, which had assured the Arab League that the agreement with Israel did not apply to the occupied territories[31]. The Cairo meetings resulted in the inclusion of trade and labour relations in the dialogue — issues the Nine had up to then been reluctant to discuss. Subject matters to be pursued now included (apart from those two) industry, agriculture, infrastructure, financial co-operation, and cultural and scientific co-operation. The compromise on trade meant that tariff concessions would be brought up — a move which would particularly worry the French and Italians, and potentially annoy the USA. Labour relations were included in the working group on cultural and social questions, while a new seventh group was created for scientific and technological co-operation[32].

The next round of experts took place in Rome from 22—24 July 1975. Again, external political events (a condemnation of terrorism by the European

Parliament) threatened the meeting; but again, the Arabs eventually acquiesced. The overall objective agreed upon for the dialogue in this meeting was a treaty between the Arab League and the European Community similar to that with Israel under the Mediterranean policy[33].

The third expert-level talks took place in Abu Dhabi, 22—27 November 1975. This meeting ended the preparatory stage of the dialogue. The Arab side put forward a series of concrete economic suggestions, aimed at the abolition of all duties and other export restrictions, the stabilization of export earnings, financial guarantees for Arab investment in Europe, and technical assistance to diversify and develop exports. Although the meeting opened with a political offensive by the Arabs, attacking the Community for not supporting a UN resolution denouncing Zionism, and asking the European Community to clarify its position with regard to the rights of the Palestinians, it quickly turned to problems of economic and social character, and substantial progress was achieved in several working groups. The Arab side presented a series of projects for infrastructural, agricultural and industrial development. In response, the European Community selected a certain number of projects in order to establish an order of priority (e.g. for port development in the Infrastructure Commission), to achieve a degree of homogeneity of industrial and agricultural projects for the region, and to avoid competition in industrial development. The scientific and technological co-operation group exchanged information on solar and geothermal energy research.

The three groups on infrastructure, agriculture and industry made enough progress to be able to charge specialists with the further pursuit of their objectives, developing programmes for port and transportation systems, and setting up task forces for such projects as Arab refineries, petrochemical plants, and meat production. Serious difficulties, however, were encountered by the groups working on trade, finance, and social affairs: in the Trade Commission, the European side refused to accept Arab demands for a preferential treatment on a non-reciprocal basis, such as the Community had accorded to the member states of the Lomé Convention — the experts did not have any mandate to concede such far-reaching benefits. In the financial working group, both sides noted the need for the protection of investment in the other side's member countries against non-commercial risks, and it was decided to set up a temporary specialized working group to examine possible measures. The Arab side did not, however, accept reciprocity of treatment. The group dealing with labour problems faced the difficult question of Arab workers in the Common Market countries at a time of general recession in the West[34].

The communiqué ending the Abu Dhabi talks officially ended the preparatory stage of the dialogue. The next step was to be the convening of the General Commission; in the meantime, specialist groups would pursue the work of three of the seven committees. The Dublin formula — the compromise solution accepting the PLO as a member of the Arab delegation, but without special reference to its status within the delegation — was accepted by both sides for the General

Commission; the Europeans also agreed that the dialogue in principle would have to have a political dimension, although the exact form of this dimension was subject to disagreement.

The meeting of the General Commission on the ambassadorial level took place on 18—21 May in Luxembourg. At the beginning of this new stage in the dialogue, the situation was as follows: the foreign ministers of the Nine had grudgingly accepted a political dimension of the dialogue, provided it would not be turned into a forum to discuss the Middle East conflict and the Palestinian problem. Nevertheless, it was quite clear that the Europeans wanted to emphasize the operational and practical aspects of the talks (not the least because of the relationship with the USA, and initial assurances that the dialogue would not become politically loaded with discussion of the Middle East conflict), while the Arab side underlined the importance of the political aspect of the dialogue — indeed, the Arab delegation was only held together by this common interest in the political aspects[35].

The first part of the General Commission meeting was devoted to the future structure of the dialogue. It was agreed to make the General Commission a permanent institution, which would meet twice annually as the supreme coordinating body of the work, to decide on project choices and to provide the impetus of the talks. This body could meet at the foreign ministerial level, but would normally be confined to the ambassadors[36]. The working parties were also to become a permanent feature. The talks then turned to matters of substance: a joint communiqué was published[37]. The political part affirmed the agreement between the two sides about the importance of the recognition of Palestinian rights for the solution of the Israeli—Arab conflict, and then proceeded to outline the positions of both sides about this conflict. The European Community position was virtually a repetition of the 6 November 1973 declaration. Nevertheless, moderate Arab delegates had prevented some more radical representatives, whose main interest in the talks were political, from using this as an excuse to discontinue the dialogue altogether. The economic section of the communiqué listed a number of projects to be studied (e.g. rural development of southern Sudan, the Polytechnical Institute in Syria, solar energy research). Possibly the most important decision in this section was the principle of joint finance of projects under the auspices of the dialogue. Finally, it was agreed to maintain a permanent exchange of views and information on economic affairs between the two sides.

The dialogue was, then, handed back to the experts who between May and November 1976 met in the seven main working groups, and in several specialized sub-groups. This structure, in mid-1976, looked as follows[38]:

(1) Industrialization group (permanent).
   Specialized groups:
   Petroleum products and petrochemical industries
   Industrial standards
   General conditions for contracts

(2) Infrastructural group (permanent).
    Specialized groups:
        Marine transport
        Overland transport
        Air transport
(3) Agricultural rural development group (permanent).
(4) Financial group (permanent).
    Specialized group:
        Investment promotion
(5) Trade and commercial group (permanent).
(6) Scientific and technological group (permanent).
    *ad hoc* meeting:
        Establishment of polytechnic institute in the Arab world
(7) Social and cultural group (permanent).

In February 1977, the General Commission of the dialogue was convened again, this time in Tunis[39].

As before, the meeting in Tunis began with an exchange of political positions of both sides, with the Arab side repeating demands for the recognition of the PLO and of the Palestinians' right to their own state, and urging the Nine to take concrete steps against Israel to ensure her acceptance of the return of the occupied territories. Apart from these fairly predictable demands, however, the political component of the Conference also contained Arab suggestions for the creation of a Euro–Arab consultation mechanism in the UN, the formation of a committee within the dialogue to discuss political aspects, and the participation of Arab countries in the Belgrade Conference (Conference for Security and Co-operation in Europe – the follow-up to Helsinki). Only the second proposal found its way into the final communiqué, with a promise by the Nine to assess the suggestion carefully. Otherwise, the communiqué only brought a reiteration of the Community's concern over the continuation of Israel's occupation of Arab territory, and a condemnation of the settlement policy in those territories and of unilateral steps to change the status of Jerusalem.

On the economic aspects of the dialogue, the Tunis meeting made some further headway. The General Commission examined the progress made by the working groups and the specialized expert groups, and approved in principle the allocation of funds for feasibility studies of specific projects. The Arab League later on made $\$15 \times 10^6$ available for this purpose, the Community $\$3.5 \times 10^6$.

One of the major areas considered by the General Commission was the *transfer of technology* to the Arab world. The European side presented a draft declaration which suggested the following specific steps:

(1) Collaboration in an assessment of the needs of Arab states and regions.
(2) Creation of education programmes in European institutions, co-operation with European and Arab institutions, programmes of sending European experts and teachers to Arab countries.

(3) Exchange of information.
(4) Adaptation of technology to Arab needs.
(5) Support and assistance in feasibility studies and in the development of research capacities in the Arab world[40].

On *trade questions*, no progress was made. The Arab side still insisted on generalized and unilateral preferences, which the Nine were unwilling to grant. They pointed out that such preferences, were already given to a number of Arab countries under the Lomé Agreement and the Mediterranean policy of the Community[41].

In the area of *industrialization*, progress was limited, the sticking point being the conflicting interests in refinery and petrochemical projects. Greatest progress was made in the standardization of technical norms and contract clauses. Here, the General Commission asked the working groups to finalize their work in mid-1977[42].

The *Infrastructure Committee* had made considerable headway. It now comprised five working groups (shipping, air transport, land transport, communications and telecommunications, and installations and buildings) with work furthest advanced on port expansion and new overland connections between Arab countries[43].

Progress was also achieved in the *Agricultural* and *Cultural* Committees[44]; the General Commission provided finance for detailed study of a development scheme in the Juba Valley in Somalia[45], and three further specific projects were approaching the point of concrete steps. In the realm of cultural co-operation, the appropriate committee was asked to prepare a list of institutions involved in the study of language, culture and civilization of the other area, to support the exchange of school books, and to develop a programme for co-operation in the information sector[46].

The next meeting of the General Committee of the dialogue took place on 26—28 October in Brussels. Once more, the pace of progress in the dialogue had slowed down considerably, and needed new momentum to be injected[47]. Political developments had made this easier : on the occasion of the European Council meeting in London in June 1977 the Nine had published a declaration on the Middle East, which largely satisfied Arab demands for a recognition of the legitimate rights of the Palestinians and a condemnation of Israel's occupation policy[48]. The meeting in Brussels made considerable progress on a number of technical issues[49], but could not resolve the more fundamental differences in position in the economic sphere — while the Arab side wanted a preferential association agreement with the Nine, the Europeans were not prepared to expand the geographic scope of preferencial association agreements beyond the already existing arrangements (which, through the Lomé agreement and the Mediterranean Policy, included all but the oil-rich Arab states, anyway). However, even on this fundamental difference, some progress was made : the Arab side had clearly relaxed its previous demands not only for free access to European markets, but also for the stabilization of export earnings and of the return of petrodollar surpluses invested in Europe; and the Nine were prepared to consider a non-preferential institutional arrangement with the Arab world as a whole (one of the key demands of the Arab side).

Specifically, the communiqué outlined the following major results of the Brussels meeting[50]: on the political level of the dialogue, the Nine accepted Arab demands to include the Middle East declaration of the European Council and the UN General Assembly resolution of 27 October 1977, which condemned the Israeli occupation policy. Each side also repeated its position on the Israeli—Arab problem. Economically, the clearing of the financial procedures for projects under the *aegis* of the dialogue allowed the start of a number of specific programmes in the area of basic infrastructure, agriculture and cultural cooperation. The projects approved by the Committee included: a symposium about relations between the two civilizations, studies for a new port in the Basrah area in Iraq and in the Tartous region in Syria, and programmes for potato cultivation in Iraq, for irrigation of the Juba valley in Somalia, and for cattle-stock farming in Sudan. Together, the nine projects approved had an estimated volume of $\$4.580 \times 10^6$, of which $\$3.580 \times 10^6$ was to be provided by the Arab side.

Had the Brussels meeting resulted in considerable additional momentum and a marked relaxation of key positions on both sides, political developments in 1978, and in particular the Egyptian—Israeli negotiation process, brought the dialogue to an almost complete standstill[51] — at least as far as the central elements were concerned. Some experts' meetings and working groups were boycotted by several Arab countries, although most of the nine projects agreed in Brussels could be started. More fundamentally, with the split of the Arab world into two camps, the mechanisms of the dialogue were no longer functioning properly, and this meant that it had become almost impossible to inject political momentum into the process[52].

On analysing the first five years of the dialogue, one is struck by two contradictory features: the very slow pace of the dialogue — even by European Community standards — the numerous difficulties encountered, the relatively minor results on the one hand, and its continuation, and constant upgrading, on the other. While the initiative for the dialogue came very much from the 1973/4 oil crisis, it is also striking that the Arab participation extends beyond the oil producers. Moreover, oil has so far played a minor role in the dialogue, and this only in the form of oil products and petrochemical industries, although OAPEC is one of the Arab organizations associated with the dialogue through its tie to the Arab League. Whether this dialogue can therefore be seen as part of the European response to the events in 1973/4, and whether, if this is accepted, it is a very significant contribution to a European international energy policy, is not entirely clear. Nevertheless, the fact that the dialogue includes the oil producers, not only opens up hopes for improved stability of supplies for the Community, but also better opportunities for economic diversification for the oil producers. In other words, the dialogue could improve the chances for balancing the positive spin-off effects in the international oil trade between the groups more evenly than has so far been the case. The involvement of all Arab states creates geographical scope for expanded development efforts which would benefit all concerned. But the Community has so far refused to give the Arab League trade preferences along the lines of those granted to the Lomé Convention members, or associated countries under the Mediterranean policy. The reason for this refusal is first that some of the Arab League members (Somalia, Sudan, Mauritania, plus the Maghreb and Mashrek countries) benefit from preferences under those agreements, anyway. Second, the Community worries about the reaction of other industrialized countries to a further expansion of trade preferences outside GATT. Economic co-operation could be improved considerably however, even without such a general system of preferences for the Arab League, say, in the form of a multilateral co-operation agreement between the Community and the oil producers. The treatment of the Arab League as an economic community would no doubt be the most effective way to enlarge the geographical scope of development efforts[53]. Such a structure could even help to alleviate some of the internal differences and conflicts within the two groups: the pressure to arrive at joint negotiating positions, and to work on joint projects could become a unifying force for both blocs.

The present achievements are rather rudimentary, but they do appear to point to the possibility of a more balanced relationship between the European Communities and the oil suppliers, and a Euro—Arab co-operation could also make a substantial contribution to the solution of the problems of underdevelopment in the Arab world. In this sense, the Euro—Arab dialogue does deal with one of the central issues of energy policy: the long-term stability of supplies[54]. However, the dialogue offers no solution to the problem of potential competition between industrialized countries for secure sources of supply, or to the problem of phasing out Arab oil exports to the Community and Arab dependence on oil revenues.

Could the Euro—Arab dialogue *become* an instrument to deal with some aspects of long-term stability of supplies? Major obstacles have not yet been overcome. It has been suggested that the main objective of the Arab side has been to change Europe's Middle East policy, with the economic benefits playing a secondary role, but the facts do not bear out this interpretation. It is remarkable that the Community has managed, largely, to maintain its position on the Middle East conflict. The only significant concession in political terms has been the inclusion of the PLO in the Arab delegation — a move which helps to improve the chances of a Middle East settlement. The behaviour of the Arab side leads to the conclusion that they are aware of the potential of the dialogue for the economic and social development of the Arab world[55]. The dependence of the Arab world on the European industrialized countries, which we have found to cut across the dependence of these industrialized countries on Arab oil exports, has made itself felt through the dialogue. The apparent vulnerability of Europe *vis-à-vis* the oil producers has largely been redressed by the channels and influence resulting from the various links between the Arab world and Europe.

The real political problem of the dialogue seems to lie elsewhere: in the differences and rifts within the Arab world. These differences, themselves mainly a function of developments in the Israeli—Arab conflict, are beyond the efforts of the Nine to shape the dialogue. Yet these differences mean that progress in the dialogue will be possible only to the extent the Arab side is reasonably united — and this, in turn, will depend to a considerable degree on the progress made towards a Middle East settlement. This is not the only matter which would disturb Euro—Arab relations since other potential and actual rifts exist in the Arab world. But for some time to come the Israeli—Arab conflict will exert a crucial influence on the EAD.

CHAPTER THIRTEEN

# The Mediterranean policy, the Lomé Convention, and trade relations with Iran

Southern Mediterranean countries traditionally have been linked to Western Europe by a variety of economic, social, political and cultural ties; these ties, established before, and vastly expanded during the period of colonial domination of the Maghreb and the Mashrek by Western European nations, remained largely intact afterwards, and the rapidly growing economic weight of Western Europe with its vast markets and its prime role as supplier of industrial exports helped to intensify those links.

The character of the relationships between the Community and the southern and eastern Mediterranean countries corresponds closely to the centre-periphery model; patterns of trade alone suggest that the Mediterranean countries are heavily dependent on their exchanges with the Community, while they are for the Community both markets and suppliers of minor importance — with one exception: oil[56]. The presence of migrant workers from the southern Mediterranean in the Community[57] and the flow of tourism in the opposite direction also establish ties and dependencies which impose constraints and specific problems for Europe's southern neighbours.

Intensity and imbalances of these economic ties eventually demanded some kind of political formalization of relationships. This demand originated in principle from the Mediterranean countries who saw formal trade or association agreements as a means to correct some of the imbalances inherent in the relationship, while from the Community's point of view the justification of such agreements was essentially political — they could serve as substitutes for the non-existing foreign policy of the Community institutions[58]. The first phase of Europe's Mediterranean policy, therefore, consisted of a patchwork of bilateral agreements. The first such agreement was with Greece in 1962, and then with Turkey in 1964[59]; both were seen as preparatory association agreements eventually leading to full membership. A non-preferential trade agreement was concluded with Iran in October 1963 (to become effective in 1964). It provided for limited reductions in the common external tariffs for imports of carpets, dried grapes, dried apricots, and caviar, as well as for a non-discriminatory tariff quota for dried grapes. The agreement was amended in 1967 to include some further concessions, and renewed annually after expiry until November 1973 when Iran did not ask for an extension[60]. A trade and technical co-operation agreement was also signed with Lebanon in 1965, to become effective in 1968; it has been renewed annually following its expiry in 1971. The importance of this treaty was mainly in the area of technical co-operation, with provisions for training programmes and research co-operation[61]. Israel also secured a trade arrangement which, however, did not constitute more than an intermediate and provisional settlement — the only substantial concession made to Israel was a 40% tariff reduction on grapefruit. Israel therefore continued to press for a closer association[62].

Trade concessions made during this first stage still largely coincided with the rules and the spirit of the General Agreement on Tariffs and Trade (GATT); they were the outcome of four sets of factors: bilateral requests from Mediterranean countries, domestic interests in member states, broader political considerations and national policies within the European Community, and the rules of GATT[63]. In the next stage, the Community concluded extensive preferential trade agreements with only a nominal commitment to full free trade — the first deviation from the rules of GATT. Agreements were concluded with Tunisia, Morocco, Lebanon and Egypt. The two former treaties were 'partial association agreements' purely relating to trade; they gave industrial products of the two countries free entry to the Community (with the exception of a few sensitive products), and provided for *ad hoc* reductions in agricultural tariffs (between 50 and 100%) on a variety of products covering about 35% of Morocco's and 70% of Tunisia's exports to the Community[64]. Negotiations between Egypt and Lebanon and the Community opened in 1970, but were interrupted over the question of the Arab boycott against Israel and the European Community's refusal to accept discrimination against European companies because of their trade with Israel. Eventually, a compromise was found ending in the inclusion of a mutually acceptable non-discrimination clause[65]. The agreements provide for a reduction of industrial import tariffs of the Community by 45%, rising to 55% from 1974, and for some special concessions for agricultural products[66]. All four agreements were reciprocal, i.e. the Mediterranean countries granted the Community concessions on quotas and tariffs for machinery, electrical equipment, chemical products and certain agricultural goods, as well as other exports from the Community[67].

The third stage of the Community's Mediterranean policy began in 1972, when the Council of Ministers asked the Commission to develop proposals for a global approach to European Community relations with the Mediterranean countries. Presumably this move was not altogether unconnected with events in the international oil market which by that time amply demonstrated the strength of the producers, particularly the Arab oil exporters. The Commission's proposals presented to the Council in September 1972 suggested broadly:

(1) New agreements removing all obstacles to free trade between the Community and each Mediterranean country in industrial goods as from 1 July 1977 (except for some sensitive goods, and for less advanced developing countries which were to be given a longer period). Agricultural products should be freed gradually from tariff restrictions, with concessions varying on a product-by-product basis.

(2) An improved programme of aid and economic co-operation, development assistance, and technical support.

The Community Council decided to initiate negotiations with six countries — Spain, Israel, Morocco, Tunisia, Algeria and Malta. In January 1975 the Commission proposed to open negotiations with Egypt, Syria, Lebanon, and Jordan with the objective of concluding agreements similar to the ones eventually reached with the southern Mediterranean countries[68].

The Middle East crisis in 1973/4 profoundly changed the situation, and intensive internal bargaining within the Community, mainly between the consumer-oriented approach of the British government and the farmer-oriented policy of the French and the Italians, further delayed the negotiations[69]. Although the Mediterranean countries repeatedly urged the continuation of negotiations, and even threatened to boycott the Euro—Arab dialogue[70], talks did not resume before October 1974. Negotiations with Israel proceeded at a faster pace, and led to the conclusion of the first agreement under the new, 'global' approach. The agreement foresaw the establishment of free access for Israeli industrialized products by 1977, and a substantial reduction of Community tariffs for agricultural products, covering about 85% of Israel's exports to the Community. In turn the European Community will be granted free access for 60% of its exports by 1980, and full access by 1985[71]. In January 1976, agreements were also signed with Tunisia, Morocco, and Algeria — after further political complications due to the Arab boycott regulations against Israel. This was resolved by a compromise including a non-discrimination clause in the agreements, but also allowing the Arab countries a qualification of this clause in separate letters to the Commission[72].

To gain a better impression of the character of these agreements, it might be useful to take a closer look at the agreement between Algeria and the European Community, which reflects the concessions made to Tunisia and Morocco. Valid for an unlimited period, the unilateral trade concessions granted to Algeria include the following points[73]:

(1) Abolition of all restrictions on free access of Algerian raw materials and industrial goods to the Community, with two exceptions. For petroleum products and cork, exports to the Community will until 1980 be subject to an annual ceiling. The quota for petroleum products covers about 20% of Algerian production. After 1980, all restrictions on these two items will be removed.
(2) Algerian agricultural products will not enjoy free access, but substantial concessions have been made. On wine, the product in which interests of European producers and Algerian farmers are most directly opposed, these concessions concern:
    (a) Wines for direct consumption: tariff reductions of 80%, with the obligation to respect the minimum import price.
    (b) Quality wines: tariff free quota rising from 250 000 hectolitres to

450 000 for the five years of the duration of this agreement, with a gradual phasing out of all exports in bulk.
(c) Wines intended for fortifying: tariff reduction of 80% within the limits of an annual quota of 500 000 hectolitres.

In addition, Algeria also secured concessions for the export of olive oil, dates, oranges, and mandarines. Overall, the agreement covers about 80—90% of Algerian products exported to the Community and the concessions vary between 20 and 100% with a certain number of provisions (quotas—import calendars for early produce, etc.)

To strengthen economic and social development in Algeria, the Community grants Algeria $114 \times 10^6$ u.a. over a five-year period, divided up as follows:

(1) $70 \times 10^6$ u.a. in the form of loans from the Euopean Investment Bank at normal market conditions.
(2) $19 \times 10^6$ u.a. in the form of loans on special terms for a period of 40 years, with a grace period of 10 years and an interest rate of 1%.
(3) $25 \times 10^6$ u.a. in the form of grants. About half of this sum would be allocated to the reconversion of Algerian vineyards.

With the agreement of the Algerian government aid given by the Community could take the form of co-financing of development projects with the credit and development institutions of Algeria, of the Nine, of international organzations, or of third countries such as other Arab oil exporters. The agreement on co-operation is to take place within the framework of Algeria's development programme, and emphasizes the importance of regional co-operation between Algeria and other Maghreb states.

Energy is explicitly mentioned in the agreement as one area of co-operation: European participation in the exploration, production and processing of Algeria's energy resources is to be favoured.

The third important area covered by the agreement is the problem of Algerian migrant workers. Here, the text sets up the principle of no discrimination against Algerians in Europe in terms of working conditions or remunerations. Algerians are to receive the same advantages in all nine countries, and will benefit from all periods of insurance, employment or residence in the Community. Pensions and similar payments will be freely transferable to Algeria.

The second set of association agreements, with the Mashrek countries, Egypt, Syria, Jordan and Lebanon, was concluded in February 1977. Overall, they are less comprehensive than the agreements concluded with the Maghreb countries; although the basic elements are similar, the former agreements do not include provisions for migrant workers and their financial endowment is smaller. The financial split between interest-subsidized loans of $165 \times 10^6$ u.a., $108 \times 10^6$ u.a. grants and $27 \times 10^6$ u.a. special loans at low interest rates mainly serves as a stimulus to mobilize funds from other sources. The trade arrangements allow for the free entry of industrial goods of the Mashrek states into the Community from July 1977, with some exceptions for refined petroleum products, cotton fabrics and yarn, fertilizers and aluminium, where the Community has or may set import quotas to be phased out gradually up to 1980.

Unlike the Maghreb countries, Mashrek countries export only a small percentage of total exports to the European Community in the form of agricultural products, and the percentage is even lower if one considers only products included in the Common Agricultural Policy (10% for Egypt, 2.5% for Jordan, and less than 1% for Syria). Here, tariffs have been cut generally by between 40 and 80%, with numerous restrictions to protect the Community farmers. All trade concessions are unilateral; the Community receives in turn only most favoured nation status; and the Mashrek countries retain the right to strengthen their tariff protection.

Iran was not included in the preferential trade co-operation approach developed by the Community since 1972[74]. Fearing to lose ground to the Arab world, and in particular countries with ambitious development programmes, such as Algeria, who now benefit from preferential access to the European market, Iran pressed for a similar special relationship with the Community. Some member governments supported the Iranian demand for preferential access along the lines of the Mediterranean policy. In addition, Iran could point to the fact that between 1963 and 1971, under the old trade agreement, Iranian imports from the Community rose by 223%, while Iranian non-oil exports rose by only 85%[75]. At the end of 1973, Iran had effectively cancelled the old agreement and in early 1974 she declared her intention to negotiate an agreement eliminating all discrimination against Iranian goods and providing free access to European markets for all goods produced by joint Iranian/European ventures[76]. This was at the peak of the energy crisis, and the Commission had considered the idea of broad co-operation agreements on a bilateral basis with the main oil producers[77].

Negotiations were slow to make progress. The Community refused to consider any preferential status for Iran; it had repeatedly committed itself to restrict such preferential concessions to the Mediterranean countries and the ACP countries of the Lomé Convention. In early 1976, the Commission suggested the use of generalized preferences, the concessions made by industrialized countries to all LDCs. This system could be adapted to suit Iran's need, possibly by information at regular intervals from Iran about her export production development, and the subsequent opening of generalized preferences by the Community, which then would be filled effectively by Iran[78]. However, there was no agreement within the Nine on such a procedure, which was thought to be unsatisfactory for a variety of reasons by some member governments. Interestingly, the UK admonished other member countries not to concede special advantages to oil producers so as not to weaken Europe's indigenous (meaning British) oil production and refining[79]. Unable to satisfy Iranian demands in the trade area, the Commission suggested a shift of the emphasis of the agreement towards co-operative aspects. It is unclear, however, whether Iran will be prepared to make an agreement which would not give her safe market outlets for her growing industrial production. And the change of regime in Teheran has suspended the whole question of institutionalized European–Iranian relations.

Apart from the Mediterranean, the Community has granted preferential access to its market to a number of African, Caribbean, and Pacific countries – the ACP members of the Lomé Convention. This agreement superseded the agreements of Yaunde and Arusha, which has associated former French and English colonies in a comprehensive trade and co-operation agreement. Most member countries are at a very early stage of industrialization, and only one of them (Nigeria) is a member of the group of main oil suppliers to the Community, although some Arab countries (Mauritania, Somalia, Sudan) and two other oil exporters, Gabon and Trinidad, are also partners in the Convention. The connections of Lomé with Europe's energy policy objections are, therefore, somewhat tenuous. Nevertheless, the Lomé Convention is of some relevance in this respect because of its inclusion of Nigeria and Gabon, of its importance as a model, and of the possibility of joint Arab/European development projects in Africa[80].

Possibly the most interesting aspect of the Convention is its innovative character in several key aspects of relations between industrialized and developing countries: the export earning stabilization scheme, the sugar agreement, the unilateral trade preferences, and the truly multilateral mode in which the Convention was negotiated – ACP countries acted together and appeared as one bloc in the discussions with the Community, thereby considerably improving their negotiating strength[81].

The most publicized part of the agreement is the export revenue stabilization scheme (Stabex). Starting from the experience that revenue fluctuations of commodity exports of LDCs have a very serious impact on their development, the Community developed in the negotiations with the 46 ACP countries a scheme which would stabilize export earnings without interfering with market forces, the free flow of international trade, or existing or future commodity agreements[82].

The Stabex scheme originally applied to a list of 12 basic commodities, or 29 products including those derived from the basic commodities. The criteria for setting up this list was a threshold dependence for any ACP country of at least 7.5% of its total export earnings on a commodity (2.5% for the least developed countries); this dependence is reassessed each year, and new commodities have been included in the list, which comprises:

(1) *Tropical agricultural products:* groundnuts, cocoa, coffee, cotton, coconut products, bananas.
(2) *Related products:* hides, skins, leather, wood, tea, raw sisal, palm products.
(3) *Processed goods* based on these commodities: groundnut oil, oil cake, cocoa paste, cocoa butter, extracts, essences or concentrates of coffee, etc.
(4) *Iron ore*, which was put on the list, only after long and difficult negotiations, on the insistence of the ACP countries. The Community refused originally to expand Stabex into minerals.

For each product, a reference revenue level is calculated on the basis of the average export earnings in the preceding four years. This average is, therefore, flexible. If the actual earnings for any year fall below the reference level by 7.5% or more (2.5% of the least developed countries), then the difference between the reference level and actual earnings is eligible for a contribution from the Stabex fund. This fund has been endowed with a total of $375 \times 10^6$ u.a. for the five-year duration of the agreement, i.e. with $75 \times 10^6$ u.a. per annum. Any remainder is carried forward, while drawings on the next year's funds can be authorized if the applications for transfer exceed the annual fund of $75 \times 10^6$ u.a. The Council of Ministers, however, has also retained the authority to order a reduction of transfers if the fund runs out of money.

The transfers are interest-free, but in principle repayable except for the LDCs. To establish whether a country has to repay some or all of previous transfers, the administrative apparatus of the Convention assesses the export value per unit of the commodity in question. Only if the unit value exceeds the reference value, and if the quantities are at least equal to the reference quantities, does the repayment have to be made.

Some examples may illustrate the Stabex scheme. Assuming a reference value per unit of the commodity in question is 100, and the reference quantity also 100, reference export earnings are 10 000. If (1) the unit value rises to 120, but quantities fall to 80 (be it for reasons of lower demand, of natural disasters, or of climatic circumstances or other reasons unrelated to discriminatory trade restrictions), then total export earnings in this year will be 9600 which makes the country eligible for a transfer. If (2) the unit value rises to 105, and the quantity remains at 100, then total income is 10 500, and the country has to pay back 500 units of currency. If (3) the unit value becomes 120, the quantity remains at 100, and the country has had one previous transfer of 1000, then only these 1000 units have to be repaid. Finally, if the value per unit remains at 100, but the export quantities rise to 110, then the country does not have to pay anything back, in spite of higher export earnings.

The sugar agreement which forms part of the Convention essentially gives the sugar producers among the ACP countries guaranteed export markets in the Community at prices which are effectively index-linked through their relation to domestic Community sugar prices. The agreement is in principle unlimited in duration, although it can be cancelled after five years, and has the Community buy sugar in specific quantities at a guaranteed minimum price negotiated annually on the basis of the Community's domestic production prices.

Trade regulations mark the area of greatest importance to the ACP countries — in 1974, 57.7% of all ACP imports, and 68.9% of all exports, were accounted for by the Community[83]. The Lomé Convention reversed the previous reciprocity principle of trade preferences, and accorded the signatory countries unilateral preference[84]. The ACP countries were only obliged to grant the Community the most favoured nation status, which effectively meant only an obligation not to discriminate against the Community. This non-discrimination clause, however, does not apply to trade between ACP countries, and between ACP countries and other developing countries — an important provision favouring intra-Third World trade. As for the preferences granted, they give 99.6% of all exports of the ACP countries free access to the Community. The only exceptions to this free access are certain agricultural products covered by the Common Agricultural Policy (CAP) of the Community. About 12.5% of all exports of the ACP countries in 1973 came under this category; 96.5% of these exports, however, could be imported into the Community without tariffs. The preferences accorded to agricultural products of the ACP countries over third countries can reach 20% or more (20% fruit juice, 22% tinned ananas and tinned grapefruit, 24% tobacco). Further important measures in the trade area are the treatment of the ACP countries as one single unit in terms of the origin of products, which should tend to encourage regional trade among ACPs, and some concessions as to the regulations concerning the origin of exports.

Industrial co-operation has a new emphasis in the Convention — an entire section is dedicated to this theme. Industrial co-operation is seen as part of the new relationship between industrialized countries and the developing world, for which the Lomé Convention intends to supply a model. The section deals with a very broad spectrum of co-operative projects and measures, ranging from the establishment of processing industries, over education and training, measures relating to technology transfer and technology adjustment, to projects designed to increase demand for industrialized goods from ACP countries. Apart from this broad range of areas for co-operation, the section established a *Committee for Industrial Co-operation* and a *Centre for Industrial Development*, both jointly managed by the two groups. The former body is a committee of the Council of Ministers of the Convention (the highest body of the agreement, comprising both sides and with the function of supervising the execution of the co-operation projects, identifying problem areas, and of suggesting solutions. The Centre has been set up to provide information, establish contacts, and to assist in co-operation between European companies and the ACP countries.

Finally, the Convention also foresees an intensification of development aid to the ACP countries. This aid was to be given on the basis of choices and priorities made by the ACP governments themselves; an amount of $3399 \times 10^6$ u.a. was budgeted for this purpose for the five years. To distribute the aid, programme missions were to be sent to all ACP countries in order to establish priorities and choices. In the first year of the new aid regime, 40% of allocations went to rural projects, 27% to communication development, 20% to economic and social development projects, and 10% to education[85]. The bulk of the total aid pledged, $2100 \times 10^6$ u.a., would be in the form of grants, and a further $430 \times 10^6$ u.a. in the form of low-interest loans.

Apart from these concrete measures and instruments to reshape the relationships between the Community and the ACP countries, the Convention sets up a number of institutions to deal with the administration of the Convention itself as well as with problems arising out of its execution. While the Convention was initially planned to last until 1980, both sides have indicated that they intend to turn the Convention into a permanent arrangement.

With agreement now reached on Lomé II, some conclusions can be drawn about the first five years of the functioning of the agreement between Europe and the ACP countries. The real importance of the Lomé Conventions probably lies in their model character, rather than in the actual impact of their provisions, which are not always of great significance: the admission of more than 99% of ACP exports to the Nine, for example, is much less of a concession than might appear on first glance — 75% would enter duty-free, anyway[86].

The most revolutionary aspect of the Lomé Convention, the Stabex scheme, seemed to work reasonably well during the first years of activities — in 1975, 19 countries profited from the scheme, in 1976 12; the total volume of payments for the two years was $109.1 \times 10^6$ u.a., of which only 37% had to be refunded[87]. During the period of the Lomé I Convention, the list of products eligible for the Stabex scheme was revised, and countries having applied to join the ACP group were included in the scheme[88]. These revisions of products eligible did not go as far as the ACP countries had wished, however, and led to additional demands for the inclusion of minerals[89].

Industrial cooperation began seriously with the institutionalisation of the Centre for Industrial Development. In its first year of operation (1977), the Centre contributed to 24 new European projects in ACP countries, resulting in the creation of 3000 jobs in ACP countries and about 1500 in the EC: a small but not insignificant beginning[90].

Aid disbursements resulted in considerable dissatisfaction among ACP countries. Some of this dissatisfaction had to do with the speed of implementation; the overall amount was also felt to be not very generous. A special problem related to aid for regional projects (to which 10% of total Lomé aid was earmarked): African countries had great problems in determining 'regions'[91]. The ACP countries therefore pressed for substantial changes on the implementation rules and the volume of aid disbursements under the Lomé Convention[92].

The evolution of the overall trade between Europe and the ACP countries also demonstrated the limitations of the Convention. EC imports from ACP countries increased by significant figures in 1972, 1973, 1974, 1976 and 1977, but fell in 1975. Raw materials and agricultural exports accounted for the bulk of ACP shipments in Europe, with manufactured products accounting for only 4%. ACP exports to Europe grew somewhat faster than exports of other developing countries, indicating the value of the preferential arrangements and other ties developed through the Convention[93]. But Nigerian oil exports dominated the exports of the more than 50 ACP countries, and without these oil exports the trade balance between the EC and the ACP would have been heavily negative for the latter. European measures to develop a new international division of labour remained somewhat ambivalent: while efforts were made to encourage industrialisation of ACP countries, there was a clear reluctance to accept the implication of this in terms of market access for the new exporters of manufactures[94].

The inclusion of the Lomé Convention in this chapter as part of a policy to achieve European objectives in the international energy system might appear far-fetched at first; yet the Convention clearly aims at improving European access to energy and mineral raw materials[95]. Several important actual or potential energy exporters are among the ACP states — Nigeria, Gabon, Niger, Angola, to name but a few. The fundamental recognition that improved access to energy and raw material supplies from the Third World will henceforth have to be linked with steps towards a better overall North–South cooperation is at the very core of Lomé — and this in itself constitutes a major advance. Some have argued that with the network of special agreements with Third World regions the Nine have begun to move down the road towards a conventional, neo-colonial 'Superpower'; and this argument finds some support in the limited benefits the associates of the Nine in the Third World have so far reaped[96]. Yet on balance this view seems too simplistic — it does not recognise the importance of the changes in the international distribution of power after 1973/4. The Lomé Convention is a transitional institution — with its share of insufficiencies and old mistakes, but also with the potential for a new deal between rich and poor countries. As part of Europe's international energy policy, it points in the right direction — but it is, in itself, quite insufficient.

CHAPTER FOURTEEN

# A global energy policy framework: Some proposals

The survey of Europe's future energy position and the international energy context within which it will have to pursue its objectives, is complete. The crucial question remains: what shape should Europe's international energy policy take in the coming years?

The answer follows largely from the conclusions of the analysis presented so far. These are:

(1) *The world energy system is undergoing a process of energy transition away from hydrocarbons.* This establishes the overriding need for vigorous national energy strategies to facilitate smooth adjustment. Although such national efforts are outside the scope of this study, their importance can only be stressed as a vital concomitant factor of the following suggestions. Such national strategies have to recognize the importance of energy decisions on social organization, values and culture. In particular:

(a) Energy objectives have to be harmonized with the overall evolution of social preferences in societies (quantitative versus qualitative growth, consumption versus creativity, distribution versus accumulation of wealth nationally and globally, etc.).

(b) Energy decisions have to be formulated democratically; politicization of energy issues has been one of the most remarkable aspects of the field over the last few years. Yet new approaches are probably necessary to overcome the inherent problems with traditional bureaucratic energy decision-making as well as pluralistic/pressure group oriented energy policy processes.

(c) Energy strategies have to incorporate elements which increase the flexibilities of energy systems and expand the range of options available, so as to be able to cope with the wide margins of uncertainty surrounding future energy supply and demand patterns.

(2) *The energy problem is global; the present world energy system is characterized by very high levels of interdependence and centralization.* Both the geographic concentration of hydrocarbons and the historic evolution of the international energy system have led to very large amounts of hydrocarbons being transported over enormous distances. The consequences have been the dependence on world markets for both exporters and importers. For the energy importers, supply shortfalls and/or drastic price increases could have very severe effects on their economies, societies and political systems. Such shortfalls could

be the result of political interferences with supplies, or prolonged interruptions of exports as a result of domestic instability and civil war in key producer countries or regional instability in the Persian/Arabian Gulf, or of a structural squeeze in which import demand exceeds export availabilities — either because of insufficient production capacity or export policies. For the exporters, a shortfall of oil revenues, caused by loss of markets or domestic/regional disturbances, and/or a sharp decline of oil prices, could equally affect the functioning of their economies, the stability of their societies, and the survival of present regimes. Each side also depends on the other: the OPEC countries have intricate links with the Western market economies, and a high stake in their economic well-being and political health. The importers depend on the economic and political stability of the exporters as a precondition of supply security. Yet here we find an important asymmetry: by and large, Western economic stability can be taken for granted, if not affected by a very severe energy crisis originating from OPEC countries. On the other hand, OPEC countries are all undergoing a process of social and economic transformation, which almost by definition destabilizes in one way or another. Obviously, there is no alternative to some form of transformation; but the challenge of 'hyper-stability', of continuity and reliability in conjunction with rapid change, is daunting. This asymmetry — the necessity to create order in an international environment prone to instability — will constitute one of the central problems of international energy politics.

The vulnerabilities indicated show that the international energy system has reached a stage of overcentralization. For the importers, the task will be to reduce levels of energy dependence on hydrocarbon imports; for the OPEC countries, overdependence on oil exports can be reduced only through the creation of alternative sources of income, and thus the transformation of their societies and economies to a point where they will be able to generate self-sustaining economic structures without oil and gas export revenues.

This economic diversification and the concomitant social transformation is an arduous process full of pitfalls, as the Iranian example has demonstrated. Crash programmes are unlikely to succeed, and would impose severe social and economic strains. While this process is underway, producers are threatened by four dangers: the loss of their control over the international energy system as a consequence of internal divisions within OPEC; the loss of markets in industrialized countries; domestic instability and upheaval as a result of the strains of the transformation process; and finally the decline of oil exports before the economic and social objectives of diversification have been achieved. In some OPEC countries, the latter danger is still almost completely hypothetical, in others it is not far over the horizon. Both importers and exporters thus face the task of reducing their dependence on international energy markets, and both have to cope with a number of risks and dangers on the way.

(3) *On the international level, the political implications of energy transition are, therefore, a transition from the present forms of interdependence and vulnerability to new, sounder forms (increased capital flows, transfer of knowledge and technologies; trade in goods which represent energy in 'frozen' form,*

such as petrochemicals or industrial products which use the energy formerly exported from OPEC countries themselves). Such an evolution also implies a considerable decentralization of actual energy structures, with higher degrees of self-sufficiency of all units in the system.

But while there seems a general agreement about the desirability of such a development, exhortations to pursue adequate national policies are plainly insufficient. One cannot ignore the difficulties in 'getting from here to there' — and the last five years have provided ample evidence for the amount of time and effort which will be needed at both ends. Thus, for a considerable number of years we shall still be faced with the risks and insecurities inherent in the present situation. This means that policies have to be found, and mechanisms to be developed, which can tackle those insecurities. Previous chapters have shown that some efforts have been made, but they have also demonstrated their shortcomings.

There is a second reason why international management is needed while national efforts at reducing dependencies continue: the reduction of dependence on energy exports/imports has to be synchronized. Imagine, for example, that OPEC rapidly achieved the ability to substitute oil exports with other exports, and oil revenues with income from other sources, while the consumers still remained heavily dependent on OPEC oil. This would obviously exacerbate the problem of energy insecurity. On the other hand, one could envisage a successful energy diversification of the importers away from OPEC oil, and a failure of OPEC development strategies. OPEC would thus be confronted with a shrinking market and probably a dramatic loss of revenues. This would most likely result in severe upheaval in OPEC countries, which could easily develop into major regional and international conflicts. It would probably also introduce a new period of 'cold war' between North and South, since the net result would be lack of progress and stalemate in global development.

Both are obviously unlikely developments in this extreme form, but we should not dismiss the possibility of problems in the synchronization of different national energy policies.

In any case, energy security will for many countries, and for the international system as a whole, remain a chimera; the real task facing the international system is the control and management of *energy insecurity*.

This takes us back to the past and present management structures of the international energy system — after all, the present state of affairs has been with us for quite a while. The international energy system, as it developed after World War II, has certainly always had a central management structure to deal with at least some of the fundamental objectives outlined above. Since this was a quasi-colonial structure, the objectives satisfied were mostly those of the industrialized importers, although the international oil companies, which ran the control structure, were flexible enough to accommodate some demands by the exporters, as well. Obviously, the profits of the companies themselves were also an important consideration in their management.

The ability of the Western international oil companies to take over these central management functions, shaping price and supply decisions and assuring security and continuity of deliveries, rested ultimately on the centralization of power in the international system as a whole.

The West, and more particularly the United States, dominated the Third World oil exporters sufficiently to support a centralized structure of trade. From the point of view of the developing exporters, this type of centralization looked much less acceptable, since it implied the subordination of their own energy priorities and general autonomy through the integration of a vital source of wealth into an international system beyond their control. These disadvantages remained secondary so long as the system expanded rapidly and profits increased for all concerned. Yet some exporters were made to feel their vulnerability as soon as they challenged the rules. Iran, Iraq and Venezuela, in the attempt to achieve national economic and political goals, all experienced the negative side of over-dependence on the international oil market.

The long and protracted boom of the world economy in the 1950s and 1960s stifled the impact of growing structural imbalances in the international energy system, as it did in the world economy as a whole. Soon after World War II, the centralization of power in the international system began, first, to fray at the edges, and then gradually to come apart. Slowly the balance of forces began to shift in favour of the oil exporters. The dominance of the majors collapsed under the two-pronged attack from the independent oil companies and from exporting governments, and in parallel the political and military ability of the West to coerce the exporters declined. At the same time, the dependence of industrialized importers on the international oil market continued to grow. By the early 1970s, the power relationship had definitely shifted; by 1973, OPEC had taken control of the crucial decision-making functions in the international oil market, and thus the international energy system. Price levels and quantities exported were now considered the exclusive affair of OPEC governments.

In essence, the history of the political economy of the international oil market has thus been a quest for control. This is hardly surprising at present levels of interdependence, since both exporters and importers easily appreciated the vital importance of such control for the realization of their fundamental economic and social objectives, and for the protection of vital economic security concerns.

(4) *Two different types of control can be distinguished (and the history of the international oil market provides examples for both): hegemony and bargaining.* The former characterizes a situation in which one group of actors, or one principal actor (e.g. the USA) controls the essential parameters of the international oil market (directly or indirectly); the latter marks a configuration where several actors, or groups of actors, exert some influence without being able to control all vital elements of the system. The former depends on a concentration of power, the latter on a diffusion. The former relies principally on coercion (although the methods of coercion can be quite subtle or even 'internalized'); the latter on bargaining.

The international energy system reflects the global diffusion of power which has been one of the characteristics of the 1970s. It might appear as if one hegemony had been replaced by another (that of OPEC). But the analysis of Part 2 has shown that in fact the present situation represents a bargaining situation — in part, because OPEC's power is extremely lopsided and one-dimensional, supplemented by important weaknesses and vulnerabilities. But part of OPEC's limitations stems from the very character of oil power: its mobilization by OPEC, through halting or reducing exports, could be enormously destructive for the West and therefore ultimately suicidal for OPEC. An important leverage of the consumers, thus, paradoxically lies in their vulnerability.

OPEC and the industrialized consumers are thus locked in a difficult, but far from unique international bargaining situation — it results ultimately from two inverse global trends: the increase of interdependence and the diffusion of power. This has meant, quite simply, that the requirements for regulatory capacity in the international system have increased while the actual capacity to control has been reduced. No single actor, no power alone, can decide on the most crucial problems of international affairs; decisions rest on the ability of several important actors to co-operate. Equally, the diffusion of power has meant an increase in the number of actors capable of blocking progress and vetoing decisions. This applies to international energy issues as much as to others. The bargaining situation at hand is analogous to industrial conflict: both sides are organized and capable of mobilizing considerable power. Confrontation could damage both severely, and ultimately destroy the basis of their existence. Like the trade unions, OPEC started out as the underdog, like them, it now has, if anything, some advantage over the other side in the bargaining process. Yet ultimately its power is destructive, and as some trade unions (and better than others), OPEC has begun to accept the responsibilities and constraints which go hand-in-hand with its power.

The present dynamics of control in the international energy system are thus the result of complex bargaining processes, involving both co-operation and conflict. Conflictual aspects are unavoidable, given the differences of interest involved; yet it is clear that only co-operation can assure a smooth international adjustment process to a new world energy structure. Attempts to establish hegemony are bound to fail, with serious consequences; the only realistic course seems the management of insecurity through co-operation.

(5) *International energy co-operation has to be organized, basically around the two most important groups: the developing energy exporters and the developed energy importers.* It would be desirable, however, to include within a framework of management other socialist countries — as was shown in the case of Eastern Europe, this region might become a significant net exporter during the coming years, and should thus be given the opportunity to participate in international energy co-operation.

More important still is the participation of the developing energy importers. Some of them are the most vulnerable countries among all importers, yet they have so far been unable to play an active role in world energy politics[97]. Besides, they can be expected to expand considerably their participation in the international energy system during the rest of the century. There are massive energy resources to be developed in those countries; and a recent World Bank study[98] predicts that the LDCs will increase their share of total world energy consumption and also expand their imports. Total energy imports (mostly oil) of the LDCs as a group (excluding OPEC) are estimated to be between 3.84 and $4 \times 10^6$ b/d in 1985, and by the year 2000 this import demand will have grown to between 7.6 and $9.37 \times 10^6$ b/d. These figures, incidentally, conceal growing oil and energy exports from some LDCs; the actual numbers of import requirements will, therefore, be considerably higher.

(6) *What are the fundamental problems to be solved by any global energy management structure?* First, a reconciliation of the 'natural interests' of exporters and importers, which are interlinked through the present situation of interdependence. These natural interests present themselves to net exporter countries (primarily of hydrocarbons) as assured markets for their exports; rates of production and export prices designed to produce revenues sufficiently large and sustained ultimately to buy independence from energy exports; and the construction of viable societies and economies based on diversified sources of wealth. The natural interests of the developed oil importers lie in supply and price levels which allow them to move smoothly and without major economic and social disruptions to an alternative energy economy. Those of the LDC importers are much the same so long as one adds their special need for measures to alleviate their specific disadvantages.

International prices of energy traded between states are of obvious relevance to all three groups and form an important mechanism for reconciling some of the contradictions inherent in these different interests. The central role of prices as the expression of the balance of forces in the international energy system ought to be reduced, and their role as mechanisms to adjust supply and demand given more emphasis. At the same time, there is little prospect that this can be done through market mechanisms, which do not reflect the long-term dimensions of the energy problem, nor the intensive politicization of international trade in energy. Oil, in particular, is too closely linked to national fate, national aspirations, and national objectives to offer any prospects of being amenable to such treatment, even were the economics to favour it.

The second basic objective any global energy order should meet is the aspiration for broadly based development. The admittedly ambitious notion of an energy transition managed through extensive international co-operation is justified by the enormous risks inherent in a failure to co-operate in phasing out oil; similarly the objective of a greater equality between North and South can draw considerable force from the implications of a failure in developing the Third World. In order to avoid the political and socio-economic consequences, which might result from major failures in development efforts, new approaches in international energy politics will have to be developed. The weakest developing countries are simply not in a position to shoulder the burden of expensive energy imports, which can only increase substantially if no specific arrangements are made to take account of these countries' situations.

The position of the LDCs also underlines the need for international energy management from another point of view: the rising energy demand of the Third World as a result of population growth and the need for socio-economic progress. In the future, the simultaneous pressures on world hydrocarbon reserves from developing and developed countries could not be sustained very long. If international trade in hydrocarbons continues to increase, extremely difficult choices will have to be faced in the not too distant future: on what basis should hydrocarbons be rationed? Access to energy could become politicized to an as yet unimaginable degree, with lines of conflict drawn between developing and developed countries, between OPEC countries and both groups, between the rich and the less rich among the developed countries, and between the less poor and the poorest among the developing countries. To prevent such disasters, the non-oil LDCs (like the industrial powers) have to be helped on the way to a different type of energy structure as soon as possible — a structure relying to the maximum extent on indigenous resources, and in particular on energy income rather than energy capital. This alternative, mainly 'soft'[99] energy path, has to be made possible by R & D, by aid to develop indigenous resources, and by the developed importers setting an example.

\* \* \*

International energy management can be considered as a *co-ordination of national energy policies* within an international framework; it is unlikely to be, and need not be, one global policy. The nation-state will, it is assumed, continue to play its crucial role. International management therefore has to be the result of bargaining and negotiations between nation-states.

Such co-ordination of energy policies will have to be institutionalized in some form. To explore possibilities, let us invent an 'International Energy Council' and its secretariat, and look at the tasks and difficulties that it will face. The first problem in the creation of such a Council would be its *membership*. Against the desirability of broad participation one has to set the need for effective decision-making. In principle, three groups of countries (OPEC, the industrialized importers, and the non-oil LDCs) should be involved. A balance between the three, by having equal numbers of members from each, would be ideal. One possibility is a limited membership (as in the North—South dialogue, which also drew on the same three groups). To make the Council more representative, it could include additional members from each group on a temporary basis, rotating the membership as is done in the UN Security Council. If such a representational system were too rigid, it could be amended by 'General Energy Assemblies', with a broader membership, at regular intervals. Membership of the Council should be potentially open to all countries, including socialist ones. If these countries turn to world markets (either as suppliers or, more likely, as importers), it will be generally desirable and probably in their own interest for them to join an international structure of management. This would increase the credibility of the organization and provide an alternative to East—West confrontation and strategic power-play over the control of the Gulf area. Of course, it could also make decision-making in the organisation more difficult — however, positive decision-making will in any case be possible only to the degree that all parties are willing to compromise.

The *function* of the Council has, as already pointed out, to be essentially political. It should be conceived as a negotiating framework with a strong set of limitations imposed on the freedom of bargaining — in order to avoid confrontation and crises. Crucial decisions would be political decisions of fairly short duration — i.e. they would be subject to correction and political approval at regular intervals. Thus, authority would remain with national governments, but it would also be limited by negotiated rules. A managerial capacity would, however, be necessary as well. For this purpose, the Council should have a permanent secretariat to supervise the actual implementation of agreements, and to conduct collection of information and independent research on the global energy situation.

The first substantive issue to be dealt with by the proposed International Energy Council would be the price of oil — more precisely, the price of internationally traded crude oil. For this purpose, the price of the marker crude would be fixed annually in the Council within a previously agreed general framework — say, a minimum increase of 5% and a maximum of 15% per annum as a yardstick for the next 10 years of the agreement. The overall idea would be to provide for a gradual, predictable and economically manageable rise in oil prices for a certain number of years to ease the energy transition away from hydrocarbons. The approximate level of the annual adjustment would be decided beforehand in such a way as to leave a certain flexibility for annual revision in line with supply/demand evolution. Such a solution seems preferable to indexation of oil prices — it would be less difficult to agree on since it does not set a general precedent, avoids the complicated issue of which yardstick to use for indexation, and allows some flexibility of management. A second element of the new price structure should be a two-tier system, with special oil prices for the oil-importing developing countries.

From a practical point of view, the two-tier price system could be confined to countries with a per capita annual income below, say, $500. The second tier would be established through a special subsidy fund (partly to be contributed by suppliers, and partly by developed importers, possibly as an element of the International Energy Fund described below). The fund would be limited in time (15 years, for example), and take into account the initial increase in import demand from the developing non-oil LDCs through an annual increase in their quotas of subsidized import rights. After a sufficient transitional period, the non-oil LDCs which benefit would be expected to switch towards domestic alternatives to hydrocarbon imports, and the quotas would be reduced, and then phased out. Payments could be made as grants, or quotas could be bought by directly drawing on the subsidies provided by the fund. Re-exports of imports would be prohibited, with the sanction of loss of preferential treatment; if limited cheating occurred, however, it would not be too serious — it could simply be considered as hidden development aid.

One aim of the two-tier price system would be to alleviate the burden of the poorest energy-deficient LDCs, which is particularly high in the early phase of industrialization and development. The other would be to give LDCs a comparative energy-cost advantage over industrialized and other developing countries so as to attract new industries into them. Price level and quotas should be set and phased so that the need for transition to differently based energy economies is emphasized in these countries, too — it would be dangerous to subsidize them to the extent that hydrocarbon imports are encouraged. On the other hand, allowance should be made for the difficulties of adjustment, and quotas should initially be increased annually for a number of years, so as to alleviate transition arrangements. The quantities involved in such a two-tier system would be limited, and the loss of revenues for supplies would be small, even if they had to bear the full loss of income themselves[100]. The subsidy could also be made on a loan basis with low interest rates and long grace periods — the crucial element is to link the financial arrangement to oil import levels.

Apart from the greater equality and justice, which a two-tier price system would introduce in the international energy system, the proposals on pricing outlined above would seem to be in the interest of both developed importers and developing exporters: it would help both to plan their energy future rationally. The degree of price stability injected by the proposals would allow energy investments to be made with a high degree of predictability about future price levels, and thereby remove a crucial element forestalling such investments. It would allow planners in OPEC countries to draw up development plans based on a fair degree of certainty about future income. Since the oil price trend is upward, it would also encourage energy conservation, and make profitable new sources of alternative supplies. And finally, the scheme would even inject an additional element of stability into the world economy, since supply interruptions and inflationary shocks from drastic price increases would become less likely.

As indicated, the finances for a two-tier price system could be established through an International Energy Fund. Such a proposal appears justified for a number of other reasons, as well: there seems to be a need for further increases in the level of international energy prices to stimulate alternative energy production and discourage demand. Yet if these price increases were made without special additional arrangements, they would further aggravate the problem of recycling petro-dollars, and the mismatch between a capital surplus on the one hand, and on the other a desperate capital need (for both energy investments and development). The Fund would recycle additional OPEC income into the two areas where investments are most needed: energy and development (the latter through subsidizing oil prices for poor LDCs, which would be directly translated into greater demand for imports from industrialized countries and other LDCs).

The income of the Fund could be provided by any one of several means:

(1) An international tax on every barrel of oil, and every cubic metre of gas consumed.
(2) The income, or some of the income, from oil price increases decided by the Council.
(3) Payments by developed importers and suppliers in equal parts.

The endowment of the Fund would be credited to the suppliers of financial resources; in the case of revenues from price increases being paid in, the credit would obviously belong to the developing exporters. The Fund would have three major functions:

(1) The financing of a two-tier price system as described above.
(2) The financing and commissioning of energy research and development, with a particular emphasis on the need of LDCs.
(3) The financing or financial backing of energy production from alternative sources in importing countries.

The last function would be the most important, since it would provide a mechanism to channel income from oil price increases directly into alternative energy investment. The Fund could in effect become an international energy resource bank, which would guarantee investments in LDCs, and itself finance some projects. Returns on such investments would go to the countries providing the funds for this enterprise. In effect, then, the Fund could combine aspects of the World Bank with those of an international resource bank, working both as a facilitator of capital investments through the provision of expertise and guarantees, and as a lender for countries which want to launch their own projects.

The advantages of such, or similar, arrangements would be multifold: for the exporters, it would be a gesture towards the non-OPEC LDCs and provide them with investment possibilities in energy production in other countries; for the LDCs, the Fund could provide a substantial subsidy on oil imports, help for the development of their own resources, and know-how through the Fund's R&D; the developed importers would profit from the demand created in LDCs, and possibly also from energy investment capital of the Fund.

The other key task of the proposed Council, apart from deciding price levels, would be the *co-ordination of import and export targets.* Clearly, some kind of world indicative planning of international energy trade will be necessary, if the risks of very serious gaps between supply and demand are to be avoided. Such planning might also increase the flexibility of the system: OPEC countries might be persuaded, for instance, to increase exports beyond levels they first considered, if they are given at the same time assurances that this will be necessary only over a limited period of time. Again, it would be impossible to arrive at such co-ordination without political negotiations at regular intervals. Two different sets of agreement would in fact appear to be necessary: some long-term guidelines (an indicative plan for energy imports and exports for, say, a period of 10 years), and actual quotas on international trade, decided within the framework of these guidelines on an annual basis. This second step could be seen as a review of the extent to which developments are in line with expectations, and of what corrective actions have to be taken. The secretariat of the Council could elicit for both purposes recommendations from independent experts.

Although some flexibility will be necessary, import/export targets should be adhered to as strictly as possible. In fact, the purpose of this proposal is to shift the burden of adjustment, the provision of flexibility in the system, away from the international markets to the nation-states. Nation-states would thus become accountable to the international system, rather than being able to externalize the failures of their energy policies and shift the burdens on to someone else. (The outstanding example is, of course, the United States, which not only expects OPEC to keep production in line with a world demand greatly increased by US imports, but also makes the European and Japanese energy position much more difficult.) If its targets are not met, it would be basically up to the nation-state to compensate the failure.

The Council would decide annually, or bi-annually, on quotas on exports and imports on the basis of recommendations by the secretariat – in effect an approval or a slight revision of the original joint commitment in the indicative plan. Although the distribution of the quotas among the groups could be left to themselves and to their organizations (OPEC and the IEA – developing importers would be in a special position), a detailed plan of export and import quotas by country might actually be possible since some national planning will be needed, and in most cases available anyway. If an importing country exceeds its quota, it ought to be taxed heavily with an additional levy (to be paid into the International Energy Fund?). If imports fall short of quotas, the difference could be carried forward. If exporters do not meet targets for reasons other than *force majeure*, the difference should be made up from stocks, and be paid by them. A serious failure to meet export targets would be equivalent to interference with supplies, and should be treated as such.

Bargaining over these quotas will obviously be very difficult; it would be desirable to have a strong commitment to an overall indicative plan of import and export evolution. However, an alternative to actual quotas, and the levies indicated above in case of failure to meet targets, is difficult to envisage. Some such mechanism will be essential if major adjustment crises are to be avoided. That something like this might not be totally impossible is suggested by the acceptance of oil import targets by the IEA member countries[101] (although this is as yet only a declaration of intent without sanctions), and by the role the IMF has begun to play in some cases as an effective instrument for imposing domestic economic discipline. Governments might not be altogether adversely disposed to have some international authorization and, indeed, pressure on them to do what is accepted as necessary but domestically unpopular.

The proposed scheme offers a number of advantages for both importers and exporters. It would in the longer run increase the security of supplies and of markets. It would make predictable the claims on suppliers' resources, thus allowing them to design their conservation policies in line with domestic responsibilities, rather than with international obligations and pressures, which are all the more insistent for being ill-defined. From the industrialized importers' point of view, it would obviously help to improve stability of supplies. It would also provide the time to develop alternative sources of energy, and to reduce demand through energy conservation without over-reliance on dubious crash-programmes. On the other hand, it would also set a time-frame within which importers would have to meet certain adjustment targets — a factor so far neglected in present policies, which recklessly externalize costs due to energy adjustment failures.

The last major aspect to be incorporated in the proposed Council would be measures to deal with the *security of imports against supply interruptions*. Here, the Council might adopt something like the IEA Emergency Allocation Scheme in an expanded version, or the IEA might be modified itself to provide such a system for a larger number of countries. One of the conditions for membership in the Council would also be a clause outlawing politically motivated supply cutbacks. Non-compliance with this clause would be penalized by expulsion. Such measures would create an additional insurance against embargoes and cutbacks: the constitution of an exporter action group would become more difficult since some exporters at least might choose not to risk expulsion. Ultimately, however, the viability of the safeguard system depends on the viability of the institution as such, and on the extent to which it will be able to attract exporters. If the majority of oil and natural gas exporters were to join, this would create sufficient momentum since exporters outside would have to worry about having to shoulder the burden of the marginal supplier (strong fluctuations in export levels, and/or in export prices). Once this concern over marginalization in the market became widespread, security might lose much of its urgency as an issue, because suppliers would be reluctant to resort to politically-motivated cutbacks. Needless to say, the amount of transfer of national sovereignty by the exporters would have to be paralleled by equally stringent guarantees by the importers about access to markets, with the same penalties attached in the case of boycott attempts and other measures designed to protect domestic supply production against competition from international supplies.

This set of proposals serves primarily to illustrate the type of problems which one must expect to have to tackle in the future international energy system, and as a challenge to draw political conclusions from present trends. In reality, progress towards international energy co-operation will be slow and interrupted by crises. Nevertheless, pressures inherent in the present situation will be felt in the way indicated. Indeed, the developments in the Conference on International Economic Co-operation and elsewhere have underlined this conclusion. Some movement in this direction is clearly recognizable. The next step now might be agreement between exporters and importers to consult on prices. Supply and demand quotas in international trade could be initiated by unilateral production ceilings, and their extension to key suppliers, in particular Saudi Arabia.

The specific mechanisms developing out of such a step-by-step approach could be very different from the ones outlined above. What is crucial, however, is the emphasis on institutionalized international co-operation and joint management of the global energy system. Some form of world indicative planning for internationally traded energy, some mechanism making international energy prices reasonably predictable, and installing a two-tier system, seem absolutely necessary to avoid the risk of serious adjustment crises. All indications show that the world is headed for a period of energy scarcity and high-cost energy. The implications of this for global economic, social and political development have hardly begun to be appreciated. To do so, one has only to consider energy input—output tables: the world food problem, for instance, is intricately linked with the energy situation, and a solution of the former depends to a considerable extent on solutions for the latter[102]. The same is true about the broader problem of underdevelopment. Any such solutions, however, can hardly avoid taxing the industrialized countries with their very high energy consumption per capita, if a more equitable world structure is to be achieved: hence the argument for a two-tier price structure.

The alternative to joint international management would be confrontation, which could take the more irrational and — given the present level of interdependence — probably self-defeating form of an attempt by one group of countries to dominate other countries. Alternatively, this confrontation could be moderated by the awareness of interdependence, but this relies essentially on the stability of a variety of factors, such as the prevalence of the conservative regimes in key exporter countries, the Middle East situation, and the reluctance of both sides to take big risks. In international energy politics there is already a growing recognition of interdependence and the need for compromise. Joint institutionalized management should move further towards this kind of awareness by increasing the stability and reducing the uncertainties of the present situation. Without such steps, adjustment crises can be avoided only if industrialized importers reduce their energy demand effectively, and OPEC introduces slow and regular increases of oil prices, without drastic changes, in line with the need to discourage demand and to stimulate alternative oil and other energy production. But this approach neglects the problem of energy-deficient LDCs (which is bound to pose more serious difficulties than is now recognized) and makes a series of crucial assumptions which cannot be taken for granted. These are:

(1) The industrialized countries will enact stringent domestic energy restraints.
(2) The rate of price adjustment needed to meet OPEC's revenue demands will not cause major economic difficulties for the industrialized importers.
(3) The internal bargaining processes within OPEC will produce solutions

acceptable simultaneously to all major exporters and the industrialized developing importers.
(4) The reduction of dependence on international trade in energy will be spontaneously synchronized in importing and exporting countries, posing problems of loss neither of supplies nor of markets, as the future supply—demand pattern evolves out of the reshaping of different national energy priorities.
(5) The excess energy producers such as Saudi Arabia will continue their policy of producing the amounts required by international demand, whatever the level of production required.
(6) The accumulation of surplus revenues will not create unmanageable problems.
(7) The social and economic development of exporter nations will evolve without major political upheavals or setbacks.

Events in 1979 have already destroyed the validity of several of these assumptions — to rely on them any longer would be foolish.

It becomes easier to envisage alternative energy perspectives if one imagines only some of these conditions prevailing at any one time. A few such scenarios seem worth outlining, in approximate order of probability. The list is far from complete, and outlines only some extreme cases.

## The likely future

The growing dependence of importers on OPEC oil might push production levels in exporting countries close to limits of capacity. Those limits could be dictated by physical factors of reserves, field characteristics, etc., but more probably would be imposed by economic and political imperatives, such as conservation, or the inability of, for example, Saudi Arabia to absorb excess earnings. The contradiction between the national priorities of some exporters and the responsibilities they are expected to assume for the well-being of the international system could become much more acute than it already is. Since the cost of this contradiction would have to be borne mainly by a small number of low absorbers, and ultimately to a large extent by Saudi Arabia, the problem would tend to become more and more pressing, the dilemma more extreme[103].

At this point, supply constraints could easily reappear in the system, possibly in the form of a definite and fairly low production ceiling set by Saudi Arabia. This would translate yet again into drastic price increases, and effectively deny supplies to certain importers and consumers. The pattern of supply would no doubt be in favour of strong countries and groups, although political allocations (e.g. in favour of the Fourth World, or of 'friendly' countries) are conceivable. The result would be an intense struggle over distribution, which would almost certainly aggravate the North—South conflict, strain the relationship between North America and Europe/Japan, and severely harm the world economy. The political ramifications of such a crisis are impossible to imagine, but they would certainly be traumatic. The adjustment towards a situation of scarcity would in this scenario be telescoped, and thereby exacerbated and fraught with risks of political miscalculation and irrational behaviour. The year 1979 brought about a mild version of such a crisis (see Postscript). Yet a re-run of this second energy crisis in the way indicated here cannot be excluded.

## OPEC breakup

Some high absorber countries might find that their ambitious gamble for rapid economic diversification away from oil and natural gas is not succeeding sufficiently quickly. In a situation of continued dependence on oil revenues and of declining hydrocarbon reserves, possibly coupled with major failures in their industrialization programmes, these countries would be strongly tempted to press for higher oil prices. OPEC might break apart. If the split turned out to be permanent, it could easily involve military conflict and/or subversion between members, with grave implications for the short- and long-term energy supplies of the importers, whose major concern ought to be stability in the main supplying region, the Persian/Arabian Gulf. The short-term risks would be of supply interruptions due to instability in key suppliers, open warfare or sabotage in the Gulf; the long-term risks would be of price war with periods of artificial energy glut at low prices followed by sharp price increases and supply constraints.

More probably, however, such pressure for higher prices within OPEC would push up the compromise level negotiated within OPEC, and this compromise might ignore the ability of the world economy to adjust to price increases. Again, the impact would be gravest for the weaker countries. This has already been confirmed by events in 1979.

## A return to consumer power: effective reduction of oil import levels

Industrialized importers might succeed in reducing their demand for OPEC hydrocarbons. This, in turn, could lead to two consequences. The more likely one would appear to be a strengthening of OPEC cohesion under pressure, the development of a rationing scheme for the shrinking market, and price increases to make up for the decline in revenues — price increases which could probably be sustained in the light of low demand elasticities in importing countries. Again, the economic adjustment could pose problems, if the exporters were to overestimate the adaptive capacity of the world economy. Alternatively, OPEC might break up, with the possible consequences outlined in the previous scenario.

In addition to these possibilities, a change of regime in Saudi Arabia would obviously be of crucial importance for the future of the international energy system. A drastic, or even a gradual, reassessment of energy policy priorities in Saudi Arabia would have dramatic consequences if the change lowered the present importance given to international responsibilities. Clearly, the present

Saudi oil policy is affected by *Realpolitik*; but the desired political leverage could be fully maintained even if the government decided to continue the production ceiling of $8.5 \times 10^6$ b/d, or even lower it somewhat.

Although the suggested scheme of international management would not totally eliminate these risks and uncertainties, it would reinforce elements of stability and reduce uncertainty for both importers and exporters. One of the elements of uncertainty in the present situation is the various linkages between energy aspects and other political issues[104]. Until 1978, these links (with the Israeli–Arab conflict, Gulf politics, North–South issues, the East–West balance through the concern of the dominating moderates in OPEC about the impact of their decisions on the global balance of power, and with the health of the world economy) made for a certain stability. Indeed, together with domestic stability in key producer countries, they were the *only* factors of stability in the 1974–1978 situation producing OPEC decisions generally tolerable, though not without problems, for importers. That a continuation of this configuration cannot be taken for granted and a fundamental change in any of the elements quoted, or even a *changed perception* of these linkages by OPEC decision-makers, could overturn this precarious stability has already been demonstrated by the Iranian revolution and the implications of the Israeli–Egyptian peace treaty.

In the long run, it therefore seems desirable to reduce those linkages, and to confine trade-offs between exporters and importers to the energy realm proper. The proposed scheme is based on the notion that negotiating decisions in the international energy system should only involve trade-offs between the interests of importers and exporters in the energy system, emphasizing long-term interests instead of linking price decisions to progress in the Israeli–Arab conflict or the North–South dialogue. Yet these links exist, and have to be taken into account. In fact, it might be possible to use them for progress towards the institutionalization of international energy co-operation. A settlement of the Israeli–Arab conflict, for example, could include a commitment by Saudi Arabia and other Arab exporters, who used oil as a weapon in the Israeli–Arab context, to join the proposed International Energy Council as a way of ensuring the permanent renunciation of supply interruptions as a political instrument. Similarly, the links between world oil issues and North–South problems could be explicitly recognized by the industrialized countries in dealing with energy issues and thus be used to secure some exporter commitment towards institutionalized co-operation. What is suggested here is a package deal, which would comprise new institutions and agreements in several areas.

Once international energy co-operation is established in a solid and institutionalized way, it will have to prove flexible. Energy will undoubtedly pose new problems, as yet scarcely broached; and many of them will be international. The siting of refineries, and the mix of crude oil and products in international trade come to mind. The scheme outlined above could absorb new questions and issues fairly easily within its co-operative machinery – it would simply be brought up within the Council, and negotiated there. One possible, indeed probable, development of international energy co-operation would seem to be the expansion from oil and natural gas to other internationally traded energy resources, namely uranium, and conceivably (at a later stage of this century) coal. Negotiations on the prevention of nuclear proliferation have been bedevilled by the problem of security of nuclear supplies – uranium, enriched uranium, etc., and reprocessing capacity. Nuclear energy security could be linked to international energy co-operation in the hydrocarbon area, leading to some kind of linkage between the two. An example could be nuclear energy supply guarantees for some OPEC countries. Such speculations, however, go beyond the 1980s.

Finally, something must be said about a crucial and highly sensitive aspect of international energy co-operation: its relationship to private enterprise. The international oil industry has long had a global management structure. In terms of day-to-day operations, private industry is still indispensable. But will it adjust to a new superstructure of governmental co-operation? The answer might be unexpectedly positive. As already noted, the tendency towards international co-operation goes against the ingrained convictions of the industry, but probably less against its actual behaviour, which always favoured organization rather than competition and stability rather than unpredictable fluctuations.

The changing relationship of forces in the international oil market has ended the period of relative freedom of manoeuvre of the private industry. A new set of rules and regulations, set essentially by producer governments, has come into existence, and private enterprise has adapted very well to these new circumstances. Adjustment to a new structure of international energy co-operation should also be possible, provided the industry finds it can continue to pursue its own objectives. The industry's interest in stability should be served by any arrangement designed to minimize the risks of confrontation between importing and exporting governments, which exposes the companies to dangerous cross-pressures. Profitability should not be impeded by any scheme of governmental energy co-operation. The objective would be to set a framework of rules, with the maximum degree of decentralization and delegation of actual management. Operating fees would presumably become increasingly important under the scheme suggested, and the need to develop alternative energy sources and the emphasis on expansion of energy production in non-oil exporting LDCs under the guarantees of an International Energy Fund should offer promising areas of investment. The crucial function of the international energy system – the reconciliation of differing interests – can no longer be assumed by private enterprise, and the industry will have to accept this fundamental change. In fact, it has already largely done so. At the same time any scheme for governmental co-operation should utilize the resources of the oil companies, enlist their co-operation, and provide them, inside the political framework, with possibilities for adequate pursuit of their objectives. Few, if any, of the aspects in the scheme suggested above, would *per se* have a negative impact on company activities and profits; overall, it should offer positive advantages and opportunities.

The specific proposals made above are obviously not immediately politically realistic; neither are they immune to detailed criticism. This should be left to others. One particular question, however, deserves to be asked: if these proposals are actually realized what could they do to avoid the kind of energy crisis triggered by a repeat of the development already experienced in Iran, possibly in worse form, in Saudi Arabia and other parts of the Persian/Arabian Gulf? The answer can hardly be satisfying, although the kind of international co-operation advocated here might help in increasing supplies from other sources, in moderating energy policies of new regimes, if and when they emerge, and even in making such a development marginally less likely. After all, it would allow the exporters to pursue their economic and social strategies with a longer time framework and less pressure. Still a fundamental asymmetry remains: the dependence of the Western world, and of Europe in particular, on a degree of stability in the Third World oil-producing regions. To ensure such stability under conditions of rapid social and economic change, and of the reassertion of traditional cultural identities against their dissolution in models and concepts developed in a Euro-centric world, will be a colossal task. New approaches, new forms of global pluralism, will certainly be needed to master it. The proposals made above might fit into such an evolution.

Undoubtedly, the dangers illustrated by the developments in Iran also demand some very fundamental rethinking of strategies of social change. Most of this, if not all, has to be done by the countries themselves — the cultural synthesis which seems to have been successful (although not without major breakdowns in the process, either) in Japan and some other East Asian countries. The search for ways in which Europe can contribute to strategies of balanced development in oil-exporting (and other) developing countries, which can avoid excessive instability and upheavals, will be difficult. But there appear to be no other convincing policy tools which could impose order on these countries.

\* \* \*

'Interdependence' and 'controlled insecurity' — these could be considered the *leitmotive* of the analysis presented here. Interdependence summarizes the structural characteristics of Europe's present energy position. As the world heads for a long and difficult period of transition from an energy capital to an energy income system, dependence and vulnerability will continue to characterize Europe's position. The consequence will have to be an organization of insecurity on the international level, until in the more distant future the parallel search for energy security is fulfilled by a restructuring of Europe's domestic energy structures. The management of interdependence will be the only choice, and the fundamental challenge, facing a European international energy policy.

The full amount of energy insecurity in the future is probably still not adequately understood — in spite of the repeated warnings of a 'second energy crisis' by the OECD, by the CIA, by oil companies and independent research assessments such as the MIT Workshop on Alternative Energy Strategies, and the events of 1979. Even the term 'energy crisis' seems to be somewhat misleading: it implies notions of a crisis similar to the one in 1973/4, while in fact a more likely development will be growing constraints on energy availability and higher energy costs in real terms, punctuated by political and economic adjustment crises.

The result of this is bound to be an intensification of conflict along the energy dimension of the international system. The greatest risk is of a confrontation between industrialized importers and OPEC countries. However, an increase in international competition for scarce energy is also likely to be a cause of conflict between industrialized nations. This competition could take the form of a direct conflict over the allocation of internationally traded oil between industrialized countries — but another important factor with potentially even more far-reaching consequences will be the problem of payment for oil imports. If, as seems likely, hydrocarbon prices continue to increase in real terms, the change in terms of trade in favour of OPEC countries will make it even more difficult for industrialized countries (let alone oil-importing LDCs) to meet their oil bills through increased exports. This might be compounded by a relative slow-down in OPEC imports; the rate of increase of the years 1974 to 1977 seems unlikely to materialize again, and a considerable slow-down should be expected as the problems of import absorption become evident in these countries. First indications of such a slow-down are already apparent.

Such a development would undoubtedly result in an increase in industrial competition not only for OPEC, but for all markets. At this point the problem begins to lead to the complex issues of international industrial change and the future of world trade. While this cannot be pursued further, the additional element of conflict induced by energy scarcity and high cost is obvious.

In such an atmosphere of heightened competition and trilateral conflict, Europe would be particularly vulnerable: the USA has the advantages of a much better resource endowment in terms of energy, and the flexibility and leverage of a superpower with a relatively high degree of economic independence. The political, military and diplomatic advantages the USA has to offer the OPEC countries, and its economic capacities in important areas (aircraft, computers, arms) are for Europe hard, or even impossible, to match. The other main competitor, Japan, is likely to improve its economic position compared with Europe; although not free from domestic strains arising from the transformation into a more pluralistic society, Japan's performance over the next decade and beyond should outpace Europe's. Moreover, the factors of strain indicated here are bound to have a different impact in different European countries, thereby exacerbating internal tensions within the European Communities. This, in turn, could combine with the general economic difficulties likely to be created by the energy transition into growing instability in Europe's East—West security balance.

Perhaps this is too gloomy a picture; even a position of weakness in international relations often permits a peculiar kind of leverage and freedom of manoeuvre. Nevertheless, it seems prudent not to rely on this hope, nor on the possibility that actual developments will be less dramatic than they now appear. What, then, is the alternative? The answer given in this chapter is: some form of close and institutionalized international co-operation. As the most vulnerable region in the trilateral world of energy importers, Europe should take the diplomatic initiative in such an effort. Such an initiative would need a common will, and some boldness — for international management of the energy situation is still far from an accepted idea. Yet there would be advantages in Europe launching such an initiative — if it came from the USA, it would be psychologically more difficult for OPEC to accept. Europe now enjoys the reputation of a non-imperial, non-hegemonial power and is widely held to be more 'forthcoming' on international North–South issues than the USA. This could be put to advantage. It would be an initiative for a global effort — there is little scope for any other European strategy. Europe has the most urgent interest in such a scheme, and its own future depends on its political will to press forward.

# Postscript

The manuscript for this book was completed in late 1978. Yet events in the international oil market and its political environment in 1979 added up to nothing less than another revolution — comparable to that of 1973/4 with the quadrupling of oil prices and the successful use of the oil weapon. The term 'second energy crisis', therefore, seems fully justified — particularly since the consequences of this new round of what clearly will be a long and drawn-out process of energy transition punctuated by repeated political flare-ups are in some respects more serious and dramatic than those of the first 'energy crisis'. It thus seemed appropriate to put the events of 1979 into perspective, and to draw some conclusions in relating the 1979 revolution (almost, but not quite yet a 'crash') to the preceding analysis.

The principal event in the international oil market in 1979 clearly was the doubling of prices. Yet this dramatic increase (which in terms of the sums involved is fully comparable to the quadrupling in 1973/4, even in real terms, i.e. after discounting inflation) was only the result of much more profound changes. As in 1973/4, the price increase reflected deep structural changes and their catalysation by political events. In 1979, the trigger of change was the parallel unfolding of the Iranian revolution and the Israeli—Egyptian peace treaty. The structural imbalance was provided by an increasingly precarious supply/demand equation, in which the inability of the industrialized consumers to get oil imports under control (by which is meant a successful turning around of the trend towards increasing oil import demand) could only be compensated through levels of oil exports from the Persian/Arabian Gulf which in the face of a growing trend towards oil conservation and the increasingly evident destabilizing impact of massive oil revenues in those countries of the Gulf region were less and less sustainable. The Iranian revolution had a profound impact on other Gulf oil exporters in at least three different ways: it provided a powerful new push towards oil conservation and lower production ceilings, it weakened the governments on the other side of the Gulf directly through the removal of a factor of regional stability, and it focused and accelerated the growing domestic instability in those countries. The Israeli—Egyptian peace treaty and the subsequent polarization and fragmentation of the Arab world compounded this effect yet further and left the conservative Arab oil exporters, and in particular Saudi Arabia, in a much weaker position. The result was the disintegration of OPEC, which retrospectively revealed itself as an important factor of stability in the international oil market in the period 1974 to 1978. The 'free for all' which resulted added up to a serious weakening of the structures of order which had kept the oil market in balance — principally the special relationship between Saudi Arabia and the USA, and OPEC cohesion under Saudi leadership. The doubling of oil prices was not the only consequence of this process of disintegration — the shift of oil exports from traditional channels into the spot market and into new contracts with a number of newcomers (bilateral deals between governments and their national oil companies and recipients of third party sales of the large international oil companies) fundamentally changed the market structure — and the implications of this change for Western oil security, for future price developments and for the long-term stability of oil supplies cannot yet be fully assessed.

The Western response to this fragmentation of the international oil order was in some respect correct but much too weak. A series of meetings in the IEA, in the European Community and on the summit level (the Tokyo Summit) aimed at a lowering of the short- and medium-term oil import targets of the industrialized countries. This was the right direction — but the targets were neither sufficient nor credible. The IEA objective of lowering oil imports in 1979 by 5% as compared with the initially projected levels was missed by a wide margin, and the Tokyo targets for 1985 were so generous as to demand little additional adjustment beyond that resulting from higher international oil prices and lower economic growth. Nevertheless, the fact that for the first time the industrialized countries could agree on national objectives did indicate a move in the direction of international rationing of available oil supplies among the industrialized countries, and a further evolution of intra-OECD co-operation in this direction seems likely — simply because the alternatives would be too damaging. Yet the industrialized consumers failed to take up the task left by a faltering OPEC — that of providing leadership in the establishment of an alternative structure of order for international oil. In that sense, the fragmentation of OPEC control was not compensated by a parallel effort of the trilateral world to recreate stability; on the contrary, the proliferation of bilateralism indicated a serious lack of co-operation on the part of the industrialized countries, as well. The Soviet invasion of Afghanistan and the growing certainty that Eastern Europe would appear as a substantial net importer of OPEC oil in the 1980s raised the spectre of an alternative international order under Soviet auspices. Although this seems on balance an unlikely future, it does underline the urgent need for new initiatives.

The major conclusion to be drawn from this is, then, that the structures of order in the international oil market have been weakened while the patterns of interdependence continue. Thus, the vulnerabilities of oil importers and oil exporters have been underlined — and this applies not only to the vulnerability of the trilateral countries, but also to OPEC: the effective removal of factors of stability in the oil price structure now makes a price development which overshoots the equilibrium point at which supply and demand elasticities will start to work effectively towards a contraction of the world oil export markets possible, even probable. Paradoxically, everybody now more or less agrees that a gradual, controlled rise in oil prices for a number of years would be the desirable solution — but the inclination of individual OPEC governments to foresake a rational long-term pricing policy in the face of mounting pressure for shortsighted decisions will probably be too strong: the weakness of governments, the threat of domestic instability, the desire to score points by living up to radical credentials or by catching up with the pace-setters simply mean that some OPEC governments will be unable to follow rational policies, and that moderating influences will be too weak to allow OPEC as an organization to reaffirm its grip on oil prices. And even if this should turn out to be too pessimistic a view in the short run, OPEC will surely be overburdened with the global responsibility of conducting the difficult policy of smooth energy transition. Domestic upheaval or regional conflicts in the Gulf region could quickly shatter any new equilibrium which might be achieved.

This implies that the type of multilateral co-operative effort outlined in Chapter 14 is still — maybe more so than ever — an urgent task of international politics. But since opportunities to arrive at such co-operation were missed in the period 1974 to 1978, it will now be more difficult to secure such agreement. Any negotiation with producer countries will now have to be preceded by a serious effort of the industrialized consumer countries to restore their bargaining power through cuts in oil imports. This, in turn, also implies a significant deepening of intra-OECD co-operation on energy. On this basis, negotiations could be initiated with the producers along the lines, and with the principal objectives, outlined in Chapter 14. This will still be difficult, given the weakening of governments in key producer countries, and it will be important to demonstrate that the bargain proposed will take into consideration the legitimate interests of the producers. A strenthening of consumer co-operation might help to make negotiations possible, however. For it would probably lead to a closing of ranks of the producers, as well — an entirely desirable development. A process of negotiation could also help governments to take political risks, as any reversal of the trend towards fragmentation and any increase of active co-operation between producers would also contribute (admittedly only marginally) to domestic authority and stability of a regime.

International co-operation will, however, contribute only little towards the major problem of instability — that originating from within producer societies, particularly in the Gulf region. Although some things could probably be done to reduce this risk and to avoid the pitfalls of uncontrolled, socially destabilizing transformation of OPEC societies, the ultimate dilemma remains: the growth path of an oil-exporting economy is highly precarious — whether growth is fast or slow. To reduce Western vulnerability towards this type of instability will be inevitable. But if efforts in this direction are not accompanied by attempts to create new structures of order between producers and consumers, we risk abandoning the Middle East to chaos and complicating the difficult and risky path of energy transition yet further. There is no real alternative to producer-consumer co-operation.

# Notes to Part four

1. P. R. Odell and L. Vallenillas, *The Pressures of Oil, a Strategy for Economic Revival,* Harper and Row, London 1978
2. An interesting analysis of international economic implications of an effective US energy policy can be found in *America's Oil and Energy Goals: The International Economic Implications,* IEPA, Washington 1977. The result of unilateral US successes in cutting imports without similar reductions by Europe and Japan would be shifts of the adjustment burden to them, and therefore a sharp deterioration of their relative economic position
3. See H. Maull, *Oelmacht: Ursachen, Perspektiven, Grenzen,* EVA, Cologne 1975, Ch. 11; C. Tugendhat and A. Hamilton, *Oil: The Biggest Business,* 2nd edn., Eyre Methuen, London 1975, pp.293–295. A useful summary of bilateral deals can be found in Bo Widerberg, *Oil and Security,* SIPRI, Stockholm 1974, Annex
4. Committee on Energy and Natural Resources, US Senate, *Access to Oil— the United States Relationship with Saudi Arabia and Iran,* GPO, Washington 1977
5. See Part III, Chapter 10
6. See Part III, Chapter 10
7. 'Solemn Declaration Adopted by the First Conference of Sovereigns and Heads of State of OPEC Member Countries on 6 March 1975, at Algiers'
8. *The Financial Times,* 7, 10, 14, 15 April 1975; *International Herald Tribune* 16 April 1975
9. *Le Monde,* 17 Oct. 1975. On the North–South dialogue (CIEC), see W. Wessels, (ed.), *Europe and the North–South Dialogue,* Atlantic Institute, Paris 1978 (Atlantic Paper No.35); J. Amuzegar, 'Requiem for the North–South Conference', *Foreign Affairs,* Oct. 1977, pp.136–159; H. Kruse, 'Die Konferenz ueber internationale wirtschaftliche Zusammenarbeit (KIWZ). Vorgeschichte, erste Ministertagung und Aufnahme der Kommissionsarbeit', *Europa Archiv,* No. 5, 1976, pp. 147–153; E. Wellenstein, 'Der Pariser Nord–Sued Dialog, Die Konferenz ueber internationale wirtschaftliche Zusammenarbeit (KIWZ)', *Europa Archiv,* No. 17, 1977, pp.561–570. Documents about CIEC in *Europa Archiv,* No. 5, 1976, pp.D 126–142 and No. 17, 1977, pp. D 469–496. The best permanent coverage was provided by *Agence Europe*
10. On the results and work of the Energy Commission, see I. Stabreit, 'Die Ergebnisse der KIWZ im Energiebereich', *Europa Archiv* No. 17, 1977, pp. 571–576; L. Turner and M. Hargreaves, 'Energy as a central factor', in W. Wessels, (ed.), *Europe and the North–South Dialogue,* Atlantic Institute, Paris 1978 (Atlantic Paper No. 35), pp.26–37; and T.A. Byer, *The end of the Paris energy dialogue and the need for an International Energy Institute,* (unpublished Ms.)
11. *The Sunday Times,* 28 March 1976; L. Turner and M. Hargreaves, 'Energy as a central factor', in W. Wessels, (ed.), *Europe and the North–South Dialogue,* Atlantic Institute, Paris 1978, (Atlantic Paper No. 35), pp. 26–37
12. L. Turner and M. Hargreaves, 'Energy as a central factor', in W. Wessels, (ed.), *Europe and the North–South Dialogue,* Atlantic Institute, Paris 1978, (Atlantic Paper No.35), pp. 26–37
13. T.A. Byer, 'The end of the Paris energy dialogue and the need for an International Energy Institute', (unpublished Ms); L. Turner and M. Hargreaves, 'Energy as a central factor', in M. Wessels, (ed.), *Europe and the North–South Dialogue,* Atlantic Institute, Paris 1978, (Atlantic Paper No. 35); see *Agence Europe,* 16 Sept. 1976, *The Financial Times,* 19 Nov. 1976
14. CCEI–EN–8 (EEC), *Proposed Conclusions on Continuing Co-operation and Consultation* (Working Paper of the CIEC Energy Commission), Paris 1977
15. *Agence Europe,* 15 June 1977
16. See W. Wessels, (ed.),*Europe and the North–South Dialogue,* Atlantic Institute, Paris 1978, (Atlantic Paper No. 35), Annex. The following analysis draws mainly on T.A. Byer, *The end*

of the Paris Energy Dialogue and the need for an International Energy Institute, (unpublished Ms.), as well as Agence Europe, 15 June 1977 and L. Turner and M. Hargreaves, 'Energy as a central factor, in W. Wessels, (ed.), *Europe and the North—South Dialogue*, Atlantic Institute, Paris 1978, (Atlantic Paper No.35)

17  See J. Amuzegar, 'Requiem for the North—South Conference', *Foreign Affairs*, Oct. 1977, pp. 136—159
18  J. Amuzegar, 'Requiem for the North—South Conference', *Foreign Affairs*, Oct. 1977, pp. 136—159 and W. Wessels, (ed.), *Europe and the North—South Dialogue*, Atlantic Institute, Paris 1978 (Atlantic Paper No.35). See also *Agence Europe*, 16 June 1977, 20/21
19  J. Amuzegar, 'Requiem for the North—South Conference', *Foreign Affairs*, Oct. 1977, pp.147—150
20  J. Amuzegar, 'Requiem for the North—South Conference', *Foreign Affairs*, Oct. 1977, pp.145—147
21  J. Amuzegar, 'Requiem for the North—South Conference', *Foreign Affairs*, Oct. 1977, pp.150—151
22  J. Amuzegar, 'Requiem for the North—South Conference', *Foreign Affairs*, Oct. 1977, pp.153 seq.
23  W. Hager, 'The Nine in the Commodity Commission', in W. Wessels, (ed.), *Europe and the North—South Dialogue*, Atlantic Institute, Paris 1978 (Atlantic Paper No.35) pp.38—49 (45)
24  W. Hager, 'The Nine in the Commodity Commission', in W. Wessels, (ed.), *Europe and the North—South Dialogue*, Atlantic Institute, Paris 1978 (Atlantic Paper No. 35), p.46
25  See also J. Carr, 'The position of the European Community', in W. Wessels, (ed.), *Europe and the North—South Dialogue*, Atlantic Institute, Paris 1978 (Atlantic Paper No.35), pp.13—25
26  H. Maull, 'The strategy of avoidance: Europe's Middle East policy after the October War', in J.C. Hurewitz, (ed.), *Oil, The Arab—Israeli Dispute, and the Industrialized World: Horizons of Crisis*, Westview, Boulder, Colorado 1976, pp.110—138
27  *Keesing's Contemporary Archives*, Vol. 1974, p.26 546. For general view of the Euro—Arab dialogue see U. Braun, 'Der europaeisch—arabische Dialog: Entwicklung und Zwischenbilanz', *Orient*, No. 1, 1977, pp.30—56; K. Meyer, 'Der europaeisch—arabische Dialog am Wendepunkt? Stand und Aussichten', *Europa Archiv*, No.10, 1978, pp.290—298
28  *Keesing's Contemporary Archives*, Vol. 1974, p.26 799
29  *Keesing's Contemporary Archives*, Vol. 1975, p.27 131; U. Braun, 'Der europaeisch—arabische Dialog: Entwicklung und Zwischenbilanz', *Orient*, No. 1, 1977, p.36
30  *The Financial Times*, 12 May 1975, 13 May 1975; *International Herald Tribune*, 13 May 1975, 21 May 1975; *Arab Report and Record*, 1—15 May 1975
31  *Le Monde*, 18—19 May 1975; *Arab Report and Record*, 16—31 May 1975
32  *Keesing's Contemporary Archives*, Vol. 1976, p.27 571; U. Braun, 'Der europaeisch—arabische Dialog: Entwicklung und Zwischenbilanz', *Orient*, No.1, 1977, p.37
33  *Arab Report and Record*,16—31 July 1975
34  *Keesing's Contemporary Archives*, Vol. 1976, p.27 571; *The Financial Times*, 28 Nov. 1975; *The Economist*, 29 Nov. 1975, p.45; *Le Monde*, 12 Dec. 1975; U. Braun, 'Der europaeisch—arabische Dialog: Entwicklung und Zwischenbilanz', *Orient*, No.1, 1977, pp.37—40
35  *Agence Europe*, 17—18 May 1976; U. Braun, 'Der europaeisch—arabische Dialog: Entwicklung und Zwischenbilanz', *Orient*, No.1, 1977, pp.40—41
36  *Agence Europe*, 21 May 1976
37  Full text in *Agence Europe*, 24 May 1976, Document; also *Europa Archiv*, No.18, 1976, pp. D 473—496
38  *Agence Europe*, 14—15 June 1976
39  *Agence Europe*, 7 April 1977; U. Braun, 'Der europaeisch—arabische Dialog: Entwicklung und Zwischenbilanz', *Orient*, No. 1, 1977, pp.41—49
40  U. Braun, 'Der europaeisch—arabische Dialog: Entwicklung und Zwischenbilanz', *Orient*, No. 1, 1977, pp.45—47
41  *Agence Europe*, 16 Feb. 1977
42  *Agence Europe*, 16 Feb. 1977
43  *Agence Europe*, 16 Feb. 1977
44  U. Braun, 'Der europaeisch—arabische Dialog: Entwicklung und Zwischenbilanz', *Orient*, No.1, 1977, pp.47—48
45  *Agence Europe*, 4—5 April 1977
46  U. Braun, 'Der europaeisch—arabische Dialog: Entwicklung und Zwischenbilanz', *Orient*, No.1, 1977, p.48

47 *Agence Europe*, 14 Sept. 1977, 24–25 Oct. 1977
48 *Agence Europe*, 24–25 Oct. 1977
49 For results see *Agence Europe*, 2–3 Nov. 1977; 7–8 Nov. 1977
50 Full text in *Agence Europe*, 4 Nov. 1977, Documents
51 *Agence Europe*, 8–9 May 1978, 20 April 1978, 13 April 1978, 12 April 1978
52 *Agence Europe*, 20 April 1978
53 K. Meyer, 'Der europaeisch–arabische Dialog am Wendepunkt? Stand und Aussichten', *Europa Archiv*, No.10, 1978
54 See the original intentions behind the dialogue as seen from the European Summit in Copenhagen 1973: *Europa Archiv*, No.2, 1974, p.D 56
55 K. Meyer, 'Der europaeisch–arabische Dialog am Wendepunkt? Stand und Aussichten', *Europa Archiv*, No.10, 1978, pp.292
56 G.-P. Papa and J. Petit-Laurent, 'Commercial relations between the EEC and the Mediterranean countries: an analysis of recent trends in trade flows', in A. Shlaim and G.N. Yannopoulos, (eds.), *The EEC and the Mediterranean Countries*, CUP, Cambridge 1976, pp.265–304
57 See G.N. Yannopoulos, 'Migrant labour and economic growth: the post-war experience of the EEC countries', in A. Shlaim and G.N. Yannopoulos, (eds.), *The EEC and the Mediterranean Countries*, CUP, Cambridge 1976, pp.99–138
58 S. Henig, 'Mediterranean policy in the context of the external relations of the European Community 1958–1973', in A. Shlaim and G.N. Yannopoulos, (eds.), *The EEC and the Mediterranean Countries*, CUP, Cambridge 1976, pp.305–324. See also W. Hager, 'Das Mittelmeer–"Mare Nostrum" Europas?', in W. Hager and M. Kohnstamm, (eds.), *Zivilmacht Europa–Supermact oder Partner?*, Suhrkamp, Frankfurt 1973 (English version under the title: *A Nation Writ Large?*, Macmillan, London 1973), pp. 233–264
59 See G.J. Kalamotousakis, 'Greece's association with the European Community: an evaluation of the first ten years', in A. Shlaim and G.N. Yannopoulos, (eds.), *The EEC and the Mediterranean Countries*, CUP, Cambridge 1976, pp.141–160; J.N. Bridge, 'The EEC and Turkey: an analysis of the association agreement and its impact on Turkish economic development', in A. Shlaim and G.N. Yannopoulos, (eds.), *The EEC and the Mediterranean Countries*, CUP, Cambridge 1976, pp.161–178
60 *The Middle East and North Africa, 1975–1976*, Europa Publications, London 1975, p.148
61 *The Middle East and North Africa, 1975–1976*, Europa Publications, London 1975, p.148; see also G.V. Vassiliou, 'Trade agreements between the EEC and the Arab countries of the Eastern Mediterranean and Cyprus', in A. Shlaim and G.N. Yannopoulos, (eds.), *The EEC and the Mediterranean Countries*, CUP, Cambridge 1976, pp. 199–216 for a detailed analysis of a second agreement with Lebanon and Egypt during the next stage of the Mediterranean Policy.
62 See S. Minerbi, 'The EEC and Israel', in A. Shlaim and G.N. Yannopoulos, (eds.), *The EEC and the Mediterranean Countries*, CUP, Cambridge 1976, pp. 243–261 (245)
63 S. Hanig, 'Mediterranean policy in the context of the external relations of the European Community 1958–1973', in A. Shlaim and G.N. Yannopoulos, (eds.), *The EEC and the Mediterranean Countries*, CUP, Cambridge 1976, p.316
64 *The Middle East and North Africa, 1975–1976*, Europa Publications, London 1975, p.146
65 *The Middle East and North Africa, 1975–1976*, Europa Publications, London 1975, p.147
66 See *The Middle East and North Africa, 1975–1976*, Europa Publications, London 1975, p.147, and G.V. Vassiliou, 'Trade agreements between the EEC and the arab countries of the Eastern Mediterranean and Cyprus', in A. Shlaim and G.N. Yannopoulos, (eds.), *The EEC and the Mediterranean Countries*, CUP, Cambridge 1976, pp.199–210
67 *The Middle East and North Africa, 1975–1976*, Europa Publications, London 1975, pp.146–147
68 A. Shlaim and G.N. Yannopoulos, 'Introduction', in A. Shlaim and G.N. Yannopoulos, (eds.), *The EEC

and the Mediterranean Countries, CUP, Cambridge 1976, pp.1—10 (4—5); The Middle East and North Africa, 1975—1976, Europa Publications, London 1975, p.144

69 A. Shlaim and G.N. Yannopoulos, 'Introduction', in A. Shlaim and G.N. Yannopoulos, (eds.), The EEC and the Mediterranean Countries, CUP, Cambridge 1976, pp. 5—6

70 A. Shlaim and G.N. Yannopoulos, 'Introduction', in A. Shlaim and G.N. Yannopoulos, (eds.), The EEC and the Mediterranean Countries, CUP, Cambridge 1976, p.5

71 See The Middle East and North Africa, 1975—1976, Europa Publications, London 1975, p.146

72 The Financial Times, 19 Jan. 1976

73 The following sections are based on the EEC Commission, Directorate General for Information, Information Co-operation and Development, 120/76, and Agence Europe, 19, 20 Jan. 1976, pp.4—5. For Tunisia's and Morocco's association agreements, see Information Co-operation and Development, 118/76 and 119/76

74 K. Kaiser, 'Iran and the Europe of the Nine: a relationship of growing interdependence', World Today, July 1976, pp.251—259 (252—253)

75 EC Commission, Directorate General of Information, Information External Relations, EEC—Iran, No.94/75

76 EC Commission, Directorate General of Information, Information External Relations, EEC—Iran, No.94/75

77 EC Commission, Directorate General of Information, Information External Relations, EEC—Iran, No.94/75

78 Agence Europe, 8 May, 1976; K. Kaiser 'Iran and the Europe of the Nine: a relationship of growing interdependence', World Today, July 1976, pp.258—259

79 Agence Europe, 8 May 1976; see also Agence Europe, 22 April 1978

80 A number of joint projects involving Arab capital and European know-how and/or capital participation have already been launched: The Financial Times, 9 July 1976; K. Meyer, 'Der europaeisch—arabische Dialog am Wendepunkt? Stand und Aussichten', Europa Archiv, No.10, 1978

81 Keesing's Contemporary Archives, Vol. 1975, p.27 051

82 The following is based on EC Commission, Directorate General of Information, Information Co-operation and Development, Stabex, No.94/75, and Keesing's Contemporary Archives, Vol. 1975, p.27 051. See also the detailed study by Frans A.M. Alting von Geusau, (ed.), The Lomé Convention and a New International Economic Order, A.W. Sijthoff, Leyden 1977

83 EC Commission, Directorate General of Information, Information Co-operation and Development, The Lomé Convention ACP—EC, No.99/75, p.1

84 The following is based on EC Commission, Directorate General of Information, Information Co-operation and Development, The Lomé Convention ACP—EC, No.99/75, and on Keesing's Contemporary Archives, Vol. 1975, pp.27 051—2

85 See Keesing's Contemporary Archives, Vol. 1975 pp.27 051 and 27 052

86 Agence Europe, 27 March 1976

87 Agence Europe, 26 Oct. 1977, 23 Dec. 1977

88 Agence Europe, 11 Nov. 1977

89 Agence Europe, 20 April 1978

90 Agence Europe, 24 Feb. 1978

91 The Economist, 17 July 1976

92 Agence Europe, 21 April 1978

93 Agence Europe, 20 May 1978, 15 March 1978

94 Agence Europe, 20, 21 April 1978

95 See The Economist, 1 Feb. 1975

96 See J. Galtung, The European Community: A Super-Power in the Making, Allen and Unwin, London 1972; for an analysis of economic benefits for associated countries during the early stages of association, see A. Shlaim and G.N. Yannopoulos, (eds.), The EEC and the Mediterranean Countries, CUP, Cambridge 1976, particularly Chs. 4, 9, 10, 13, 15

97 See G. Foley, The Energy Question, Penguin, Harmondsworth 1976, p.92; P.R. Odell, Oil and World Power, Background to the Oil Crisis, 3rd edn., Penguin, Harmondsworth 1974, pp.142—144; World Energy Outlook, OECD, Paris 1977. From 1970—1974, the coefficient between growth in GDP and growth of energy consumption for oil-importing LDCs was 1.13, compared with a coefficient for the OECD (1960—1974) of 0.99. For the ten least industrialized countries in the Third

World, the coefficient was as high as 1.464—for a 1% increase in GDP, energy input had to be increased by almost 1.5%

98 *Energy: Global Prospects 1985—2000,* McGraw-Hill, New York 1977 (Report of the Workshop on Alternative Energy Strategies), App. 1

99 See A.B. Lovins, *Soft Energy Paths: Towards a Durable Peace,* Penguin, Harmondsworth 1977

100 According to the WAES study (see note 98) import demand of oil-importing developing countries in 1985 would be about $4 \times 10^6$ b/d. Accepting this figure, the cost of a two-tier structure with a price difference of $5/b would be in 1985 something like $7.2 bill., if the scheme applied to *all* LDCs. The scheme suggested here should be substantially cheaper—probably lower than present OPEC development aid. Compared to this aid, the suggested two-tier system would spread support in line with the burden; at present, most of OPEC's aid goes to Egypt and Syria, both actually small *exporters* of oil

101 *The Financial Times,* 6 Oct. 1977

102 G. Foley, *The Energy Question,* Penguin, Harmondsworth 1976, pp. 105—109 contains some interesting information on this

103 Conceivably, this could lead to tension within OPEC. If all the other suppliers followed conservation policies, and the West turned to Saudi Arabia for additional exports, Riyad might be inclined to push within OPEC for a fairer distribution of the burden of surplus production to meet Western demand

104 On this point, see L. Turner, 'Die rolle des oels im Nord—Sud dialog', in *Europa Archiv,* No.4, 1977, pp. 102—112; D. Rüstow, 'US—Saudi relations and the oil crisis of the 1980s', *Foreign Affairs,* April 1977, pp.494—516; H. Maull, *Oil and Influence: The Oil Weapon Examined,* IISS, London 1975,(Adelphi Paper No.117)

# Index

Abu Dhabi, development plans, 132
Afghanistan, Soviet invasion of, 324
Africa, refinery capacity, 119
Agriculture, 127, 179, 180, 184, 186, 188, 191, 194
Algeria, 196
  agriculture in, 179, 180
  assessment of, 179
  Community exports to, 109
  development plans, 132, 179
  economic and social development, 296
  export policy, 180
  future oil exports, 196
  in North–South dialogue, 268
  mineral wealth, 179
  natural gas exports, 54, 66
  trade agreements with, 295, 296
  trade with European Community, 102, 104, 105, 106, 107, 108
  value of oil exports, 134
Arab–Israeli conflict, 111, 112, 114, 137, 141, 153, 244, 245, 292, 318
Arab League, relations with Community, *see under* Euro–Arab dialogue
Atomic reactors, *see* Nuclear reactors
Australia,
  coal imports from, 53, 60
  uranium ore exports, 69

Belgium,
  coal industry, 21
  nuclear power in, 32
Brazil,
  uranium exports, 56

Canada,
  coal exports, 53, 60
  investment in oil industry, 121
  uranium oil export policy, 71
Chemical industry, 122
Coal,
  a major energy asset, 9
  British development of, 26
  efficiency of, 25
  European balance sheet, 22
  fall in consumption, 21
  for power stations, 24

Coal (*cont.*)
  future of, 21, 25
  imports of, 51, 52, 58
  industrial use of, 24, 25
  in place of other fuels, 24
  investment in equipment, 24
  new markets for, 23
  oil company interest in, 226, 227, 228, 229
  price of, 61
  production of, 84
    costs, 23
    environmental effects, 23
    increase in, 17, 23
    labour costs, 61
  production capacity, 25
  reserves of, 25
  synthetic natural gas from, 25
  trade in, 21
  transport of, 24
  USA exports, 51
  utilization of, 53
  world market prices of, 62
  world production of, 17
  world reserves of, 58
Coal gasification, 17, 25, 51
  nuclear power for, 34
Coal liquefaction, 51
Coke ovens, 24
  coal in, 25
Coking coal, demand for, 61
Common Agricultural Policy, 300
Conference on International Economic Co-operation, 242, 268–283
  achievements and failures, 277
  conceptual differences, 279
  development plans, 278
  Energy Commission, 269–277
    action-oriented phase of, 271
    agreements, 283
    analytical phase of, 270
    continuing consultation, 276
    disagreements in, 275
    price discussions, 273, 275
    results of, 274
    sea bed resources, 276
    expectations of, 279

334    Index

Conference on International Economic
    Co-operation
  Energy Commission (*cont.*)
    financial questions, 278
    general outcome of, 277
    organizational aspects, 279
    results, 282
    tactics among members, 280
    working groups, 269
Conservation of energy, 49, 174, 240, 270
Co-operation, 205, 267, 307
  alternatives to, 264, 315
  between oil companies and producers,
    225, 229
  bilateral and unilateral approach to, 265
  character of, 264
  CIEC attempts, 272
  concept of, 263
  flexibility of, 318
  need for, 263, 307, 325
  private enterprise and, 319
  within OPEC, 233
Co-ordination and national policies, 309

Decision-making,
  nuclear power and, 41
Denmark,
  natural gas from, 18, 30
  natural gas reserves, 29
  oil imports, value as percentage of GDP,
    113
  oil reserves, 44
Dependence on imports, 10, 153, 263
Development aid, 300
Domestic energy supplies, 17–51
Dyson–Tsiolkovskii shell, 3

East European energy balance, 80
East–West relations, 310, 318
  role of, 156
Economic decisions, 5
Economic security, 6
Egypt,
  oil exports, 82
  trade agreements with, 294, 296
Electricity,
  costs of, 35
  demand for, 34
  distribution, 35
  for space heating, 34
  future demand, 35
  nuclear generation of, expansion in, 67
  problems of, 41
  production of, wasteful nature of, 18
Emergency Allocation Scheme, *see under*
  International Energy Agency
Energy balance, 16, 85
Energy insecurity, 305
Energy policies, timing by suppliers, 139
Energy strategy, evolution of, 5

Energy transition, 4, 274, 303, 320
  international organization of, 7
  political implications, 304
Environment,
  constraints on fuel production, 50
  damage to, 7
  nuclear power and, 37
  protection of, 23
Euro–Arab dialogue, 284–292
  achievements of, 291
  agricultural questions, 286, 288, 289
  financial questions, 286, 288
  industrial questions, 286, 287, 289
  infrastructural questions, 286, 288, 289
  scientific and technological questions,
    288
  social affairs, 286, 288, 289, 292
  structure of, 287
  subject matters, 285
  trade questions, 286, 288, 289, 291

Ford Foundation/Mitre study of nuclear
    safety, 38
France
  coal industry, 21
  national oil companies, 212
  natural gas reserves, 29
  nuclear energy policy, 74
  nuclear power in, 32
  oil imports, value as percentage of GDP,
    113
  oil reserves, 44, 164
  opposition to nuclear power in, 41
  uranium reprocessing, 73
Fuel costs, comparative, 35

Gas, natural *see* Natural gas
Germany,
  coal industry, 21
    production increase, 23
  natural gas reserves, 29
  nuclear power in, 32
  oil imports, value as percentage of GDP,
    113
  oil stocks, 164
  opposition to nuclear power in, 41
  uranium imports, 56
  uranium reprocessing, 73
Global energy policy, 7, 8, 303
  consumer power, 317
  development, 308
  emergency schemes, 314
  fundamental problems, 308
  future of, 316
  need for, 10
  price fixing, 308, 310
Global rationality, 6
Government,
  expenditure in oil suppliers, 128

Index 335

Government (cont.)
  interference, 6
  policies for natural gas, 30
  revenues from oil, 47
Greece, 293

Imports of energy, 51–86
  dependence on, 153
  level of, 156
Industrialization,
  from oil revenues, 127
  difficulties, 127
Industry,
  coal consumption in, 24, 25
  control of energy use, 6
  co-operation, 300
  in Iran, 182
  nuclear power in, 34
  producing manufactured goods, 118
Insecurity, controlled, 320
Instability, 323
Interactions between importers and exporters, 265
Interdependence, 320
International distribution system, impact of interruption of supply on, 157
International Energy Action Group, 238
International Energy Agency, 153
  confrontation strategy, 247
  emergency allocation scheme, 158, 162, 247, 248, 314
    consumption constraints, 165
    decision-making in, 162
    divisions in, 171
    failure of programme, 163
  formation of, 239
  import/export targets, 311
  international oil market and, 249
  lowering imports, 322
  relations with OPEC, 246
  role in price levels, 246
International Energy Council, 309
  advantages of, 314
  co-ordinating import/export targets, 312
  function of, 310
  oil prices and, 310
International Energy Fund, 311, 319
International Energy Institute, 273
International Energy Resource Bank, 312
International Maritime Industry Forum, 167
International Resource Bank, 273
Iran,
  agriculture in, 180, 182
  assessment of, 180–184
  Community exports to, 109
  economic failure in, 182
  estimated exports from, 184
  future oil exports, 196
  industry in, 182

Iran (cont.)
  marketing involvement, 176
  mineral wealth, 183
  natural gas exports, 53, 66
  nuclear programme, 76
  oil exports, 82
  political development of, 183
  relations with USA, 244
  relations with USSR, 158, 183
  revolutionary change in, 138, 181, 196, 304, 323
  social distortions in, 183
  tanker fleet, 167
  trade relations with, 293, 297
  trade with European Community, 104, 105, 106, 107, 108
  underdevelopment, 181
  value of oil exports, 134
Iraq,
  assessment of, 184–185
  development plans, 132
  future oil exports, 196
  investment, 184
  mineral wealth of, 184
  oil exports, 82, 185
  oil potential, 185
  political stability of, 185
  trade with European Community, 104, 105, 106, 108
  value of oil exports, 134
Ireland,
  natural gas reserves, 29
  oil, imports value as percentage of GDP, 113
Israel, 294
Israeli–Arab conflict, 111, 112, 114, 137, 141, 153, 244, 245, 292, 318
Israeli–Egyptian peace treaty, 322
Italy,
  national oil companies, 212
  natural gas reserves, 29
  nuclear power in, 32
  oil imports, value as percentage of GDP, 113
  oil reserves, 44

Japan,
  national oil companies, 212
  oil stocks, 164
Jordan, trade agreements with, 296

Kuwait,
  assessment of, 189–190
  Community exports to, 109
  conservation policy, 190
  development plans, 132
  import requirements, 190
  industry, 189
  involvement in marketing of oil, 176

Kuwait (cont.)
  natural gas production in, 176
  tanker fleet, 168
  trade with European Community, 104, 105, 106, 108
  value of oil exports, 134

Lebanon, 293
  trade agreements with, 293, 294, 296
Lesser developed countries, 267, 310
  Community exports to, 109
  effect of oil embargoes on, 162
  energy-deficient, 311
  energy imports, 308
  industrialization, 126
  international energy fund and, 319
  need for energy management, 309
  oil exports, 108, 120, 122, 125
  problems of, 315
  resource bank and, 312
  Stabex and, 298
Libya,
  assessment of, 188—189
  attack on oil companies, 215
  Community exports to, 109
  development plans, 132, 188
  industry, 188
  oil export policy, 189
  trade with European Community, 103, 104, 105, 106, 108
  value of oil exports, 134
Lifestyles, new, 7
Lomé Convention, 76, 281, 286, 289, 293, 297, 298—301
  development aid, 300
  industrial co-operation and, 300

Management of energy problem, 5
Manufacturers, 118
Marketing, involvement of producers in, 175
Martinique summit, 239
Mediterranean countries, trade agreements with, 293, 294, 295, 296
Middle East,
  investment in oil industry, 121
  natural gas imports from, 63
  oil exports to USSR, 82
  operating and investment expenditure, 123
Middle East crises, 295
Middle East settlement, 284
Military action, to secure supplies, 7
Military expenditure, 196
Mineral wealth,
  Algeria, 179
  Iran, 181, 183
  Iraq, 184
  Saudi Arabia, 191

Morocco,
  trade agreements with, 295

Nairobi, UNCTAD conference, 242
Natural gas, 26—32, 317
  benefits to producers, 176
  centralization of international market, 66
  consumption, expansion of, 86
  depletion rates, 29
  development of, 270
  European balance sheet, 16, 27
  from coal, 25
  from Libya, 188
  gap between demand and supply, 18
  government policies, 29, 30
  imports of, 53, 63
    levels of dependence, 66
    political factors, 54
    projects, 65
    technical and political risks of, 64
  liquefied, problems of, 64
  oil companies' role in production, 226, 227, 228, 229
  political aspects of, 31
  price of, 200
  pricing policy, 30
  production of, 4, 50, 62, 63
    by USA, 243
    growth of, 62
    political factors, 49
    targets, 31
  relations between suppliers and consumers, 101
  reserves of, 18, 28
  security of supply, 156, 157
  sources of, 17
  storage costs, 28
  supply factors, 30
  tanker fleet, 158
  total resources, 31
  transport and distribution, 28, 63
    political factors in, 157
  uses of, 26
  value of, 18
  versatility of, 28
  world reserves of, 62
Netherlands,
  natural gas production, 28, 49, 63
    exports, 53, 66
    government policy, 30
    reserves, 17, 18, 29, 62
    targets, 31
  nuclear power in, 32
  oil imports, value as percentage of GDP, 113
  oil stocks, 164
Nigeria,
  agriculture in, 186

Nigeria (cont.)
    assessment of, 186–188
    Community exports to, 109
    development potential, 186
    future oil exports, 196
    industry, 125, 187
    oil industry, economics of, 124
    oil production, prospects, 187
    problems of, 186
    results of development, 187
    trade with European Community, 103, 104, 105, 106, 107, 108
    value of oil exports, 134
North Africa, natural gas imports from, 63
North Sea,
    natural gas in, 18
    oil production, 9, 20, 221, 224
        costs of, 47
        development of, 44
        economic restraints on, 47
        exploration, 46
        new sources, 46
        political factors, 48
        rate of extraction, 46
    oil reserves in, 45, 46
North–South dialogue, 241, 242, 267, 305, 309, 318, 322
    *See also* Conference on International Economic Co-operation
    conflict, 316
    role of USA, 281, 282
Norway,
    natural gas production, 53, 63
    natural gas reserves, 62
    oil exports, 77
    oil production, 57
        effects on economy, 77
Nuclear energy, 18–20, 32–42, 319
    accidents and malfunctions, 38
    comparative costs, 35
    decision making in, 41
    development of, 33
    disposal of wastes, 39, 40
    environmental impact, 37
    expansion of, 32, 42
    exports from Community, 57
    fast breeder reactors, 37
    for Third World, 42
    growth of, 86
    imports of, 54
    industrial use of, 34
    oil company interest in, 227, 228, 229
    power stations, 73
    problems of, 18, 19
    public opposition to, 19, 37, 40
    radiation risks, 39
    risk of disaster, 39
    safety record of industry, 38

Nuclear energy (cont.)
    scope for, 34
    security of supplies, 37
    social issues of, 42
    supply and demand structures, 19, 32
    timing of strategy, 41
Nuclear fuel cycle, 55, 67
Nuclear proliferation, 101
    dangers of, 57, 69
Nuclear reactors, 37
    export of, 76
    fast breeders, 73
    fuel costs of, 69
    in developing countries, 76
    light water, 69
    programme,
        political implications, 20$n$
Nuclear trade, 67–76
Nuclear weapons,
    production of, 76
    proliferation of, 69, 72

Oil, 20–21, 42–49
    as weapon, 112, 318
    average stocks held, 164
    breakdown of cost of, 201
    conflict between companies and producers, 210, 215
    confrontation between producers and USA, 238
    conservation of, 174, 272
    consumption of, 21
    creating export earnings, 102
    dependence on, 15, 60, 153, 200
    depletion rates and financial requirements, 177
    development and production costs, 47
    discovery of new sources, 46
    distribution of, company control of, 223
    distribution of resources, 217
    distribution of surplus profits, 211
    distribution under crisis conditions, 165
    diversification away from, 305
    embargoes by exporters, 8
    employment in industry, 125
    energy balance, 16
    European balance sheet, 43
    European reserves, 44
    exploration in North Sea, *see under* North Sea
    exports in developing countries' 'economies', 118
    'golden age of', 15
    Government revenues from, 47, 176
        depletion rates and, 177
        factors in, 178
        increasing, 176
        in Norway, 78

Oil (*cont.*)
  importance of, 42
  imports of, 77
    by USA, 8
    from OPEC, 84
    sources, 81
    value as percentage of GDP, 113
  industry,
    infrastructural development, 126
    spin-off effects, 122
  international market, 203
    control of, 240, 304
    IEA and, 249
    importers controlling, 209
    interaction and distribution, 214, 217
    OPEC control of, 233
    power aspects of, 111
    relationship of forces in, 205, 317
    restructuring of, 213
  interruption in supply, effects of, 114
  investment in, 120, 121
  long-term supply of, 173–199
  minimum safeguard price, 247
  operating and investment expenditure, 123
  phasing out of, 173
  political economy of, 118
  price of, 47, 271, 308, 310
    bargaining processes, 203
    capability of influencing decisions, 202
    conflict in, 210
    consultation on, 273
    control over resources, 204
    decisions, 202
    distribution of profits, 210
    economic aspects, 204
    elements in, 200
    evolution of, 207, 208
    external influences, 205
    formation and evolution, 200
    hegemonial factors, 205, 209
    increase in, 17, 323
    influence in decision making, 202
    influence on strategies, 206
    long-term stability, 229
    long-term trends, 204, 220
    OPEC bargaining, 218
    OPEC control of, 215, 267, 277, 306
    OPEC internal dispute, 236
    patterns of interaction, 204
    political aspects, 203, 204, 206
    'rent' aspects, 200, 201, 202
    role of companies, 207, 220
    role of International Energy Agency, 246
    role of OPEC, 230

Oil
  price of (*cont.*)
    role of Saudi Arabia, 231, 235, 236
    role of USA, 237
    stability of, 117, 229
    strategies, 206
    suppliers influencing, 117
    two-tier system, 310
  production of, 4, 44, 50
    by USA, 243
    company control over, 209
    control by producers, 213
    costs of, 202
    decline in, 174
    depletion measures, 48
    depletion rates, 193
    economic restraints, 47
    growth rates, 79
    in North Sea, 20
    in USSR, 79
    increase in, 86
    political aspects, 48
    rates, 140
    revenue and, 117
    stability of, 117
  refining,
    benefits to supplier, 120
    capacity, 119, 120
    concentration in consumer areas, 120
    OPEC capacity, 175
    role of companies, 223
  relation between suppliers and consumers, 101
  requirements of, 76
  reserves,
    North Sea, 45, 46
    of OPEC, 83
  restraint in consumption of, 165
  revenues,
    allocation of, 129
    industrialization and, 127
    use of, 126
  role in trade relationships, 103
  royalties, 201
  security of supply, 153, 291, 292
    consumption restraints, 165
    embargoes, 159, 160
    emergency scheme, 158, 159, 160, 162
  strategic importance of, 102, 103
  surplus profits, conflict over, 215
  tanker fleet,
    capacity, 165
    developments in, 166
    oil producers in, 167, 216
  vital role of, 4
  world spare capacity, 112
  world supply of, 173

Index 339

Oil companies, 220—230
  access to markets, 225
  and the Third World, 223
  attack on, 215
  becoming energy companies, 226, 230
  capital of, 221
  conflict with exporters, 210, 215, 226
  control of markets and distribution, 223
  control over production and reserves, 209
  control over resources, 221
  co-operation with producers, 225, 229
  diversification away from oil, 226
  diversification of sources of supply, 224
  dominance of, 304
  functions for producers, 225
  independents, 221
  in refining, 223
  investments, 222
  management and administration, 223
  management functions, 305
  maximization of profits, 207
  national, 212
  production figures, 222
  profits, 230
  proliferation of, 212
  reduction in share, 213
  relations with OPEC, 224, 226, 229
  role before 1959, 207
  self-preservation of, 207
  strategies, 223
    withdrawal, 224
  technical know-how, 221, 225
  USA, 211
Oil importers,
  conflict with producers, 217
  controlling markets, 209
  losses of, 216
Oil producers, *see also* OPEC countries
  agriculture and, 127
  balance of trade surpluses, 129
  Community exports to, 107
  companies' functions for, 225
  conflict with importers, 217
  conflict with oil companies, 210, 226
  control over international energy system, 240
  control over production, 213
  controlling prices, 117, 141
  co-operation with companies, 225, 229
  depletion of reserves, 140
  diversification of power and strength, 137
  economic benefits, 174
  economic development, 133
  economic diversification by, 139
  economic self-interest, 151, 177
  embargoes by, 137, 159, 160
    results for themselves, 160
    selective, 169

Oil producers (*cont.*)
  excess earnings, 115
  expansion of power and influence, 135
  foreign investments, 129, 133, 137, 138, 140, 177, 197
  forward linkages in economy, 122
  government expenditure, 128
  government revenues
    depletion rates and, 177
    factors in, 178
    increase in, 176
  impact of oil exports on, 128
  industrialization of, 110, 127, 130, 139
    difficulties, 127
  influencing international affairs, 133
  infrastructural development, 126
  involvement in marketing, 175
  level of exports, 156
  national objectives and international responsibilities, 178
  need for export markets, 127
  non-oil exports, 107
  operating and investment expenditure, 123
  problems of development of, 130
  rates of production, 140
  relationship to Europe, 101, 136
    confrontation, 138, 139, 140, 171, 172
    co-operation, 151
    future evolution of, 141
  revenue requirements and oil demand, 197
  revolutionary changes in, 138
  shortage of labour, 132
  social and cultural impact of oil industry, 128
  synchronization between import demand and export requirements, 177
  tanker involvement, 167, 216
  technical linkages with companies, 124
  timing of policy by, 139
  trade with European Community, 103
    ability to interrupt, 111
    as means of exerting influence, 112
    effect of interruptions, 114
    power aspects of, 111
  USA policy towards, 239
OPEC *see also* Oil producers
  as structured oligopoly, 234
  breakup of, 317
  burdens of, 325
  cohesion of, 216, 267, 324
  conflict with importers, 309, 321
  controlling prices, 215, 230, 231, 267, 274, 277, 306
  co-operation among, 233
  co-operation of Community with, 264
  decision making by, 218, 306

OPEC (*cont.*)
  development of, 233, 311
  dialogue with users, 268
  differences of opinion within, 218
  disintegration of co-operation, 236
  diversification by, 230, 272, 304
  embargoes imposed by, 8, 156
  economic development, 133
  expansion of imports, 194
  financial assets of, 116
  formation of, 213
  future export rates, 196, 197
  future of, 235
  future strategy, 232
  government policies, 177
  import/export quotas and, 313
  imports from Community, 110
  industrial competition with, 321
  industrialization of, 110, 138
  internal disagreements, 232, 236
  in transport, 230
  market control by, 233
  natural gas exports, 53, 66
  natural gas production policies, 66
  oil reserves, 83
  oil revenues, 116
  power of, 307
  price increases by, 202
  production rates in, 232
  recycling of income, 311
  refining capacity of, 175, 230
  relations with oil companies, 224, 226, 229
  relation with users, 307
  relations with IEA, 246
  relations with USA, 240, 241, 245
  resources and political stability, 195
  revolutionary change in, 138
  solidarity, 235
  strategy of confrontation with Europe, 137
  strength of, 230
  tanker fleets, 167
  trade balances, 102
  transformation period of, 305
  value of imports, 114
  vulnerability of, 232, 307, 324
Overcentralization of energy system, 302

Palestinian question, 286, 287
Persian Gulf, regional instability in, 139
Plutonium, 73
  for reactors, 37
  hazards of, 40
  processing of, 73
    risks of, 72
  risks of, 72, 76
  transportation of, 39, 73

Policy objectives, 151
  price formation, 200
Poland, coal exports, 53, 58, 60
Political aspects of natural gas production, 31
Political cost of energy policy, 20$n$
Political decisions, 5
Politics, technology and, 4
Pollution, natural gas and, 28
Power stations,
  coal-fired, 24
  comparative costs of, 36
  nuclear, 24
    fuel for, 73
  security of supply to, 37
  substitution of coal for oil, 24
Prices,
  as cause of conflict, 138
  control of, 141 *See also under* Oil
Radiation, risk of from nuclear plants, 39
Radioactive wastes, 40
  disposal problem, 40
  transport of, 40
Rasmussen Report, 38
Raw material prices, 278
Refining, oil companies in, 223
Reserves, strategic, 164

Saudi Arabia, 275, 280
  absorbing excess earnings, 316
  agriculture, 191
  assessment, 190–194
  Community exports to, 109
  development opportunities, 191
  development plans, 132, 191, 245
  future oil exports, 192
  industry, 191
  mineral wealth, 191
  oil production, 231
  policies, 177
  relations with, 265
  relations with USA, 235, 239, 244, 280
  risks of social change, 192
  role in price fixing, 231, 235, 236
  tanker fleet, 168
  trade with European Community, 104, 105, 106, 108
Scottish Nationalism, oil production and, 48
Sea bed resources, 276
Security, economic, 6
Shale oil, oil companies and, 226
Siberia, oil production in, 79
Social cost of energy policy, 20$n$
Solar energy, oil companies interest in, 227, 228
Solar system, reconstruction of, 3
Solid fuels, *See also* Coal and Coke
  additional supplies of, 60

Solid fuels (cont.)
  comparative costs, 35
  dependence on, 10
  energy balance, 16
  fluidized bed combustion of, 26
  imports of, 51
  scope for expansion, 58
  survey of, 21
South Africa,
  coal exports, 53, 60
Soviet Union,
  coal exports, 53, 58
  coal production, 61
  invasion of Afghanistan, 324
  natural gas exports, 53, 63, 66
  oil exports, 57, 77, 78, 79
  oil imports, 82
  oil production in, 79
    difficulties, 82
    future problems, 79
    levels of, 78
  relations with Iran, 158, 183
  relations with USA, 212
  uranium ore exports, 74, 75
Stabex, 298, 301
Standby capacity of Community, 157
Strategic reserves, 164
Sugar agreement, 299
Supplies, average stocks, 164
  long-term availability of, 173–199
  long-term stability of, 264
  security of, 153–172, 238, 290, 292
    factors, 194
    IEA emergency scheme, 159, 160, 162, 247, 248, 314
    tanker fleets as factor, 165, 166
  standby capacity of, 157
Syria, trade agreements with, 296

Technology,
  concentration of know-how in developed countries, 122
  social and political change and, 4
Third World, 306
  demand for energy, 270
  development of, 308
  nuclear power for, 42
  oil companies and, 223
  relations with Community, 282
Trade agreements with Mediterranean countries, 293, 294
Trade balances, between suppliers and consumers, 102
Trade patterns,
  between European community and energy suppliers, 103
  effect of interruption, 114
  as means of exerting influence, 112
  power aspects of, 111

Trade regulations with ACP countries, 299
Transport,
  of coal, 24
  of natural gas, 28, 63
  of nuclear waste, 39
Tunisia, trade relations with, 295
Turkey, 292

UNCTAD, 242, 278
United Kingdom,
  coal development in, 26
  coal industry, 21
    production, 28
    production increase, 23
    reserves, 17, 18, 62
  nuclear power in, 32
    delays in construction, 41
  oil imports, value as percentage of GDP, 113
  oil production by, 44
    build-up of, 45
    factors, 45
    political factors, 48
    Scottish Nationalism and, 48
  oil reserves, 46
United States of America,
  challenge to, 237
  coal exports, 51, 60
  confrontation with producers of oil, 238
  demand for electricity, 34
  economic security and, 284
  investment in oil industry, 121
  natural gas production, 243
  nuclear policy of, 74
  oil companies, 211
  oil imports, 8, 243, 245
  oil production, 243
  oil stocks, 164
  policy towards oil producers, 239
  relations with Iran, 244
  relations with OPEC, 240, 241, 245
  relations with Saudi Arabia, 235, 239, 244, 265, 280
  relations with USSR, 212
  role in North–South dialogue, 268, 281, 282
  role in oil price stability, 237
  uranium enrichment plants, 74
  uranium exports, 54
    restrictions on, 69
U.S. Atomic Energy Agency, accident study, 38
Uranium,
  enriched, 70, 73, 74
  exploration for, 56
  exports, 54
  import of, 54
    political problems, 56

Uranium (cont.)
   international market in, 56
   political difficulties, 101
   prices of, 71
   requirements of, 37
   world requirements of, 72
Uranium Institute, 70
Uranium ore,
   exports of, 74
   international market, evolution of, 71
   multinational corporations, 70
   potential suppliers, 67
   processing of, 71
      interaction between government and companies, 70

Uranium ore (cont.)
   reprocessing, 72
   requirements, 56, 67
   reserves, 67, 68
   sources of supply, 75
   world production of, 68
Uranium oxide, conversion of, 71

Venezuela, investment on oil industry, 121
Venezuela, value of oil exports, 134

Washington Energy Conference, 239
World, energy consumption in, 6